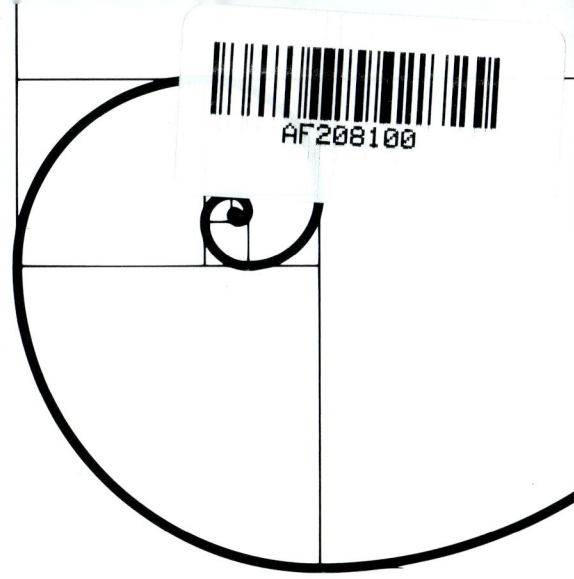

SECOND EDITION

Teaching Secondary Mathematics

Secondary mathematics teachers working in the Australian education sector are required to plan lessons that engage with students of different genders, cultures and levels of literacy and numeracy. Written by experts in mathematics education, *Teaching Secondary Mathematics* engages directly with the Australian Curriculum: Mathematics and the Australian Professional Standards for Teachers to help preservice teachers develop lesson plans that resonate with students. Part 1 covers contemporary issues in learning and teaching mathematics, such as engaging male and female students, and utilising new technologies and clear language in the classroom. Part 2 explores key mathematics content, including statistics and probability, measurement and geometry, and number and algebra.

This edition has been thoroughly revised and features a new chapter on supporting Aboriginal and Torres Strait Islander students by incorporating Aboriginal and Torres Strait Islander cultures and ways of knowing into lessons. Chapter content is supported by new features including short-answer questions, opportunities for reflection and in-class activities. Further resources, additional activities, and audio and visual recordings of mathematical problems are also available for students in the book's online resources.

Teaching Secondary Mathematics is the essential guide for preservice mathematics teachers who want to understand the complex and ever-changing Australian education landscape.

Gregory Hine is a Senior Lecturer in the School of Education at the University of Notre Dame, Australia.

Judy Anderson is Associate Professor of Mathematics Education at the University of Sydney.

Robyn Reaburn is a Lecturer of Mathematics Education at the University of Tasmania.

Michael Cavanagh is Associate Professor and Director of Learning and Teaching in the School of Education at Macquarie University.

Linda Galligan is Associate Professor and Head of the School of Sciences at the University of Southern Queensland.

Bing H. Ngu is a Senior Lecturer of Mathematics Education at the University of New England.

Bruce White is a Lecturer of Mathematics, Science and IT Education at the University of South Australia.

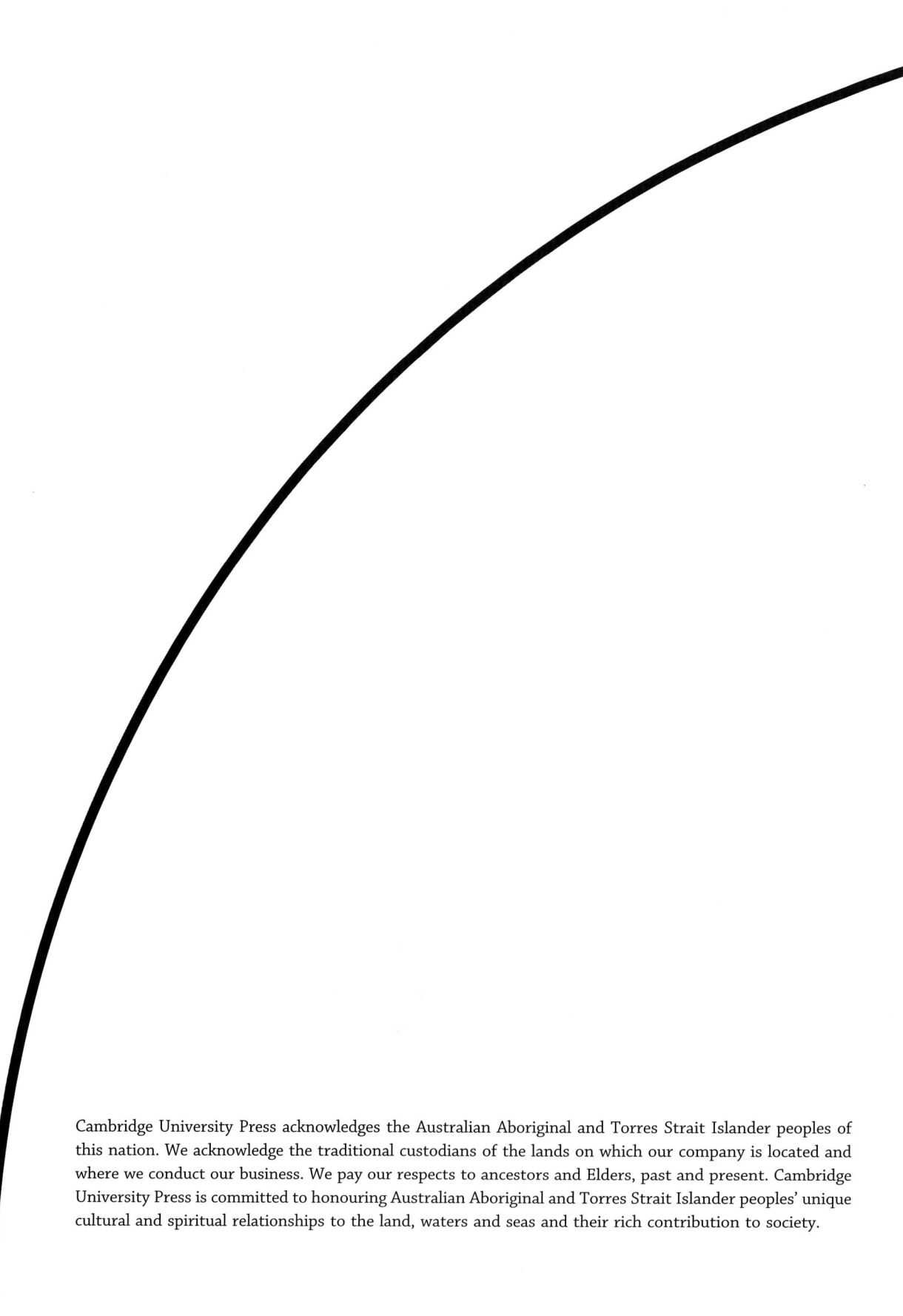

Cambridge University Press acknowledges the Australian Aboriginal and Torres Strait Islander peoples of this nation. We acknowledge the traditional custodians of the lands on which our company is located and where we conduct our business. We pay our respects to ancestors and Elders, past and present. Cambridge University Press is committed to honouring Australian Aboriginal and Torres Strait Islander peoples' unique cultural and spiritual relationships to the land, waters and seas and their rich contribution to society.

SECOND EDITION

Teaching
Secondary
Mathematics

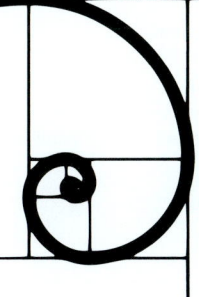

Gregory Hine
Judy Anderson
Robyn Reaburn
Michael Cavanagh
Linda Galligan
Bing H. Ngu
& Bruce White

CAMBRIDGE
UNIVERSITY PRESS

CAMBRIDGE
UNIVERSITY PRESS

University Printing House, Cambridge CB2 8BS, United Kingdom

One Liberty Plaza, 20th Floor, New York, NY 10006, USA

477 Williamstown Road, Port Melbourne, VIC 3207, Australia

314–321, 3rd Floor, Plot 3, Splendor Forum, Jasola District Centre, New Delhi – 110025, India

103 Penang Road, #05–06/07, Visioncrest Commercial, Singapore 238467

Cambridge University Press is part of the University of Cambridge.

It furthers the University's mission by disseminating knowledge in the pursuit of education, learning and research at the highest international levels of excellence.

www.cambridge.org
Information on this title: www.cambridge.org/9781108984683

First published 2016
Second edition 2021

Cover designed by Tanya De Silva-McKay
Typeset by Straive
Printed in China by C & C Offset Printing Co., Ltd, July 2021

A catalogue record for this publication is available from the British Library

A catalogue record for this book is available from the National Library of Australia

ISBN 978-1-108-98468-3 Paperback

Additional resources for this publication at
www.cambridge.org/highereducation/isbn/9781108984683/resources

Contents

About the authors

Gregory Hine is a Senior Lecturer at the University of Notre Dame Australia in Fremantle. He previously taught in Australia and the United States, predominantly in the areas of mathematics and science to middle and high school students, for 14 years. Greg completed both of his Master of Education degrees at the University of Notre Dame, South Bend, Indiana. In 2011 he completed his doctoral studies at Notre Dame's Fremantle Campus and was appointed to a full-time academic faculty position in the School of Education. Greg teaches in the undergraduate and postgraduate degree programs, mostly in secondary mathematics education, and educational action research. His areas of scholarly interest are professional noticing in the mathematics classroom, and the training of preservice and in-service mathematics teachers. In 2019 Greg was awarded the Vice Chancellor's Award for Excellence in Undergraduate Education (Secondary Mathematics).

Judy Anderson is Associate Professor in mathematics education and former Director of the STEM Teacher Enrichment Academy at the University of Sydney. In her role as secondary mathematics curriculum coordinator, Judy has been teaching and researching at the University of Sydney for 18 years. Prior to that, she worked at the Board of Studies NSW as a Senior Curriculum Officer (K–12), responsible for the development of the mathematics syllabuses for NSW schools. As a past President of the Australian Association of Mathematics Teachers (AAMT) and a member of the Executive Committee from 2007 to 2010, she provided leadership and ongoing support for teachers of mathematics throughout Australia. This was a critical role at a time of national curriculum development in Australia and the development of national testing regimes. She is currently the Secretary of the International Group for the Psychology of Mathematics Education.

Robyn Reaburn is a Lecturer in mathematics education at the University of Tasmania and has been joint recipient of a federal government award for Teaching Excellence from the Office for Learning and Teaching. Prior to her current role at the university, she taught mathematics and science at secondary schools and at TAFE. She has also taught statistics at the University of Tasmania for 13 years. Robyn's main research interests include students' and instructors' understanding of probability and statistics and the preparation of future mathematics teachers.

Michael Cavanagh is a former secondary mathematics teacher and is currently the Director of Learning and Teaching in the School of Education at Macquarie University. He is an Associate Professor in mathematics education and a Senior Fellow of the Higher Education Academy. Michael is a recipient of an Australian Learning and Teaching Council Citation for Outstanding Contribution to Student Learning and an Outstanding Professional Service Award from the Professional Teachers' Council of NSW for his work as the editor of *Reflections*, the journal of the Mathematical Association of NSW.

Linda Galligan is an Associate Professor with the School of Sciences at the University of Southern Queensland and Head of School. She teaches courses on mathematics for teachers and has strong links with schools providing mathematics programs and activities for students and teachers. Her research includes language and mathematics, and students' and lecturers' perceptions of student preparation for numeracy demands of university. Recently her focus on research has been on

student engagement in online learning, the use of tablet technology to effectively teach mathematics, and the use of modelling to improve preservice teachers' deep understanding of mathematics.

Bing H. Ngu is a Senior Lecturer in mathematics education at the University of New England. She has over 15 years of mathematics and science teaching experience in secondary schools in Australia as well as abroad. Her current research is mainly shaped by her previous mathematics teaching experience. Specifically, based on cognitive load theory and learning by analogy theory, she has conducted experimental studies with secondary students to enhance learning to solve linear equations as well as percentage problems. She has also conducted cross-cultural mathematics education research with secondary students between Asian countries and Australia. Her research has made a strong impact on pedagogical approaches, informing the development of various preservice teacher education units that she currently teaches at the University of New England.

Bruce White has been lecturing at the University of South Australia in mathematics and science teacher education since 1990. His teaching background is in secondary (Years 8–12) mathematics and science. He is a member of the Mathematics Education Research Group of Australasia and researches the use of technology in the teaching and learning of mathematics.

Acknowledgements

The authors would like to acknowledge and thank all individuals who provided feedback on the second edition of *Teaching Secondary Mathematics,* specifically Kevin Lowe and Bronwyn Ewing for their generous support and guidance.

The authors and Cambridge University Press would like to thank the following for permission to reproduce material in this book.

Figure 1.1: © Getty Images/courtneyk. **Figures 1.3** and **8.7**: © Getty Images/SolStock. **Figure 2.1**: reproduced under the Creative Commons Attribution 3.0 Australia licence (CC BY 3.0 AU). https://creativecommons.org/licenses/by/3.0/au/. **Figures 3.1** and **5.1**: reprinted from Stacey, K. (2005). The place of problem solving in contemporary mathematics curriculum documents. *Journal of Mathematical Behaviour*, 24, 341–50, © 2005 with permission from Elsevier. **Figures 3.3**, **3.4**, **3.5** and **3.6**: © Russell Kincaid 2015. **Figure 4.1**: reproduced by permission of the publisher, © 2012 by tpack.org. **Figure 4.6**: © Learn Troop 2016. **Figure 6.1**: © Getty Images/PhotoAlto/ Frederic Cirou. **Figure 7.1**: reproduced with permission of Tyson Yunkaporta. **Figures 8.1**, **8.2** and **8.3**: © J. Minstrell, 2001. **Figure 10.1**: reprinted by permission from Springer: Springer *Mathematics Education Research Journal* from Wilkie, K. & Clarke, D. (2016). Developing students' functional thinking in algebra through different visualisations of a growing pattern's structure. *Mathematics Education Research Journal*, 28, 223–43, © 2016. **Figure 10.10**: reprinted from Ngu, B.H., Phan, H.P., Hong, K.S. & Usop, H. (2016). Reducing intrinsic cognitive load in percentage change problems: The equation approach. *Learning and Individual Differences*, 51, 81–90, © 2016 with permission from Elsevier. **Figure 12.2**: house image used © Getty Images/tatarnikova. **Figure 13.2**: matches image used © Getty Images/laymul.

Table 3.1: TIMSS 2011 Assessment. © 2009 International Association for the Evaluation of Educational Achievement (IEA). Publisher: TIMSS & PIRLS International Study Center, Lynch School of Education, Boston College. **Table 3.3**: reproduced under the Creative Commons Attribution 4.0 Australia licence (CC BY 4.0 AU). https://creativecommons.org/licenses/by/4.0/. **Table 8.1**: © H.M.G. Watt, 2005, Springer Nature. **Table 10.4**: reprinted by permission from Springer: Springer Science+Business Media from Ngu, B.H. & Phan, H.P. (2016). Unpacking the complexity of linear equations from a cognitive load theory perspective. *Educational Psychology Review*, 28, 95–118, © 2016. **Table 13.2**: reprinted by permission from Springer: Kluwer Academic Publishers from Mamona-Downs, J. (2001). Letting the intuitive bear on the formal: A didactical approach for the understanding of the limit of a sequence. *Educational Studies in Mathematics*, 48(2), 259–88, © 2001.

Table from MacGregor, M. & Moore, R. (1991). *Teaching mathematics in the multilingual classroom: A resource for teachers and teacher educators*. Melbourne: Institute of Education, University of Melbourne: reproduced with permission from the University of Melbourne.

Extracts from Matthews, C. (2015). Maths as storytelling: Maths is beautiful. In K. Price (ed.), *Aboriginal and Torres Strait Islander education: An introduction for the teacher profession* (2nd end, 102–20). Port Melbourne: Cambridge University Press: © Cambridge University Press 2015.

Guide to online resources

The student online resources for *Teaching Secondary Mathematics* are freely available online at www.cambridge.org/highereducation/isbn/9781108984683/resources. Visit the site to explore a variety of resources including downloadable additional content, audio and video recordings, weblinks and further resources.

 This margin icon is used throughout the book to indicate that a resource relating to the content under discussion is available online. The descriptor and chapter number can be used to help you easily identify the corresponding resource.

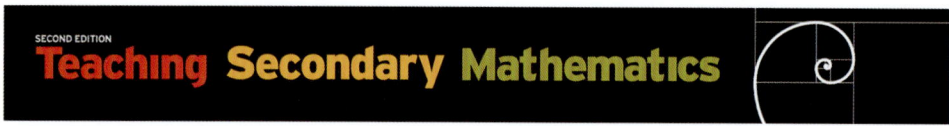

Chapter 1

The learning and teaching of mathematics

Further resources

Statistical literacy resources

The Australian Bureau of Statistics provides statistical information about Australia and the Australian Census. It is a repository of relevant and contemporary statistical data about Australia and its people.

Financial literacy resources

Moneysmart, developed by the Australian Securities and Investments Commission (ASIC), is designed to promote financial literacy. It includes a range of classroom resources aligned to the Australian Curriculum.

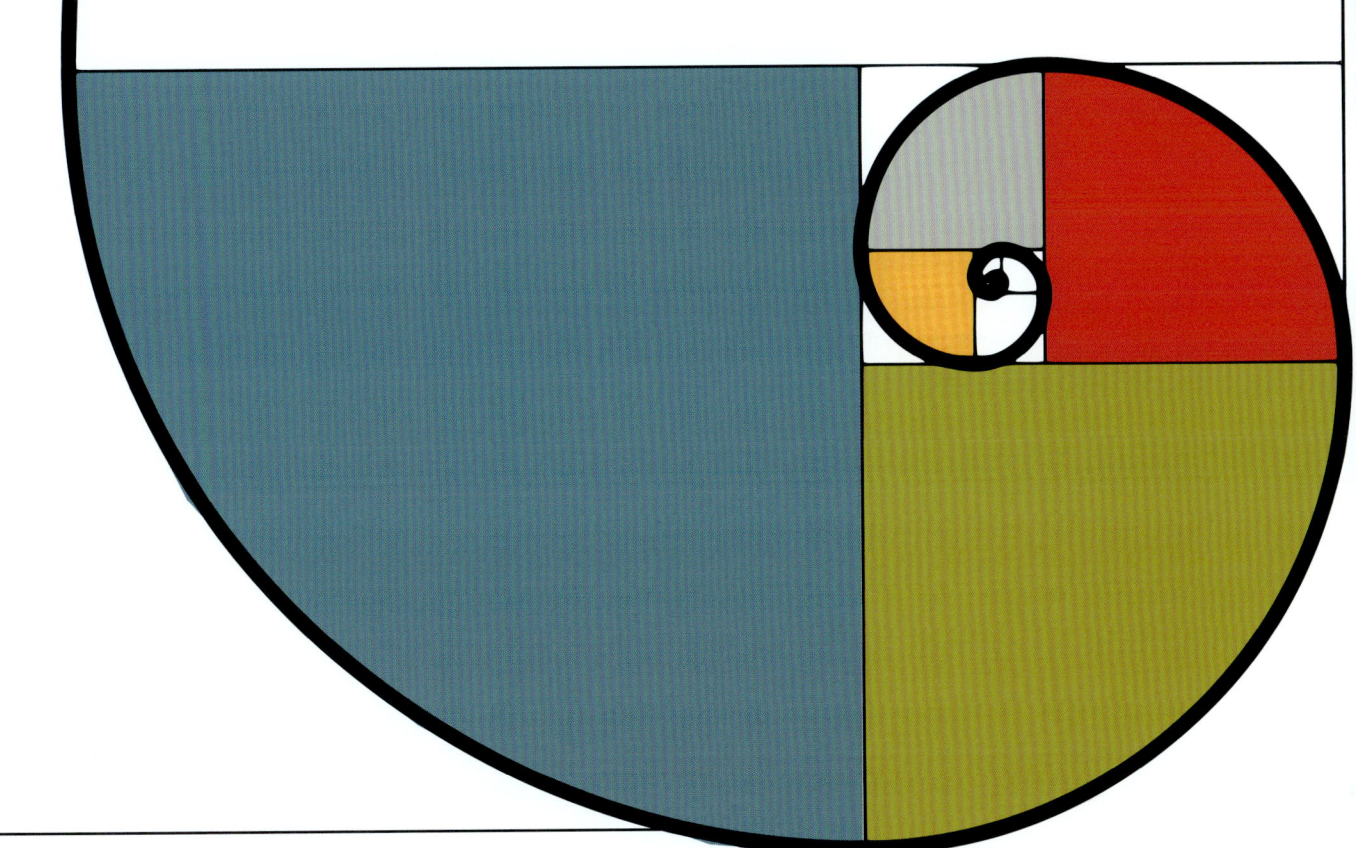

Introduction
What is mathematics?

GREGORY HINE

So, you're going to be a secondary mathematics teacher . . . well, this book has been written for you. It might not surprise you to read that mathematics teachers need to know a lot about mathematics, but this is not the only type of knowledge they need to do their job well. Scholarly work over the past few decades has highlighted that in addition to possessing excellent content knowledge, mathematics teachers need to know about the curriculum, lesson planning, selecting the best strategies to teach students, teaching diverse groups of learners, designing appropriate assessment tasks . . . the list goes on! Teaching mathematics effectively to secondary students is indeed a difficult task, and this book has been designed to help you become a successful teacher. To get us started, we look at the key differences between mathematics both as a discipline and a subject taught in schools. Then we look briefly at one empirically supported way to create a mathematically powerful classroom, before exploring the structure and additional features of this revised edition of *Teaching Secondary Mathematics*.

1

Mathematics as a discipline field

Mathematics as a discipline has evolved across many cultures over thousands of years, and mathematical ideas continue to be developed to this day. But what is understood by the term *mathematics*? There is an abundance of explanations and views to support a definition, and probably too many to mention in this book. For instance, in the late 1960s, a definition of mathematics was given as

> the study of quantity, form, arrangement, and magnitude; especially, the methods and processes for disclosing, by rigorous concepts and self-consistent symbols, the properties and relations of quantities and magnitudes, whether in the abstract, pure mathematics, or in their practical connections, applied mathematics. (Funk & Wagnalls, 1968, p. 835)

Compare that definition with one offered more recently, whereby the authors define not only what mathematics is but describe what it can used for:

> Mathematics reveals hidden patterns that help us understand the world around us. Now much more than arithmetic and geometry, mathematics today is a diverse discipline that deals with data, measurements, and observations from science; with inference, deduction, and proof; and with mathematical models of natural phenomena, of human behaviour, and of social systems ... mathematics is a science of pattern and order. (Mathematical Sciences Education Board, 1989, p. 31)

Ultimately these and other definitions provide insight into how we use mathematics to view and make sense of our world. To indicate how vast the discipline of mathematics is, the Mathematics Subject Classification (American Mathematical Society [AMS], 2020) currently partitions mathematics into 64 broad topics. According to Lawson (2016), the scope of each of these topics is so large that 'a mathematician could work their (sic) entire professional life in just one of these topics' (p. 7). Additionally, any topic could be divided into sub-classifications which could each become the focus of a doctoral thesis.

Teaching mathematics within secondary school education

It has been suggested that how we view mathematics influences how we learn and teach mathematics (Siemon et al., 2015). For example, some have questioned whether teaching mathematics is an art or a science. In answering this question, Posamentier, Smith and Stepelman (2015) offered that teaching mathematics is both an art and a science, whereby successful teachers possess an innate ability to learn, use intuition and exercise creativity, which can be 'buttressed in varying amounts with sociological, psychological, philosophical, and commonsense principles' (p. vii). As a preservice, secondary mathematics teacher (PSMT), you will spend much time thinking about the mathematics you teach as well as the way(s) that you will best help secondary students to learn this mathematics.

Creating mathematically powerful classrooms

There are plenty of ideas in this book to assist you in both regards, although we would like to present one view of teaching mathematics to secondary school students here. Schoenfeld presented an empirically driven framework focused on 'what it takes for students to develop proficiency working with contextual algebraic tasks and crafted materials to support algebra teachers' (2014, p. 410). This framework comprises five dimensions of what Schoenfeld (p. 407) referred to as characteristics of 'mathematically powerful classrooms', which all mathematics teachers can strive towards. The materials are components of the *Teaching for Robust Understanding of Mathematics* (TRU Math) suite, which contains a number of key documents designed for mathematics teachers (e.g. a TRU Math rubric), their teaching of students, and their own professional development (see http://map.mathshell.org/materials/trumath.php). These dimensions and elaborations are summarised in Table 0.1.

Table 0.1 The five dimensions of mathematically powerful classrooms

The mathematics	The extent to which the mathematics discussed is focused and coherent and to which connections between procedures, concepts and contexts (where appropriate) are addressed and explained. Students should have opportunities to learn important mathematical content and practices and to develop productive mathematical habits of mind.
Cognitive demand	The extent to which classroom interactions create and maintain an environment of productive intellectual challenge that is conducive to students' mathematical development. There is a happy medium between spoon-feeding mathematics in bite-sized pieces and having the challenges so large that students are lost at sea.
Access to mathematical content	The extent to which classroom activity structures invite and support the active engagement of all of the students in the classroom with the core mathematics being addressed by the class. No matter how rich the mathematics being discussed, a classroom in which a small number of students get most of the 'airtime' is not equitable.
Agency, authority and identity	The extent to which students have opportunities to conjecture, explain, make mathematical arguments, and build on one another's ideas in ways that contribute to their development of agency (the capacity and willingness to engage mathematically) and authority (recognition for being mathematically solid), resulting in positive identities as doers of mathematics.
Uses of assessment	The extent to which the teacher solicits student thinking and subsequent instruction responds to those ideas by building on productive beginnings or addressing emerging misunderstandings. Powerful instruction 'meets students where they are' and gives them opportunities to move forward.

Source: Schoenfeld, 2014, p. 407.

What this second edition offers preservice secondary mathematics teachers

Teaching Secondary Mathematics has been written to guide and inform the professional lives of PSMTs at the early stage of their careers. Throughout the text, readers have access to key theoretical and philosophical perspectives that have been the focus of much empirical research and that underpin best instructional practices within the secondary mathematics classroom. Additionally, various contemporary issues influencing the planning, teaching and evaluation of mathematical learning experiences have been included for reader consideration. Within each chapter, readers are provided with various activities to consolidate and extend their learning.

First, reflective questions that link closely to the topics presented are offered for individual or collaborative response (Pause and Think). Second, a variety of new and engaging pedagogical activities provides mathematics teachers of lower secondary and senior secondary students with resources to use immediately in the classroom. These pedagogical activities include Scenario boxes, Connection boxes, Short-answer questions and In Practice boxes which have been designed with PSMTs' needs in mind. At the time of publication, *Teaching Secondary Mathematics* reflects the thinking and content available in the most current version of the Australian Curriculum.

In this second edition, the authorship team have paid an increased attention to the Australian Curriculum: Mathematics (ACM) (ACARA, 2020) and the Australian Professional Standards for Teachers (APSTs) (Australian Institute for Teaching and School Leadership [AITSL], 2011). Curriculum authorities in each Australian state and territory (e.g. New South Wales Education Standards Authority, School Curriculum Standards Authority in Western Australia) are responsible for developing and implementing their own mathematics curriculum, using the ACM as an overarching and guiding document. Consequently, PSMTs need to be familiar and conversant with both the ACM and the legislated curriculum documenting their respective state or territory. Provided in Chapter 9 is a comprehensive and useful summary of what the ACM offers PSMTs, especially with regard to the overall design, aims, rationale, content strands, proficiency strands, general capabilities and cross-curriculum capabilities. Also outlined in Chapter 9 are some important considerations for PSMTs who are learning to plan, teach and evaluate mathematics lessons in preparation for a teaching internship or practicum experience.

From a design perspective, this book has been arranged into two key parts. Chapters 1–8 encompass Part 1: Contemporary issues in learning and teaching mathematics, which collectively present topics such as curriculum, student diversity, technology, planning and assessment. Chapters 9–13 comprise Part 2: Learning and teaching key mathematics content, where content strands from the Australian Curriculum are explored at depth. Within each of the 13 chapters, the needs of adolescent learners have been considered carefully, together with pedagogical approaches PSMTs can use to respond effectively to these needs. Each chapter within this second edition commences with a series of learning objectives for the audience, followed by an introductory summary. These learning objectives are tied directly to the APSTs, which the AITSL has devised to inform teachers what they should be aiming to achieve at every stage of their career. The seven standards are grouped into three domains of teaching: Professional Knowledge, Professional Practice and Professional Engagement. Generally speaking, these standards inform teachers of how to improve their practice inside and outside of the classroom. The learning outcomes in each chapter signal to the audience how the content, skills and dispositions located within support preservice teachers in their development and progression towards graduate teacher status.

The authors of the chapters are notable academics hailing from a broad range of Australian universities, and who in their current roles prepare the next generation of mathematics teachers in Australia. Their professional classroom experience and engagement with cutting-edge research combine to produce a clear mathematical voice which resonates in the summaries, insights, teachings, questions and reflections presented. Although the text has been authored primarily for an Australian audience, it is hoped that this voice will inspire and challenge PSMTs everywhere for years to come.

REFERENCES

American Mathematics Society (2020). *Classifications of mathematics*. Retrieved from https://mathscinet .ams.org/mathscinet/freeTools.html?version=2

Australian Curriculum, Assessment and Reporting Authority (ACARA) (2020). *The Australian Curriculum: Mathematics.* Sydney: ACARA. Retrieved from https://www.australiancurriculum.edu.au/f-10-curriculum/mathematics/

Australian Institute for Teaching and School Leadership (AITSL) (2011). *Australian Professional Standards for Teachers.* Melbourne: AITSL. Retrieved from https://www.aitsl.edu.au/teach/standards

Funk & Wagnalls (1968). *Funk & Wagnalls standard college dictionary* (text ed.). New York: Harcourt Brace.

Lawson, M.V. (2016). *Algebra and geometry: A guide to university mathematics*. Boca Raton, FL: Taylor and Francis.

Mathematical Sciences Education Board, National Research Council (1989). *Everybody Counts: A report to the nation on the future of mathematics education.* Washington, DC: National Academy of Sciences Press.

Posamentier, A.S., Smith, B.S. & Stepelman, J. (2015). *Teaching secondary mathematics: Techniques and enrichment units* (8th edn). Boston, MA: Pearson.

Schoenfeld, A.H. (2014). What makes for powerful classrooms, and how can we support teachers in creating them? A story of research and practice, productively intertwined. *Educational Researcher*, 43(8), 404–12.

Siemon, D., Beswick, K., Brady, K., Clark, J., Faragher, R. & Warren, E. (2015). *Teaching mathematics: Foundations to middle years* (2nd edn). South Melbourne: Oxford University Press.

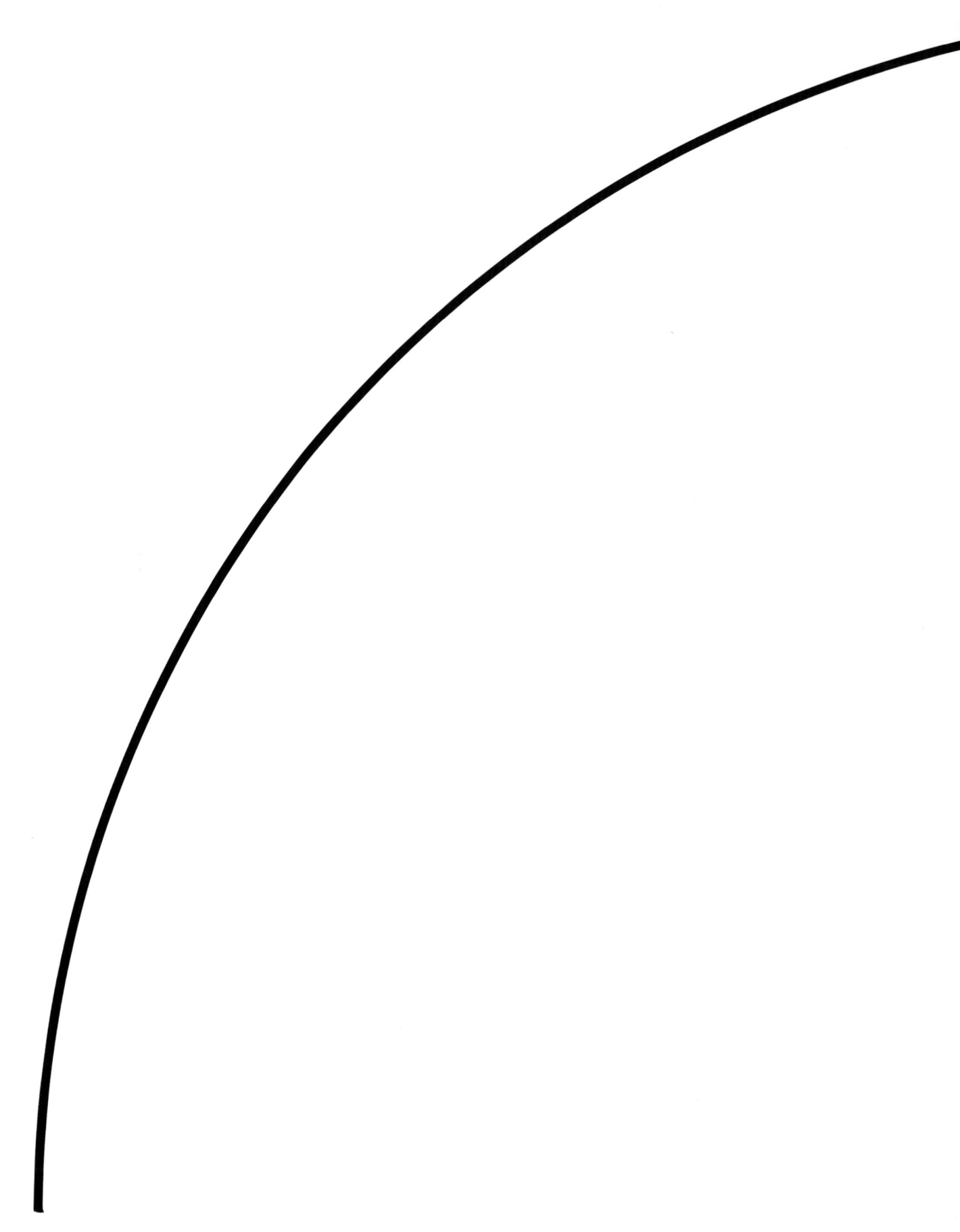

Part

1

CONTEMPORARY
ISSUES IN
LEARNING AND
TEACHING
MATHEMATICS

CHAPTER 1

The learning and teaching of mathematics

Michael Cavanagh

LEARNING OBJECTIVES

The learning objectives of this chapter are directly linked to the Australian Professional Standards for Teachers (APST) (Australian Institute for Teaching and School Leadership [AITSL], 2011). After studying this chapter, you should be able to:

- discuss the importance of mathematics in modern society, including fundamental aspects of mathematical literacy and numeracy (APST 2.1)
- summarise the key features of some important theories about student learning (APST 1.1)
- examine some recent advances in the field of neuroscience and their implications for mathematics learning (APST 1.2)
- identify some ways in which teachers can promote student engagement in mathematics lessons (APST 4.1).

Introduction

The purpose of this chapter is to consider the nature of mathematics and its importance in the lives of twenty-first-century students, and to discuss some key learning theories. The chapter begins with a description of some of the important knowledge, skills and understandings that today's students need in order to participate fully in modern society – we consider what it means to be mathematically literate in today's world. In doing so, we emphasise the importance of helping students develop a basic understanding of statistics and financial mathematics, which are important life skills. Next, we discuss some of the significant developments in theories about how students learn mathematics and some implications of these theoretical approaches for your work as a teacher. We then consider how some recent advances in neuroscience can inform our understanding of mathematics learning. The chapter concludes with a look at how we might engage students more productively in mathematics lessons in order to cultivate positive attitudes to the subject among students and to improve their learning outcomes. At various points in the chapter, we discuss some relevant scholarship and research that may provide guidance for your work of teaching mathematics in secondary classrooms.

Mathematics for the twenty-first century

The Alice Springs (Mparntwe) Education Declaration (Council of Australian Governments Education Council, 2019) is the framework document that sets out the national vision for education and the commitment to improve educational outcomes for all students. One of the central themes in the Declaration is the need to promote excellence and equity in education so that all students become 'confident and creative . . . successful lifelong learners [and] active and informed members of the community' (2019, p. 4).

We live in a world of rapid change and development, especially in terms of advances in technology. As information and communication technologies become more sophisticated, they exert an ever-increasing influence on how we conduct our daily lives. Many of these technological tools, such as databases and spreadsheets, are underpinned by mathematics (Siemon et al., 2011). Mathematical understanding is crucial in laying a strong foundation for study beyond secondary school in a range of disciplines, such as engineering, business and finance. In addition, developing mathematical understanding and becoming more numerate helps students to interpret a range of practical situations, allowing them to make more informed decisions in their everyday lives; for example, reading timetables, interpreting graphs and converting fractional amounts when cooking.

The Shape of the Australian Curriculum: Mathematics (National Curriculum Board, 2009) also notes the need for a mathematically literate workforce:

> Successful mathematics learning also provides a workforce that is appropriately educated in mathematics to contribute productively in an ever-changing global economy, with both rapid revolutions in technology and global and local social challenges. An economy competing globally requires substantial numbers of proficient workers able to learn, adapt, create, interpret and analyse mathematical information. (p. 4)

Mathematics education plays a crucial role in driving national productivity and prosperity. School mathematics lays a foundation on which to build the mathematical and scientific literacy every citizen needs to thrive in an increasingly technology-dependent world. But what kinds of mathematical knowledge, skills and understandings do today's students require if they are to be full and active members of our modern, highly technological age? To answer this question, we first need to consider the very nature of mathematics and what it means to do mathematics.

The German-born, Dutch mathematician Hans Freudenthal developed an approach to teaching known as realistic mathematics education, which he based on learning mathematics by solving well-chosen and carefully sequenced problems taken from daily life. Freudenthal believed that by attempting to solve such problems, students would gradually develop and enrich their mathematical understanding. Although he died in 1990, his work has continued to be very influential in mathematics education reform, not only in The Netherlands, but throughout the world. He once described mathematics as:

> an activity of solving problems, of looking for problems, but it is also an activity of organizing a subject matter. This can be a matter from reality which has to be organized according to mathematical patterns if problems from reality have to be solved. It can also be a mathematical matter, new or old results, of your own or others, which have to be organized according to new ideas, to be better understood, in a broader context, or by an axiomatic approach. (Freudenthal, 1971, pp. 413–14)

As you read this quotation, and perhaps you need to do so more than once, note how Freudenthal begins with the many different kinds of activities people do in their everyday lives that involve mathematics. He also refers to more abstract mathematics, but in doing so he emphasises mathematics as a *process* of inquiry, particularly through his reference to finding and solving problems, taking a systematic approach in organising one's thinking about these problems, and making generalisations from their solutions to develop mathematical understanding. Another important feature of Freudenthal's description is the underlying assumption that mathematics is an essentially human endeavour because it is always contextualised as a social and cultural activity. From Freudenthal's point of view, mathematics education has its beginnings in the sociocultural world of each student and the essential aim of school mathematics is to strive to raise students' awareness of the essentially mathematical nature of this everyday human activity (Ryan & Williams, 2007).

If we take Freudenthal's description as our starting point, we can begin to identify some of the particular mathematical knowledge and skills that are important for today's students. These are discussed in the following sections.

Statistical literacy

Data are all around us and twenty-first-century students need considerable skill in order to comprehend the large quantity of information they are presented with each day through the media and other sources. Indeed, to be an active participant in our democratic society requires the ability to deal with ever-increasing amounts of data. Many occupations, such as actuaries, business analysts and meteorologists, also require an ever-greater appreciation of data in far more sophisticated ways than has previously been the case. These jobs require a sound grasp of statistics – a powerful tool that enables us to make sense of our data-driven world. We can use statistics to identify trends over time and make meaningful comparisons among and between data sets. We can also extrapolate from data in order to hypothesise and predict what might occur in the future for the purposes of planning and decision making.

Every day, people use numbers and comparisons of numbers to justify decisions they make about their lives. Statistics is a way of making sense of a number of factors or costs when comparing these activities and providing evidence to support our choices. In statistical literacy, students learn how to add rigour to numerical claims and comparisons and how to examine other people's claims using statistical evidence and reasoning. Statistical literacy is the ability to make sense of data; it is one's ability to understand, interpret and evaluate data from a variety of sources. Callingham and Watson (2017) developed a six-level hierarchy of statistical literacy. The highest level is critical mathematical, which is characterised by a critical and questioning mindset, using proportional reasoning to interpret statistics in the media, an acknowledgement of uncertainty in making predictions, and interpreting the subtleties of statistical language. As with many other areas of mathematics, we would also want our students to develop positive attitudes and confidence in their statistical abilities so that they can be informed and engaged participants in society.

Some of the activities that could be used to promote statistical literacy include the development of surveys and questionnaires and a discussion about how the wording and ordering of the items can influence responses. Students could consider if they need to survey the entire population (e.g. all of the students in a school) or if a sample would be sufficient. This could lead to discussions about the sample size and the need for the sample to be representative of the population. Raw data often needs to be 'cleaned up' by, for example, disregarding anomalous results. In doing so, students could consider the impact of outliers on measures such as the mean of a set of scores. When the data are collected, decisions need to be made about how it could be represented (in tabular form or graphically). This could lead to discussion about the kinds of graphs that are best suited to representing data that are categorical, numerically discrete or continuous. Then the data must be analysed so that trends or other patterns can be identified. When dealing with data from other sources, students need to develop a healthy scepticism for claims made by considering how the data were collected and by whom, and for what purpose. This could lead to a discussion of biased or misleading data.

Statistical literacy: the ability to evaluate, understand and interpret data from a variety of sources.

PAUSE AND THINK

Why is it important for all students to develop statistical literacy?

Financial literacy

Financial literacy: the ability to manage money and financial risks effectively and responsibly in order to achieve one's financial goals.

Another important aspect of mathematics in the twenty-first century is the area of financial literacy. Financial literacy is significant because the ability to manage personal finances 'is a core skill in today's world. It affects quality of life, the opportunities individuals and families can pursue, their sense of security and the overall economic health of Australian society' (National Consumer and Financial Literacy Framework, 2011, p. 5). A report by the Australian Securities & Investments Commission (ASIC) defines financial literacy as 'being able to understand and negotiate the financial landscape, manage money and financial risks effectively and responsibly, and pursue and attain financial and lifestyle goals' (ASIC, 2014, p. 6). Financial literacy is therefore seen as a core life skill for twenty-first-century students so that they can fully participate in modern society. The increasingly complex nature of the world today and the greater number of alternatives for dealing with discretionary income mean that students need to develop the knowledge and skills to take charge of their own financial future. Helping students understand financial issues is particularly important nowadays since they can be expected to deal with more and more sophisticated financial products and services. Levels of financial risk are also rising, especially in areas such as managing savings and investments, purchasing a home, planning for retirement and ensuring the ability to pay for future healthcare needs (Programme for International Student Assessment [PISA], 2019).

Surveys and assessment data show that young adults typically have very low levels of financial literacy. For example, in 2012 PISA conducted the first financial literacy assessment of approximately 29 000 secondary school students aged 15 years. The survey was conducted in 18 countries and covered financial issues such as understanding a bank statement, calculating the long-term cost of a loan and knowing how insurance works. The results, which were released in July 2014, showed no significant difference between the performance of boys and girls, but they did indicate that students from relatively high socioeconomic backgrounds tended to do better in financial literacy than less advantaged students (PISA, 2012). Importantly, the results also revealed that skills in mathematics and reading were very closely related to financial literacy (PISA, 2012).

The range of skills that students need to develop in order to make appropriate financial decisions include adaptability, initiative, communication, problem solving, planning and organising, analysing issues and managing identified risks. Many of these skills are closely aligned with some of the central themes of the Australian

Figure 1.1 Financial literacy is an important life skill

Curriculum: Mathematics (Australian Curriculum, Assessment and Reporting Authority [ACARA], 2018). This is especially the case with the Proficiency Strands of the Australian Curriculum: Mathematics.

Numeracy

Today's students must achieve basic numeracy skills such as the ability to count, measure, compare and sequence. Students need to develop competence in many everyday tasks, such as budgeting, shopping, travel and leisure activities. In some respects, these skills can be learned and practised through simple exercises such as those typically found in school textbooks. However, most mathematics educators argue that to be fully numerate, students must do more than acquire and master basic mathematical routines and algorithms. These skills are necessary and essential, but not sufficient for numeracy (Australian Association of Mathematics Teachers [AAMT], 1997).

The numeracy demands placed on twenty-first-century learners cover much more than just number skills and the ability to recall basic facts such as the multiplication table (Goos et al., 2019). Numeracy involves a deep understanding of mathematical concepts and skills from across the discipline, including numerical, spatial, graphical, statistical and algebraic topics. Students need to develop their mathematical thinking strategies as well as more general thinking skills through problem-solving activities which are grounded in realistic contexts. Thelma Perso is a leading Australian mathematics education researcher with a particular interest in numeracy development and promoting literacy and numeracy for Indigenous students. She identified the following as fundamental characteristics of students who are numerate:

- knowing how to read situations and determine if or what mathematics is needed
- knowing how to choose the methods and tools they will use and explain and justify their reasons based on context
- knowing how to apply the methods, models and tools they have chosen
- knowing how to critique their own (and others') methods and determine whether they make sense in context
- knowing how to communicate their process and results appropriately depending on purpose and audience (Perso, 2011).

The 'Australia's children' report (Australian Institute of Health and Welfare, 2020) notes that numeracy is a key foundation for children's educational development and their capacity to engage fully in life beyond the classroom. The report also identifies the importance of numeracy for all students, particularly those from regional and remote schools, Indigenous children and students with a language background other than English (LBOTE). An example of a numeracy intervention is the QuickSmart program developed by John Pegg and colleagues in the National Centre of Science, ICT, and Mathematics Education for Regional and Rural Australia (SiMERR) at the University of New England. QuickSmart draws on research evidence (e.g. Cumming & Elkins, 2010) that students need to become fluent in recalling basic number facts, referred to as automaticity, in order to learn mathematics effectively. The intervention uses practice activities such as flash cards to promote automaticity and aims to improve students' mathematical understanding through pattern recognition and investigating number operation relationships. An evaluation study

Numeracy: the ability to understand and work flexibly and efficiently with numbers and other mathematical concepts to solve a variety of problems across a range of contexts.

(Drew et al., 2019) found a small increase in student achievement from QuickSmart, particularly in the primary years.

To be numerate, students must not only develop disciplinary knowledge of basic mathematical processes and skills but also establish a deep conceptual understanding, which will enable them to engage in problem solving and critical analysis, apply their knowledge creatively in unfamiliar contexts and communicate their ideas using appropriate mathematical language and symbols (Office of the Chief Scientist, 2014). Merrilyn Goos, Shelley Dole and Vince Geiger (2012) are Australian researchers who developed a model for numeracy in the twenty-first century. The model includes five dimensions:

1. *Contexts* are real-world situations where numeracy skills are applied
2. *Mathematical knowledge* highlights the mathematical basis for numeracy, particularly in problem solving and estimation
3. *Dispositions* include confidence and a willingness to use mathematical knowledge flexibly and adaptively
4. *Tools* can be physical, such as rulers and models, representational, such as graphs and symbols, and digital, such as computers and the internet
5. *Critical orientation* emphasises thinking skills, such as analysis and evaluation, and the importance of justifying arguments to support an argument or challenge a position.

Numeracy is also one of the general capabilities in the Australian Curriculum and hence the responsibility of all teachers. It is likely that some of your colleagues from other subject areas might seek your advice as a mathematics teacher on how best to incorporate numeracy into their lessons.

PAUSE AND THINK

Do you see yourself as a teacher of numeracy as well as a mathematics teacher? How can you ensure that your students develop sound personal numeracy skills?

SCENARIO   The importance of context

You have been asked to prepare a revision lesson on percentages for a mixed-ability Year 8 class. You initially consider a lesson in which you assign some exercises from the textbook for the students including multiple practice questions such as 'find 20% of 35' and 'increase 250 by 15%' using their calculators.

Instead, you decide to contextualise the revision lesson by focusing on percentage discounts, profit and loss, and the goods and services tax (GST). You also include questions where students choose and use mental, written and calculator methods.

Questions

1. Why is it important to contextualise the activity using relevant examples?
2. Why is it important to emphasise mental computation and estimation?
3. How does the revised lesson promote financial literacy?

SHORT-ANSWER QUESTIONS

1. What do you see as the most important mathematical skills for secondary students to be active participants in society today?
2. How could you, as a mathematics teacher, support your colleagues from other subject areas to implement numeracy in their lessons?

Theories about how students learn mathematics

In general, the common thread running through all learning theories is that they are based on the premise that learning involves some kind of change. However, different theoretical approaches to learning vary in quite fundamental ways in describing the nature of this change and how it might be identified or measured. This section of the chapter focuses on learning theories that have had a significant impact on the practice of mathematics education in recent decades. As a teacher, it is important for you to know and understand these learning theories since, 'For teaching to be effective, it must be grounded in what we know about how students learn' (Goos, Stillman & Vale, 2017, p. 28). We begin our discussion with behaviourism.

Behaviourism

Behaviourism was the archetype approach used in educational psychology from the 1920s until the 1950s. For behaviourists, learning is evidenced in changes in a person's behaviour, and so behaviourists are concerned with actions that are observable and measurable through empirical data. In general, behaviourist theories suggest that learners are shaped by a range of environmental influences or stimuli to which they respond (Siemon et al., 2011). Behaviourism has its origins in observational studies of animals in laboratory settings. These studies showed that animal behaviour could be influenced through reward and punishment to reinforce desirable actions. For example, rats and pigeons could be trained to perform certain actions, such as pressing levers in order to receive food, or to avoid negative consequences, such as mild electric shocks. Later studies showed that behavioural reinforcement could accelerate learning for adults and children as well.

Behaviourism: a theory of learning that is mainly concerned with observable behaviour and assumes that learners respond passively to external stimuli.

One of the early proponents of behaviourism was Edward Thorndike. He developed a theory of operant conditioning, which is sometimes referred to as instrumental learning. Thorndike developed his 'law of effect' in which he postulated that an association could be strengthened or weakened as a result of its consequences; behaviours which led to positive outcomes were more likely to be repeated than those which produced disagreeable results. In other words, using positive outcomes to promote desirable behaviours is a much more effective strategy than punishment for unwanted actions. Thorndike's work also examined the bonds that can be formed between a stimulus and its response. He demonstrated that when an association is established between a stimulus and its response and this stimulus–response association is repeated, the response is likely to continue to occur even after the stimulus is reduced or withdrawn.

Thorndike (1922) believed that elementary stimulus–response associations were the foundation of learning and applied his theory to the learning and teaching of arithmetic through his use of number bonds. He identified 390 such bonds which were based on the four operations of arithmetic and the numerals 0 to 9. He based this work on three key assumptions: first, that learning is essentially the construction of bonds or connections between situations and responses to them; second, that habitual drill and practice leads to improved memory and better learning outcomes; and third, that such improvement could be best measured by standardised testing.

Later, B. F. Skinner built on Thorndike's work to propose the notion of operant conditioning, in which behaviour is affected and altered by the events that precede and follow it. In collaboration with Charles Ferster, Skinner also introduced a new idea into the law of effect – the concept of reinforcement (Ferster & Skinner, 1957). They suggested that behaviours which are continually reinforced are more likely to be strengthened and repeated, while those which are not reinforced tend to be weakened and diminish over time. In essence, Skinner conceptualised the learning of mathematics as the formation and strengthening of associations between stimuli and responses to them. He focused particularly on observable skills since he believed that learning was evidenced in terms of changes in behaviour.

The work of behaviourists such as Thorndike and Skinner has had a profound impact on mathematics education. Although behaviourist methods are rarely advocated nowadays, it is still possible to see their influence in many mathematics classrooms and syllabuses. The hold of behaviourism is evident in the ways that student learning outcomes are often phrased in syllabus documents, the emphasis on rote learning, the abundance of graded examples and exercises like those often found in school textbooks, a narrow focus on basic skills in assessment tasks, and an approach to teaching firmly grounded in explicit instruction (Ryan & Williams, 2007). As Brumbaugh and Rock (2001) note, 'The behaviourist approach treats mathematics as a collection of skills. Learn the skills and learn mathematics' (p. 27).

Another legacy of behaviourism can be seen in the work of Robert Gagné and his hierarchy of learning. Gagné's (1965) model classifies various types of learning activities according to the complexity of the mental processes involved in completing them. The higher levels, such as concept learning and problem solving, deal mainly with more cognitive aspects of learning, while the lower-level learning tasks, such as stimulus–

response learning, are related to more behavioural features. The hierarchical nature of the model is reflected in Gagné's assertion that higher orders of learning require increasingly greater amounts of prior learning at the lower levels. Gagné proposed that students could achieve learning outcomes by following a predetermined sequence of tasks, each one building on its predecessor. Students could master the tasks at each subordinate level through repetition and practice and then proceed to the next stage in the learning continuum.

Gagné argued that particular kinds of teaching methods were also crucial to ensure students' mastery of prerequisite skills. These included clear objectives for lessons, regular reminders to students of previously learned material and guided instruction. The teacher should then demonstrate worked examples prior to students completing exercises that were very similar to those modelled. Regular feedback on students' progress towards the learning goal is also provided, along with systematic reinforcement of correct procedures. Gagné's learning hierarchies were often represented as flow charts which provided teachers and students with a step-by-step program of graded tasks and activities leading to the desired learning outcome. This approach to learning and teaching is reductionist in style because it attempts to reduce learning to a series of discrete tasks and focuses primarily on the mathematical content to be learned rather than the processes by which learning takes place.

Learning outcomes

Your choice of verbs is crucial when writing student learning outcomes for your lessons. It is generally better to avoid verbs like 'understand' or 'appreciate' since they are subjective and difficult to assess. Instead, follow the syllabus outcomes by using *behaviourist verbs* that relate to observable and measurable tasks such as 'solve', 'calculate' and 'draw'.

PAUSE AND THINK

What aspects of behaviourist learning theories could you usefully adopt in your teaching?

Gestaltism

Gestaltism takes its name from a German word meaning shape or form. It is based on the belief that the mind tends to perceive the whole form (the gestalt) first and only afterwards are the constituent parts analysed. Gestaltism can therefore be seen as in opposition to the reductionist approach taken by behaviourists such as Gagné, who emphasised a step-by-step approach to learning and teaching. For the gestaltists, learning was far more complex in nature and the sensory whole was always viewed as other than just the sum of its sensory parts. The gestalt theorists were especially critical of the rote learning methods espoused by

Gestaltism: a theory that suggests that the mind processes wholes first, before attending to constituent parts.

behaviourists and argued against the behaviourists' insistence on drill and practice as the most efficient method to learn mathematics. The gestaltists based their views on the potentially harmful effects of stimulus–response approaches following experiments which suggested that students who had learned by rote often struggled to solve problems which were presented in unfamiliar settings or varied even slightly from those which they had practised.

An example of one such experiment comes from the work of Max Wertheimer (as cited in Schoenfeld, 1988). Wertheimer asked children in primary school grades to solve arithmetic problems such as:

$$\frac{357 + 357 + 357}{3}$$

He noticed that many children promptly began the somewhat laborious task of calculating the sum of the numerator and then performing the division by the denominator. Those children who were competent in the arithmetic operations were able to solve the task correctly; however, they often did so without apparently noticing, even after obtaining the right answer, how the structure of the problem might have been useful in saving all that time and effort. Wertheimer concluded that although these students had mastered the procedures required for addition and division, they had failed to recognise the repeated addition as equivalent to multiplication and division as the inverse of multiplication. They seemed to be so conditioned to focus on the individual parts of the problem that they failed to interpret it as a whole. In other words, the majority of children had clearly not developed a deeper understanding of the underlying arithmetical structure. As Schoenfeld (1988, p. 148) notes:

> This example illustrates that being able to perform the appropriate algorithmic proced-ures, although important, does not necessarily indicate any depth of understanding.

Schoenfeld then proceeds to discuss a more widely known example from Wertheimer's work which concerns his observations of some mathematics lessons in which students were learning about the area of a parallelogram. As in the previous case from arithmetic, the students had been taught the standard procedure – in this case, how to apply the formula 'area equals base length multiplied by perpendicular height'. Wertheimer observed the teacher illustrate the formula by cutting away a triangular section from one side of the parallelogram and repositioning it on the other edge to produce a rectangle whose base and altitude were equal to that of the original figure (see Figure 1.2). He noted how the students were able to reproduce the teacher's method and explain it adequately during the class. However, when Wertheimer later asked some of the students to calculate the areas of parallelograms given in non-standard orien-tations, the students quickly became confused and were unable to provide the correct answers. Again, Wertheimer deduced that even though the students had memorised the proof and correctly followed the procedures they had learned to complete exercises during the lesson, their lack of appreciation as to how or why their

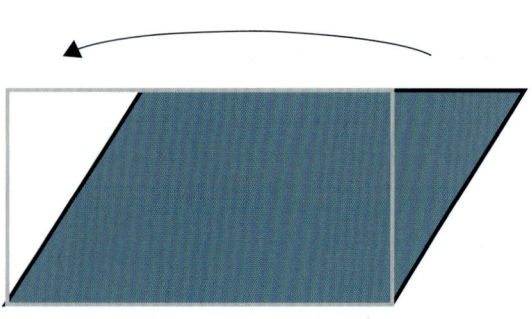

Figure 1.2 Transforming a rectangle into a parallelogram

methods worked exposed their lack of conceptual understanding. Instead, their superficial understanding limited what they could do, especially when confronted with tasks that did not closely conform to those they had previously learned to solve.

Richard Skemp's often-cited distinction between instrumental and relational understanding in mathematics learning is apt here. Skemp defined the former as 'rules without reasons' and the latter as 'knowing both what to do and why' (Skemp, 1976, p. 20). He also listed some advantages for each type of understanding. For instrumental understanding, Skemp noted that it is usually easier to understand, the rewards are more immediate and correct answers can often be obtained more quickly. As Skemp (1976, p. 23) notes: 'If what is wanted is a page of right answers, instrumental mathematics can provide this more quickly and easily.'

However, as Wertheimer's observations remind us, simply being able to obtain the right answer is not always sufficient. Skemp identified the advantage of relational understanding as its adaptability to new tasks: relational understanding is easier to remember, it can provide intrinsic motivation for learners, and it is organic in nature (by which Skemp meant that the development of relational understanding in one area of mathematics encourages a search for similar levels of understanding in others). On balance, Skemp believed that the long-term advantages of relational learning far outweigh any immediate benefits associated with instrumental understanding.

Skemp also acknowledged that despite the obvious advantages of a relational approach to learning and teaching mathematics, instrumental learning remains prevalent in the majority of mathematics classrooms and he provides some important reasons why this might be the case. These include: first, the powerful influence of high-stakes external examinations (which tend to focus on testing understanding at a more instrumental level of recall of basic rules and procedures); second, overcrowded mathematics curricula (which discourage teachers from allocating the time that is often necessary, at least in the initial stages, to teach for relational understanding); third, the difficulties associated with assessing students' relational understanding (which, to be valid and reliable, typically requires interviewing individual pupils about their work); and fourth, the challenges associated with encouraging teachers to discard long-held views about the nature of mathematics and mathematics instruction (which can be difficult to shift).

PAUSE AND THINK

How can you promote students' relational understanding of mathematical concepts in your teaching?

Constructivism

Skemp's advice on the long-term benefits of a relational understanding of mathematical concepts fits nicely within a cognitivist outlook on learning. Cognitivism is the paradigm in psychology and learning sciences that overtook behaviourism in the 1960s during the so-called Cognitive Revolution (for example, see Miller, 2003). One of the pillars of cognitivism is the notion that knowledge is not simply received by learners but rather that it is actively constructed by them. While there are now many forms of constructivism, they all share this fundamental premise (Bobis, Mulligan & Lowrie, 2013).

Constructivism: a cognitivist theory of learning that views the learner as an active constructor of knowledge.

Constructivism regards learning as a cognitive process and is derived principally from the work of Jean Piaget, who viewed learners as individuals who interact with their environment in order to make sense of it. Indeed, constructivism sees the learner as the central agent in the process of knowledge construction, through assimilating one's experience in the world (Ryan & Williams, 2007). Assimilation is a process by which we interpret input from the world around us through our senses. This information is interpreted in terms of our existing cognitive structures or schemata (Siemon et al., 2011). For constructivists, to understand a concept is essentially to assimilate it into an appropriate schema.

At times, individuals may have an experience which cannot be adequately explained by their existing knowledge schemata. In these situations, a conflict or perturbation may arise which is unable to be resolved simply through assimilation of the new knowledge or experience (von Glasersfeld, 1995). Constructivists believe that the desire to maintain an equilibrium – by which we continue to make sense of our experiences – leads to an accommodation or structural reorganisation of existing schemas and the construction of new ones as well (Peled & Suzan, 2011). As Davis and Tall (2002) remark, the assimilation and accommodation processes are very much dependent on the experience of each individual:

> Essential to an understanding of schemes is the focus of attention of a learner. What it is a person focuses on in an action scheme determines the consequent structure of that scheme for them. (p. 133)

Ben-Hur (2006) suggests that there are some fundamental characteristics of the structural changes in schemas that arise as a result of accommodation. The schematic restructuring is *pervasive*, since profoundly new experiences that require accommodation lead to rearrangements of the entire cognitive structure. Accommodation promotes *centrality* because when new learning brings about a structural change in cognitive structures it places the learner in the position to learn something more advanced in the same area. And the reorganisation is regarded as *permanent* because the kind of structural changes that result from the need to accommodate new experiences are neither reversible nor can they be forgotten.

Given that new learning is seen to arise as a consequence of the individual's desire to maintain equilibrium, constructivism proposes that teachers deliberately create situations that are likely to engender a cognitive conflict between the student's current way of thinking and the new knowledge to be learned. This approach is often referred to as conflict teaching and it can produce significant moments in the learning process, as it is designed 'both as a source for conflict and instability, and at the same time as an opportunity to reflect, reorganize one's knowledge and construct new knowledge' (Peled & Suzan, 2011, p. 75). For example, it has been observed that students who learn instrumentally are likely to over-generalise rules, formulas or procedures and use them in contexts where they do not apply. On such occasions, the use of non-examples can be a profoundly important means of generating disequilibrium. If the teacher can encourage students to make sense of the unexpected situation for themselves, exploring those non-examples may lead to deeper insights into the underlying structure of the problem.

Wertheimer's use of parallelograms in non-standard positions referred to previously is a case in point. Students who did not recognise that the area formula requires that the height of the parallelogram is perpendicular to its base could be helped to understand the formula by attempting to find the areas of non-standard figures and explaining their results.

Constructivists view the learner as an active participant in the learning process because they hold that knowledge is not passively received but instead is actively constructed. The role of the teacher also changes from that of transmitting knowledge to students (who are viewed as a 'blank slate' or 'tabula rasa') to facilitating knowledge construction by the students themselves. Therefore, the teacher's task in a lesson is to select meaningful learning experiences, to encourage students to engage productively in the learning activities and to guide students to develop their understanding of fundamental concepts.

A key aspect of constructivism is the notion that learning is an active process. Jerome Bruner, a leading constructivist theorist, proposed that lessons should provide experiences and contexts that promote engagement – a readiness to learn. He recommended that instruction should be designed to encourage student exploration of concepts, and he introduced a way of organising content in a spiral curriculum (Bruner, 1960) where complex ideas are introduced in a simplified form and then revisited at more complex levels – an idea that is evident in the structure of the Australian Curriculum. Bruner also conceived three modes of representation, based on actions (enactive representation), images (iconic representation) and language (symbolic representation). Bruner suggested that learning should progress sequentially through these three modes, again emphasising his insistence on careful sequencing of learning activities and materials with the aim of helping learners develop symbolic thinking.

Another important aspect of constructivist approaches to learning is the need for students to reflect on what they have done. Reflection allows students to make sense of their learning and abstract the underlying properties of a mathematical concept. Hence reflection is regarded as an essential step in triggering the process of accommodation (Boaler, 2000). Piaget makes an important distinction in his work between what he refers to as *empirical abstraction* and *reflective abstraction*. The former is based on our experiences and observations, which allow us to derive the common properties of objects and make generalisations from specific instances or occurrences. For example, a child is likely to have often experienced fair sharing, where each person receives an equal part of the whole. From many such experiences in a variety of different contexts the child is able to empirically abstract the concept of 'half' and then generalise its meaning in new and unfamiliar situations.

Reflective abstraction, on the other hand, is internalised and leads to different kinds of generalisations. These generalisations are more concerned with interrelationships among actions and allow new meanings to be constructed. For example, when learning the commutative property of addition (that the total is independent of the order in which the numbers are added together), a child might work with concrete materials to count and rearrange a group of objects several times. Each of these physical actions is represented internally in such a way that the child can reflect on them and realise that they give the same result each time. These reflections then lead to an abstraction of the concept of commutativity.

Piaget (1972, p. 70) refers to the *encapsulation* of a process as a mental object in such a way that:

> the whole of mathematics may therefore be thought of in terms of the construction of structures, . . . mathematical entities move from one level to another; an operation on such 'entities' becomes in its turn an object of the theory, and this process is repeated until we reach structures that are alternately structuring or being structured by 'stronger' structures.

Reflective abstraction, therefore, is activating the restructuring of one's knowledge schema/ models such that processes that were implicit at one level become the objects on which the learner acts in the next (Boaler, 2000). In a similar vein, Sfard (1991) uses the term *reification* to describe 'a sudden ability to see something familiar in a totally new light . . . an instantaneous quantum leap: a process solidifies into object, into a static structure' (p. 20).

Why, then, is this restructuring of processes as objects so important? The answer lies in the fact that when mathematical ideas and processes are reified as objects they can be stored in long-term memory in a compressed form and promptly reactivated when required. The results of research by Gray (1991) may help to illustrate this important point. In the study, Gray asked teachers to select children whom they identified as above average, average and below average in terms of their ability to perform simple arithmetic. He interviewed 72 children between the ages of seven and 12 from two schools and asked the children to solve simple addition and subtraction questions while he recorded the methods they used. He identified four main strategies which the children used: counting all (where all of the numbers are counted), counting on (where the child starts counting from one of the given numbers), using known facts (results which the child has committed to memory) and using derived facts (a flexibility in using the numbers to make the process of calculating easier).

The results of Gray's study showed clearly that lower performing children used complicated methods, such as counting all of the numbers from one or counting on from the smaller number. These methods involved additional cognitive load and were more likely to result in errors. Higher performing children, on the other hand, made greater use of derived facts and employed more efficient strategies. Gray concluded that these two groups were, in fact, doing a different kind of mathematics. High performers were able to retrieve a range of facts and procedures with ease and adapt them to derive new facts, which could then be accumulated into their memory and added to their stored repertoire. Low performers struggled with memory-intensive procedures that constrained their ability to move beyond the most laborious and error-prone of methods. In other words, the process of reflective abstraction enabled the higher performing students to develop a relational understanding of the mathematics.

Class discussions

A strategy based on constructivist learning principles that you can use to promote whole class discussions in your lessons is *think-ink-pair-share*. This is where students take time to quietly think on their own about a question and write down their ideas before talking about them with a partner. As they chat, you can move among the pairs and note any important mathematical ideas that emerge. Then you can call the class together and invite those students to share their insights during the discussion.

PAUSE AND THINK

What are some practical strategies that you can use in your lessons to ensure that students are active participants in their learning?

Socioculturalism

If constructivists view learning as a process of knowledge construction in the mind of the individual, sociocultural theorists regard learning as 'a collective process of enculturation into the practices of the mathematical community' (Goos, Stillman & Vale, 2017, p. 28). Learning therefore leads to changes in participation and identity (Lave & Wenger, 1991) as students grow in confidence and improve their ability to make sense of mathematical ideas. According to Sullivan, Clarke and Clarke (2013, p. 9), the social constructivist view of knowledge is:

> not fixed, rather it is socially negotiated, and is sought and expressed through language. In other words, students do not learn only by listening but also by engaging in experiences that contribute to their learning, and having the opportunity to share approaches to tasks with others.

Much of socioculturalism's approach to learning builds on the work of Vygotsky. He proposed that knowledge construction develops from social interactions and that the context in which these interactions take place acts as a mediator for the learning that results. Vygotsky's belief that social interaction preceded development is in contrast to the constructivist view proposed by Piaget, who regarded development as a necessary precursor for learning. The importance of mediated learning is summarised in Vygotsky's concept of *internalisation*, which he described as 'internal reconstruction of an external operation' (Vygotsky, 1978, p. 56). That is, learning first occurs on the social plane between individuals, and then within individuals on the psychological plane.

Crucial to Vygotsky's insistence on the primacy of social interaction is what he termed the *zone of proximal development*, or ZPD, which Vygotsky (1978, p. 86) described as:

Socioculturalism: a view of learning that emphasises how learning does not take place solely within the mind of the individual, but also through one's interactions with others.

the distance between the actual developmental level as determined by independent problem solving and the level of potential development as determined through problem solving under adult guidance or in collaboration with more capable peers.

The ZPD therefore describes the region between what students can do independently and that which can be done only with the assistance of others. Hence the ZPD is the realm of knowledge and skills which have not yet become apparent but could emerge as the student interacts with more knowledgeable peers, parents, texts or teachers (Lerman, 2000). In this respect, Vygotsky believed that teachers should pitch their instruction within the ZPD rather than at the level of the student's current level of independent performance. By carefully selecting activities beyond what the student can do on his/her own, the teacher would therefore assist in promoting knowledge growth to produce developmental gains.

The ZPD has other implications for mathematics teaching, particularly in terms of the provision of *scaffolding* by which social support from the teacher or the peer group is used to promote new learning. Scaffolding enables students to explore new concepts in a more sophisticated manner and more efficiently than if they were attempting to learn them on their own. It is especially important in the initial stages of learning about a new concept or process and can be gradually withdrawn as the student becomes more skilled in the new activity (Ryan & Williams, 2007). Scaffolding can take many forms in mathematics lessons, such as:

- using assessment *for* learning to check prerequisite knowledge and understanding
- allowing students to work collaboratively in small groups
- illustrating mathematical concepts using examples and non-examples
- presenting new material incrementally in small, manageable steps
- ensuring that new knowledge and skills progress from the simple to the more complex
- providing regular feedback on students' work
- questioning by the teacher
- using visual aids
- highlighting key information and using procedural prompts.

In recent years, there has been a growing emphasis on the social and cultural aspects of mathematics education and how these can potentially influence the learning and teaching of mathematics in the classroom. In particular, researchers have increasingly considered issues related to equity in mathematics education and focused their attention on the relative under-performance of disadvantaged groups (Jorgensen, 2010). A sociocultural perspective on learning views the classroom as an environment in which students and teachers from a range of cultural and social settings come together for the purpose of learning and teaching mathematics. Each individual brings their own considerable knowledge and prior experiences shaped by personal social and cultural backgrounds. This is significant for the teacher, who must consider how the classroom norms and practices may impact on students from a diverse range of circumstances.

SCENARIO

What does 'understanding' look like in mathematics?

A preservice teacher taught a lesson on linear equations to a top-streamed class of high-achieving Year 9 students. The lesson began with a worked example, $3x - 5 = 7x + 3$, on the board for students to copy into their books, followed by an exercise of similar equations from the textbook. The preservice teacher moved around the room while students were working and noticed that all students were completing the activity and correctly solving the equations.

In the self-evaluation of the lesson, the preservice teacher noted that 'students correctly completed the textbook exercises and have learned to solve linear equations well'.

Next lesson, following the supervising teacher's suggestion, the preservice teacher asked students to solve the equation $3x + 2 = 2 + 3x$. All of the students immediately attempted to solve the equation using the procedure they had learned in the previous lesson, reducing the equation to $0 = 0$, and concluding that its solution was $x = 0$. One student later said that 'x could be anything' but he was not able to explain why.

Questions

1. Why did the preservice teacher believe that the students had learned to solve linear equations well?
2. Do you agree with the preservice teacher's assessment? Explain your reasoning.
3. How could some of the ideas discussed in this chapter help the preservice teacher to make sense of this incident?
4. What are the implications of this incident for your own classroom practice?

Neuroscience

Neuroscience is a relatively new area of scientific endeavour which involves a study of the brain and the nervous system using brain imaging techniques. Most commonly, this involves functional magnetic resonance imaging (fMRI) to detect changes in blood flow as a measure of brain activity. An fMRI machine measures which parts of the brain are active when a person is thinking particular thoughts or performing certain actions so scientists can determine how the brain functions during these different activities. However, fMRI must be conducted in rather restricted and unnatural settings due to the equipment that is used. Other brain imaging techniques are less invasive and allow for data to be collected via sensors within a helmet or cap (de Smedt et al., 2011). These methods include electro-encephalography to record electrical activity in the brain and near-infrared spectroscopy, which uses infrared light to measure blood haemoglobin levels.

In education studies, neuroscientific research typically seeks to describe cognition as a set of tasks and then map these to individual parts of the brain. The aim of these 'area-focused' studies is to identify the particular brain area that activates most selectively for each task (Varma & Schwartz, 2008). Applications of neuroscience to mathematics education have focused mainly within the topic of early number learning, though some

Neuroscience: an interdisciplinary field which includes biology, psychology and computer science, and is the study of the brain and the nervous system.

researchers have also examined rational number arithmetic (e.g. Obersteiner & Tumpek, 2016).

Neuroscience has been used to investigate the condition of dyscalculia, a learning disorder associated with difficulty in learning arithmetic and performing basic number calculations. Kaufmann (2008) conducted an fMRI study of how eight-year-old children and young adults compared numbers. The researcher showed the participants pictures of two hands representing different finger patterns and asked them to indicate which hand displayed more fingers by pressing a button. Results indicated that both groups (children and young adults) performed well, achieving 99.3 per cent correct response rate. Interestingly, the brain activation patterns of the two groups were quite different, indicating that these patterns are not necessarily identical across different age groups. The author suggested that teachers take advantage of the fact that fingers may serve as concrete representations of number magnitude and the acquisition of calculation skills.

The attention to early number learning is based on the premise that children's abilities in basic aspects of number work, such as whole number arithmetic and number magnitude, are predictive for mathematical achievement in later years (Verschaffel, Lehtinen & van Dooren, 2016). However, a more generally accepted view of mathematics, such as that described in the Australian Curriculum: Mathematics, as a search for patterns and relationships suggests that the focus on basic number skills should be broadened to encompass more advanced mathematics.

Delazer et al. (2005) studied the effects of memorisation compared to learning a computation algorithm. The first group memorised the answers to problems but did not learn how to use the operation, while the other group was taught the algorithm. Both groups were then scanned using fMRI imaging as they solved familiar and unfamiliar problems related to the original tasks they had done. Results from the study showed that participants in each group activated different brain areas to perform the tasks. The brain network used by participants in the learning group supported transfer to novel problems with 78 per cent accuracy, but the network organised by participants in the memorisation group was not as successful, with just 15 per cent accuracy.

Another area that has received particular attention in neuroscientific studies is mathematics anxiety, a negative emotional reaction to doing mathematics that induces feelings of apprehension and fear. Neuroimaging studies have shown increased activity in parts of the brain associated with the detection and experience of pain for people who have high levels of mathematics anxiety (Lyons & Beilock, 2012). Other neuroimaging studies, such as Bishop (2007), have reported decreased activity in one area of the brain (the amygdala) alongside increased activity in another (the prefrontal cortex) for people who successfully overcome their mathematics anxiety. The contribution of such studies is that they suggest there could be value in teaching emotional self-regulation techniques to students who experience mathematics anxiety (Buckley et al., 2016).

There has been some critique of neuroscience as a research methodology applied to the study of mathematics learning. De Smedt et al. (2011) note that neuroscientific studies often average their results over multiple scans and for several individuals, thus failing to detect individual differences. The authors also highlight the limited generalisability of neuroscience studies to classroom contexts. This is mainly due to the reliance on invasive techniques, such as fMRI, since brain activity observed in the magnetic resonance imaging may not occur in the same way when the student is sitting at a desk during a lesson. There is also the issue of compartmentalisation in 'area-focused' neuroscientific studies – most mathematics educators discourage a view of the discipline as a set of seemingly unrelated domains and prefer to emphasise the relational aspects of mathematics learning.

The application of neuroscience to educational research is still a relatively young field and, despite criticisms such as those outlined above, the increasing number of interdisciplinary studies from neuroscience and mathematics education researchers suggests that more fruitful results might be discovered in the future.

SHORT-ANSWER QUESTIONS

1. What are the main principles of behaviourism?
2. What are the main principles of socioculturalism?
3. What do you see as the main contributions of neuroscience to mathematics education?

Student engagement in learning mathematics

Engagement typically refers to how actively involved students are in their learning. Engagement is an important consideration for mathematics teachers since the quality and the intensity of student involvement in classroom activities can have a profound impact on the rate at which they learn, the depth of understanding they develop and their ability to generalise new concepts so that they can apply them in unfamiliar contexts (Helme & Clarke, 2001). Stipek (1996) suggests that students who are actively engaged in their learning will be excited and enthusiastic when undertaking challenging tasks and be more willing to make a concerted effort to persist, even in the face of difficulties.

Engagement: the level of attentiveness, interest and curiosity that students exhibit when they are learning.

The notion of engagement is different to that of motivation, which is linked to students' self-efficacy (or how certain they are that they can perform specific tasks), their ability to persevere and overcome difficulties and their capacity to bounce back after experiencing a disappointment of some kind (Attard, 2012). Motivation is also related to one's attitudes to and beliefs about what is important, which can have a major impact on how willing a student is to engage in a particular task or activity.

Poor student engagement is a significant issue in Australian secondary mathematics classrooms (McPhan et al., 2008) and research suggests that many students are disengaging from mathematics (Sullivan, Tobias & McDonough, 2006). They are not opting to study higher levels of mathematics in the final years of high school (Kennedy, Lyons & Quinn, 2014), nor are they choosing to pursue their study of mathematics at the tertiary level (Office of the Chief Scientist, 2014). Boaler (2009) suggests that a major contributing factor leading to disengagement in mathematics classrooms is the mistaken belief that success in mathematics is a sign of general intelligence and that some people can do mathematics while others simply cannot. Boaler regards this belief as so pervasive and widely held that it is rarely mentioned, and she refers to it as the 'elephant in the classroom'.

Figure 1.3 All students are capable of 'doing' mathematics if they are engaged in their lessons

The incorrect notion that some students just cannot do mathematics is reminiscent of the work of Dweck (2000), who distinguished between two quite different perspectives on ability and intelligence. The first, which she called an *entity view*, treats intelligence as something which is fixed and stable. People who hold an entity view are likely to believe that their intelligence was predetermined at birth and will remain so throughout their lives. Dweck proposed that students who believe in the entity view require easy success in order to maintain motivation and they typically regard challenging work as a threat. These students are unlikely to remain engaged in mathematical problem-solving tasks, preferring instead a more repetitive, drill and practice style of instrumental learning where they can obtain correct answers quickly with a minimum of effort. The *incremental view*, on the other hand, regards intelligence as malleable and people who view intelligence in this way believe that they have some control to change their level of intelligence or achievement. Students with incremental beliefs gain satisfaction from learning something new and are not bothered that they might appear foolish in their initial attempts. These students enjoy a challenge and will persevere in a problem-solving activity as long as they feel they are gaining some valuable new insights or skills from doing so.

How students establish their personal learning goals can be linked to the ideas about intelligence. Since an entity view is concerned with looking smart, students with this orientation are concerned primarily with performance goals. They can easily lose confidence in themselves as soon as any obstacles arise and are susceptible to a kind of learned helplessness because they believe that they have little control over their work. So, they tend to avoid situations that they perceive to be challenging, perhaps by procrastinating or even denigrating the activity. In contrast, the incremental view is more concerned with learning new things and so these students are mainly interested in mastery or task goals. When faced with a challenge, they tend not to give up but rather to seek out alternative approaches and are likely to redouble their efforts until they can progress or achieve success.

The distinction between entity and incremental approaches can therefore lead to quite different levels of student engagement, so it is worth considering some of the implications of Dweck's ideas for teaching. For instance, performance goals might well be motivated in students whose teachers have tended to place too great an emphasis on their past achievements. These teachers may focus on performance but protect the student from negative feedback, perhaps by not discussing errors or simply reducing the requirements of the lesson activities. Dweck claimed that these teachers are likely to promote a belief in students that they can achieve success without effort and, though well intentioned, the practice of reducing the demands of tasks may simply serve to create a situation where difficult challenges are avoided rather than confronted.

Attard (2011) conducted a longitudinal case study of the engagement levels of 20 students over three years. She tracked the students from their final year of primary school into their first two years of secondary school. Her aim was to identify factors which encouraged or inhibited student engagement in mathematics lessons. Data for the study were drawn from student interviews, focus group discussions and classroom observations. A major finding from the study concerns the role of what Attard referred to as 'positive pedagogical relationships' between teachers and their students. Attard found that these relationships must be developed as a foundation for sustained engagement. Essential ingredients in positive pedagogical relationships include implementation of a range of teaching styles on a regular basis to ensure sufficient variety of learning tasks while still maintaining a sense of order and structure within lessons. Teachers should encourage active student participation in lessons by providing appropriate academic challenges. There also need to be opportunities for social interaction through activities such as group work. In addition, the relevance of the mathematics being studied must be emphasised so that students can appreciate how the work they are doing is applicable to their current and future lives. On the other hand, pedagogical practices can also serve to decrease student engagement; for example, if they are largely based around individual tasks that do not promote interaction and dialogue, or where tasks are perceived to have little or no relevance to students' lives. Attard also noted that while students are generally happy to work from textbooks, an over-reliance on these resources is counterproductive.

The Australian Association of Mathematics Teachers revised and republished its *Standards for Excellence in Teaching Mathematics in Australian Schools* (AAMT, 2006). The document describes some of the key knowledge, skills and dispositions, or positive attitudes, required for good teaching of mathematics. These are arranged in three domains: professional knowledge, professional attributes and professional practice. In terms of student engagement, the domain of professional practice is most relevant. The Standards suggest that teachers purposefully establish a classroom environment that encourages active student participation to maximise opportunities for student learning. Students are 'empowered to become independent learners … motivated to improve their understanding … and develop enthusiasm for, enjoyment of, and interest in mathematics' (AAMT, 2006).

When planning lessons, teachers should be mindful to include opportunities for self-directed learning of substantive mathematical content based on a variety of resources and teaching styles, including the use of a range of technological tools and devices where appropriate. Teachers should also draw on students' backgrounds and prior experiences to develop activities in which students can explore and apply mathematics in a range of contexts. According to the Standards, good teaching 'promotes, expects and supports creative thinking [and] mathematical risk-taking' (AAMT, 2006).

IN PRACTICE

Student autonomy

One way to motivate and engage students in learning mathematics is to provide them with some autonomy over how they complete their work. *Hint cards*, which contain enabling prompts or learning scaffolds, can be prepared before the lesson and made available to any students who choose to use them.

SHORT-ANSWER QUESTIONS

1. What are some indicators of student engagement in the mathematics classroom?
2. What are some ways that teachers can promote high levels of engagement for all students in their lessons?
3. How can you use the AAMT Standards for Excellence to guide your teaching practice?

Conclusion

The chapter began with an overview of some fundamental aspects of mathematical learning which are central to the lives of today's students and will allow them to fully participate in the nation's social and economic prosperity. In particular, we considered the importance of financial and statistical literacy for students. We also discussed some of the most significant theories about how learning takes place. We explored some recent advances in neuroscience and examined some implications of this work for helping students learn fundamental mathematics concepts and skills. Finally, we explored the notion of student engagement as it applies in mathematics lessons and we looked at some of the ways in which teachers can promote student engagement through their positive pedagogical relationships, which include implementing a range of teaching styles to provide sufficient variety of learning tasks for students while maintaining a degree of structure within lessons.

BRINGING IT TOGETHER

1. Look back over the definition of mathematics by Freudenthal at the start of this chapter. How would you describe the nature of mathematics?
2. What is meant by the terms 'financial literacy' and 'statistical literacy'? Why are these important skills for all students to develop?
3. What are some advantages of collaborative group work activities in mathematics lessons?

Use the following information to answer questions 4–6.

Minh is currently completing his final professional experience placement. Prior to starting at the school, Minh was learning about constructivism and the importance of engaging students in their learning of mathematics. Although he was taught in a fairly traditional way when he was studying mathematics in high school and at university, Minh has become enthusiastic about providing opportunities in his lessons for students to learn through problem-solving activities.

Minh's mentor teacher, Julie, tends to teach mainly from the textbook but she has allowed Minh to try out some more student-centred approaches in his lessons. Minh prepared a lesson for his Year 8 class where students work in groups to investigate the most cost-effective way to design a cylindrical drink can using the least amount of material. He designed a worksheet for the class and intended that the students would solve the problem numerically and graphically using a spreadsheet.

The lesson began well, and the students were interested in the problem. They were also keen to start the spreadsheet activity. However, things soon began to fall apart as students were unaccustomed to working in groups and wrote rather simplistic responses to the task. They lacked the ability to think about the problem for an extended time period and did not persevere in finding multiple solutions.

After the lesson, Julie's main feedback to Minh was that he should revert to using the textbook because the class was falling behind the other Year 8 classes.

4. What could Minh have done differently to help students become more accustomed to a teaching approach based on constructivist learning principles?
5. Do you agree with Julie that Minh should use the textbook as his main teaching resource from now on?
6. How do you think Minh should respond to Julie's feedback on his lesson?

REFERENCES

Further resources

Attard, C. (2011). 'If I had to pick any subject, it wouldn't be maths': Foundations for engagement with mathematics during the middle years. *Mathematics Education Research Journal*, 25, 569–87.

Attard, C. (2012). Engagement with mathematics: What does it mean and what does it look like? *Australian Primary Mathematics Classroom*, 17(1), 9–13.

Australian Association of Mathematics Teachers (AAMT) (1997). *Numeracy = everybody's business. The report of the Numeracy Education Strategy Development Conference, May 1997*. Adelaide: AAMT.

Australian Association of Mathematics Teachers (AAMT) (2006). *Standards for excellence in teaching mathematics in Australian schools*. Adelaide: AAMT.

Australian Curriculum, Assessment and Reporting Authority (ACARA) (2018). *Australian Curriculum: Mathematics*. Sydney: ACARA. Retrieved from http://www.australiancurriculum.edu.au/ mathematics/content-structure

Australian Institute for Teaching and School Leadership (AITSL) (2011). *Australian Professional Standards for Teachers*. Melbourne: AITSL.

Australian Institute of Health and Welfare (AIHW) (2020). *Australia's children*. Cat. no. CWS 69. Canberra: AIHW.

Australian Securities & Investments Commission (2014). *National financial literacy strategy 2014–17*. Retrieved from https://www.financialcapability.gov.au/files/national-financial-literacy-strategy-2014-17.pdf

Ben-Hur, M. (2006). *Concept-rich mathematics instruction*. Alexandria, VA: Association for Supervision and Curriculum Development.

Bishop, S.J. (2007). Neurocognitive mechanisms of anxiety: An integrative account. *Trends in Cognitive Science*, 11, 307–16.

Boaler, J. (2000). Introduction: Intricacies of knowledge, practice, and theory. In J. Boaler (ed.), *Multiple perspectives on mathematics teaching and learning* (1–18). Westport, CT: Ablex Publishing.

Boaler, J. (2009). *The elephant in the classroom: Helping children learn and love maths*. London: Souvenir Press.

Bobis, J., Mulligan, J. & Lowrie, T. (2013). *Mathematics for children: Challenging children to think mathematically* (4th edn). Frenchs Forest: Pearson.

Brumbaugh, D.K. & Rock, D. (2001). *Teaching secondary mathematics* (2nd edn). Mahwah, NJ: Lawrence Erlbaum.

Bruner, J.S. (1960). *The process of education*. Cambridge, MA: Harvard University Press.

Buckley, S., Reid, K., Goos, M., Lipp, O. & Thomson, S. (2016). Understanding and addressing mathematics anxiety using perspectives from education, psychology and neuroscience. *Australian Journal of Education*, 60, 157–70.

Callingham, R. & Watson, J.M. (2017). The development of statistical literacy at school. *Statistics Education Research Journal*, 16, 181–201.

Council of Australian Governments Education Council (2019). *Alice Springs (Mparntwe) Education Declaration*. Carlton South: Education Services Australia.

Cumming, J. & Elkins, J. (2010). Lack of automaticity in the basic addition facts as a characteristic of arithmetic learning problems and instructional needs. *Mathematical Cognition*, 5. https://doi: 10 .1080/135467999387289

Davis, G.E. & Tall, D. (2002). What is a scheme? In D. Tall & M. Thomas (eds), *Intelligence, learning and understanding in mathematics: A tribute to Richard Skemp* (131–50). Flaxton: Post Pressed.

De Smedt, B., Ansari, D., Grabner, R.H., Hannula-Sormunen, M., Schneider, M. & Verschaffel, L. (2011). Cognitive neuroscience meets mathematics education. *Educational Research Review*, 6, 232–7.

Delazer, M., Ischebeck, A., Domahs, F., Zamarian, L., Koppelstaetter, F., Siedentopf, C.M., Kaufmann, L., Benke, T. & Felber, S. (2005). Learning by strategies and learning by drill: Evidence from an fMRI study. *NeuroImage*, 25, 838–49.

Drew, A., Gore, J., Harris, J., Prieto-Rodriguez, E., Fray, L., Lloyd, A. & Taggart, W. (2019). *QuickSmart numeracy: Learning impact fund evaluation report*. Independent report prepared by the University of Newcastle for Evidence for Learning. Retrieved from https://evidenceforlearning.org.au/lif/our-projects/quicksmart-numeracy/

Dweck, C.S. (2000). *Self-theories: Their role in motivation, personality, and development*. Philadelphia, PA: Psychology Press.

Ferster, C.B. & Skinner, B.F. (1957). *Schedules of reinforcement*. New York: Appleton-Century-Crofts.

Freudenthal, H. (1971). Geometry between the devil and the deep sea. *Educational Studies in Mathematics*, 3, 413–35.

Gagné, R.M. (1965). *The conditions of learning.* New York: Holt, Rinehart and Winston.

Goos, M., Dole, S. & Geiger, V. (2012). Auditing the numeracy demands of the Australian Curriculum. In J. Dindyal, L. Chen & S.F. Ng (eds), *Mathematics education: Expanding horizons. Proceedings of the 35th Annual Conference of the Mathematics Education Research Group of Australasia (MERGA)* (314–21). Singapore: MERGA.

Goos, M., Geiger, V., Dole, S., Forgasz, H. & Bennison, A. (2019). *Numeracy across the curriculum: Research-based strategies for enhancing teaching and learning.* Crows Nest: Allen & Unwin.

Goos, M., Stillman, G. & Vale, C. (2017). *Teaching secondary school mathematics: Research and practice for the 21st century* (2nd edn). Crows Nest: Allen & Unwin.

Gray, E.M. (1991). An analysis of diverging approaches to simple arithmetic: Preference and its consequences. *Educational Studies in Mathematics*, 22, 551–74.

Helme, S. & Clarke, D. (2001). Identifying cognitive engagement in the mathematics classroom. *Mathematics Education Research Journal*, 13, 133–53.

Jorgensen, R. (2010). Structured failing: Reshaping a mathematical future for marginalised learners. In L. Sparrow, B. Kissane & C. Hurst (eds), *Shaping the future of mathematics education. Proceedings of the 33rd annual conference of the Mathematics Education Research Group of Australasia (MERGA)* (vol. 1, 26–35). Fremantle: MERGA.

Kaufmann, L. (2008). Dyscalculia: Neuroscience and education. *Educational Research*, 50, 163–75.

Kennedy, J., Lyons, T. & Quinn, F. (2014). The continuing decline of science and mathematics enrolments in Australian high schools. *Teaching Science*, 60(2), 34–46.

Lave, J. & Wenger, E. (1991). *Situated learning: Legitimate peripheral participation.* Cambridge: Cambridge University Press.

Lerman, S. (2000). The social turn in mathematics education research. In J. Boaler (ed.), *Multiple perspectives on mathematics teaching and learning* (19–44). Westport, CT: Ablex Publishing.

Lyons, I.M. & Beilock, S.L. (2012). Mathematics anxiety: Separating the math from the anxiety. *Cerebral Cortex*, 22, 2102–10.

McPhan, G., Morony, W., Pegg, J., Cooksey, R. & Lynch, T. (2008). *Maths? Why not? Final report prepared for the Department of Education, Employment and Workplace Relations (DEEWR).* Canberra: DEEWR.

Miller, G.A. (2003). The cognitive revolution: A historical perspective. *TRENDS in Cognitive Sciences*, 7, 141–4.

National Consumer and Financial Literacy Framework (2011, September). Retrieved from http://scseec .edu.au/site/DefaultSite/filesystem/documents/Reports%20and%20publications/Publications/ Miscellaneous/National%20Consumer%20and%20Financial%20Literacy%20Framework-2011.pdf

National Curriculum Board (2009). *Shape of the Australian Curriculum: Mathematics.* Canberra: Commonwealth of Australia.

Obersteiner, A. & Tumpek, C. (2016). Measuring fraction comparison strategies with eye-tracking. *ZDM Mathematics Education*, 48, 255–66.

Office of the Chief Scientist (2014). *Science, technology, engineering and mathematics: Australia's future.* Canberra: Commonwealth of Australia.

Peled, I. & Suzan, A. (2011). Pedagogical, mathematical, and epistemological goals in designing cognitive conflict tasks for teacher education. In O. Zaslavsky & P. Sullivan (eds), *Constructing knowledge for teaching secondary mathematics: Tasks to enhance prospective and practicing teacher learning* (73–88). New York: Springer.

Perso, T. (2011). Assessing numeracy and NAPLAN. *The Australian Mathematics Teacher*, 67(4), 32–5.

Piaget, J. (1972). *The principles of genetic epistemology* (W. Mays trans.). London: Routledge & Kegan Paul.

Programme for International Student Assessment (PISA) (2012). *Data base – PISA 2012.* Paris: Organisation for Economic Co-operation and Development (OECD). Retrieved from https://www .oecd.org/pisa/data/pisa2012database-downloadabledata.htm#

Programme for International Student Assessment (PISA) (2019). *PISA 2018 Assessment and Analytical Framework*. Paris: OECD.

Ryan, J. & Williams, J. (2007). *Children's mathematics 4–15: Learning from errors and misconceptions*. Berkshire: Open University Press.

Schoenfeld, A.H. (1988). When good teaching leads to bad results: The disasters of 'well-taught' mathematics courses. *Educational Psychologist*, 23, 145–66.

Sfard, A. (1991). On the dual nature of mathematical conceptions: Reflections on processes and objects as different sides of the same coin. *Educational Studies in Mathematics*, 22, 1–36.

Siemon, D., Beswick, K., Brady, K., Clark, J., Faragher, R. & Warren, E. (2011). *Teaching mathematics: Foundations to middle years*. South Melbourne: Oxford University Press.

Skemp, R. (1976). Relational understanding and instrumental understanding. *Mathematics Teaching*, 77, 20–6.

Stipek, D.J. (1996). Motivation and instruction. In D.C. Berliner & R.C. Calfee (eds), *Handbook of educational psychology* (85–113). New York: Macmillan.

Sullivan, P., Clarke, D. & Clarke, B. (2013). *Teaching with tasks for effective mathematics learning*. New York: Springer.

Sullivan, P., Tobias, S. & McDonough, A. (2006). Perhaps the decision of some students not to engage in learning mathematics in school is deliberate. *Educational Studies in Mathematics*, 62, 81–99.

Thorndike, E.L. (1922). *The psychology of arithmetic*. New York: MacMillan.

Varma, S. & Schwartz, D.L. (2008). How should educational neuroscience conceptualise the relation between cognition and brain function? Mathematical reasoning as a network process. *Educational Research*, 50, 149–61.

Verschaffel, L., Lehtinen, E. & van Dooren, W. (2016). Neuroscientific studies of mathematical thinking and learning: A critical look from a mathematics education viewpoint. *ZDM Mathematics Education*, 48, 385–91.

von Glasersfeld, E. (1995). *Radical constructivism: A way of knowing and learning*. London: Falmer.

Vygotsky, L.S. (1978). *Mind in society: The development of higher psychological processes*. Cambridge, MA: Harvard University Press.

CHAPTER 2

Language and mathematics

Linda Galligan

The learning objectives of this chapter are directly linked to the Australian Professional Standards for Teachers (APST) (Australian Institute for Teaching and School Leadership [AITSL], 2011). After studying this chapter, you should be able to:

- recognise the importance of language in the teaching and learning of mathematics (APST 1.2)
- identify the elements of the mathematics register in English and other languages and discuss the impact on learning, enabling you to be responsive to the learning strengths and needs of EAL (English as an additional language) students (APST 1.2, 1.3)
- reflect on the appropriate use of language by the student and the teacher in the mathematics classroom (APST 2.5)
- demonstrate a range of verbal and non-verbal communication strategies to support student engagement (APST 1.3).

Introduction

The Australian Curriculum states that students are to be 'confident, creative users and communicators of mathematics' (Australian Curriculum, Assessment and Reporting Authority [ACARA], 2018). To be able to communicate in this way they need suitable language tools (Riccomini et al., 2015). This chapter investigates why language is important and reviews the mathematics register. It highlights the use of language as a resource to teach mathematics in the classroom and discusses its importance for second language learners. The chapter will ask you to reflect on particular problems and offer activities that can be used in the secondary classroom. It will also ask you to reflect on your mathematical language fluency, which is needed to communicate to students in written, oral, symbolic and graphical form.

In a typical mathematics classroom, students read, write, listen, draw and speak mathematics. In many secondary mathematics classrooms, listening to the teacher speaking the mathematics or watching the teacher write mathematics is commonplace, but is this the best way for students to learn? In these language modes there are also particular issues in the context of mathematics:

1. The written language can be dense. In the statement 'there are just four numbers, after unity, which are the sums of the cubes of their digits', the phrase 'the sums of the cubes of the digits' takes time to unpack.
2. Words have precise meanings. The word 'integrate' in everyday English is related to but not the same as 'integrate' in mathematics.
3. The spoken language is quite different from the written. We use everyday English to teach mathematics. A preservice teacher commented: 'I think at this stage I'd make a poor maths teacher because I've never spoken this language. I read it at school, but have never "said it aloud" – BIG difference' (Galligan & Hobohm, 2013, p. 328).
4. The symbols in mathematics have their own grammar which are often in conflict with ordinary English. A student says: 'I can see this bit here has to be timesed by 3 outside the brackets, and then take away the bits in the brackets ...'
5. The written mathematics lends itself to efficiency and maximising logical thinking without ambiguity. In solving the simultaneous equations,

$$x + y = 4 \tag{1}$$

$$x + 3y = 10 \tag{2}$$

a student wrote as an answer:

$$x = 1; \ y = 3$$
$$1 + 3 \times 3 = 10$$
$$10 = 10$$

Each of these scenarios highlight some of the language issues in teaching and learning mathematics. These issues fall into several categories:

1. The mathematics register (i.e. the words, phrases and associated meanings used to express mathematical ideas). This includes the etymological view of the words of mathematics as well as the syntax, semantics, orthography and phonology of the language itself and the impact these have on understanding mathematics. An extended definition of the register (O'Halloran, 2015) also includes the symbolic and the visual forms of communication.

2. Language in the classroom. This is the use of language by teachers to communicate ideas and the dialogue used by students to communicate and learn mathematics, which is quite different to the written technical communication.

3. Technical communication. This is the accepted standard use of language, symbols and images to communicate mathematical ideas, both orally and in the written form.

> **Syntax:** the sentence pattern in language.
>
> **Semantics:** the study of meaning in language.
>
> **Orthography:** the representation of the sounds of language by written symbols.
>
> **Phonology:** the system of sound patterns.

Critiquing your mathematical language

SCENARIO

As this chapter is about language, we suggest that you have access to a recording device. The best device would be a tablet with a stylus for writing, but using a video-recording device and a whiteboard would also work well. We want you to get used to listening to your own voice and being critical of what you say. Compare what you say with what is written.

This is what one student said of the experience:

> I first must say that I was absolutely terrified at doing this, not the math writing part but the talking part. For a couple of days, I would sit down to do it but lost all my nerve. A few days went by and I decided that maybe if I practiced that would help, . . . so I sat down today, . . . and started to talk into the microphone and just kept going blank after a couple of seconds. That went on for a couple of hours; ok it was more like five hours. Then all of a sudden, I just started talking and didn't stop. I must have forgotten all about the microphone until about the last minute or so when I thought about it again and got a bit panicky but was able to finish it. Besides all that I have actually enjoyed doing this and getting results. (Galligan & Hobohm, 2013, p. 329)

The online resources for this text include three recordings of a teacher explaining fractions after she had spent a semester using recording technology. There is an error in one of the recordings – see if you can find it.

Audio & video recordings

Why language is important

The Australian Curriculum, as well as national reports on education (e.g. the National Numeracy Review Report), recognises the importance of language in mathematics. In the Australian Curriculum: general capabilities (ACARA, 2018), you will see literacy is highlighted as an important tool in the teaching and learning of mathematics, from interpreting word problems to discussing mathematics in the classroom. The National

Numeracy Review Report (Council of Australian Governments, 2008) recommended that the language and literacies of mathematics be explicitly taught since language can be a significant barrier to understanding mathematics. Poor use of the language of mathematics can also lead to misconceptions. As teachers routinely assess students' understanding of mathematics through literacy, an important question to ask yourself when setting assessment tasks is: Are your students struggling to understand the mathematics or are there specific language difficulties associated with assessment tasks you have set?

In addition to literacy in mathematics, mathematical literacy has an important role in understanding mathematics. In Chapter 1, we spoke about mathematical, scientific, statistical and financial literacy. Mathematical literacy is related to the role mathematics plays in understanding life, whether it is school, work or everyday activities. In addition to this awareness, mathematical literacy is also related to a person's confidence and competence in using the mathematics in the appropriate context.

> **Literacy in mathematics:** students being able to access the mathematics in words, and to make sense of the context and clarify what is required (Council of Australian Governments, 2008). It is different to mathematical literacy (Numeracy) defined in Chapter 1.

PAUSE AND THINK

Imagine a teacher running her fingers across the pages of the textbook and telling her students 'a sample is a subset of a population, chosen freely without any rule or any obvious bias'. The students listen quietly, but one of them is thinking, 'I thought a sample was a small amount of something like blood to test whether a person has some disease or not'. (example from: Jourdain & Sharma, 2016)

This example illustrates why language is important. It shows that the language of mathematics can sometimes be confusing and challenging, especially when students bring their prior knowledge to it (Jourdain & Sharma, 2016). Have you considered that you will often have a dual role of mathematics teacher and language teacher, especially when you are teaching EAL students?

Mathematics

The Australian Curriculum: Mathematics outlines several key elements that students should learn through their studies. These include the:

> ... vocabulary associated with number, space, measurement, and mathematical concepts and processes. This vocabulary includes synonyms, technical terminology, passive voice, and common words with specific meanings in a mathematical context. Students develop the ability to create and interpret a range of texts typical of mathematics, ranging from calendars and maps to complex data displays. Students use literacy to understand and interpret word problems and instructions that contain the particular language features of mathematics. They use literacy to pose and answer questions, engage in mathematical problem-solving, and to discuss, produce, and explain solutions. (ACARA, 2018)

PAUSE AND THINK

Create a list of the different aspects of literacy in mathematics from the statement above. Next to this list give an example of each that would be relevant in high school. We will be visiting these aspects later in the chapter.

Can you see that there is a strong relationship between mathematics achievement and literacy for high school students? Perhaps you have seen this in your own experience as a student, a preservice teacher or in other settings. What sorts of activities would you develop in your mathematics classroom to improve literacy? What could you do to ensure that students carefully read mathematical word problems?

The mathematics register

In everyday English, numbers are usually adjectives (i.e. one book, three days), but in mathematics 'one' and 'three' are nouns. In everyday English we speak of the 'function of government' or 'factors that influence climate change'. The use of 'function' and 'factor' is quite different in mathematics. In a classroom, teachers and students talk about mathematics differently from a written text. In a class, the language is often informal and natural, but more formal mathematical language 'makes varied use of a complex, rule-governed writing system that is mainly separate from that of the natural language into which it is read' (Pimm, 1991).

This collection of terms is often called the mathematics register. The mathematics register is more than the words and phrases in mathematics, it is the 'set of meanings that is appropriate to a particular function of language *together with* the words and structures which express these meanings' (Halliday, 1978, p. 195). It can also include the symbols that capture the relationship between mathematical entities and processes (e.g. $x + y > q$), as well as related mathematical conventions (O'Halloran, 2015).

As teachers, you need to be aware of the subtle differences in language. In some instances, words used in a mathematics classroom may have altered meaning and grammatical function. Adams, Thangata and King (2005) suggest that teachers need to support students to use technical language when talking about concepts. You should encourage students to make the connections between everyday meanings and the mathematical meaning of words such as function, mean or rational. Ensure you provide time for students to 'talk about mathematics as they solve problems, encouraging them to articulate patterns and generalisations' (Adams, Thangata & King, 2005, p. 446).

Mathematics register: 'the set of meanings that is appropriate to a particular function of language *together with* the words and structures which express these meanings' (Halliday, 1978, p. 195).

Words used

Words are the gateway to understanding and processing concepts in mathematics. The way in which words are assembled and presented, and the way they sound, may contribute to the ease (or difficulty) with which we can understand related mathematical concepts. In cases where that gateway is difficult, it is a teacher's role to open the gateway for better understanding.

Etymology: the study of the origins of mathematical words.

Characteristics of the English language can have a negative effect on the way students process mathematical text (Ellerton & Clements, 1996). Take the word 'diameter', for example. While the word may only supply the reader with access to the concept definition, to improve student understanding, it may be helpful to access the total cognitive structure (i.e. all the words which associate with the word 'diameter', such as 'radius'). Discussion of the origins of mathematical words (their etymology) may also help students to make connections between mathematical terms by offering them insights into the construction and meaning of the word (Jourdain & Sharma, 2016). For instance, in Chinese, the literal meanings of diameter (直径 [zhíjìng] 'straight path') and radius (半径 [bànjìng] 'half path') allow more direct access to the meaning of both words.

In English the names of number words may also hinder early learning. For example, twelve and thirteen are not part of a regular named-value system. By contrast, in Asian languages (e.g. Burmese, Japanese, Korean and Thai) number words are said and then the value of that number is named (5726 – five thousand seven hundred two ten six) (Fuson & Kwon, 1991). Here, the orthography of the language may assist in the reading and processing of text. For example, the word thirteen in Chinese is 十三 (ten three); the word for March is 三月 and for triangle it is 三角形 (Galligan, 2001). If a teacher can highlight these structures to students who struggle in high school, this may assist their understanding of the subject, particularly around the concept of place value.

Invention of words

The words that are used in mathematics are constantly being invented. An interesting project in New Zealand in the 1990s applied Māori words to mathematics (Barton, Fairhall & Trinick, 1998). As Māori is the first language and the language of instruction in some schools, the construction of more meaningful terminology was thought to have motivational and bilingual benefits for students who identify with the Māori culture. For example, the word 'hōkai' is a brace or stay used to keep an eel-pot open and it was an early choice for the word 'diagonal' (Barton, Fairhall & Trinick, 1998, p. 5).

PAUSE AND THINK

The choice of word can have an impact on the learner. When introducing a new concept, spend some time with students talking about the origins of the words and what images these words bring to mind. For example, look up the etymological meaning of the two words 'integration' and 'differentiation'. How do these words connect to the mathematical concepts related to the words?

There are at least 11 categories of difficulties associated with learning the words of mathematics (Riccomini et al., 2015). Can you think of other examples to add to the ones below?

1. meanings are context dependent (e.g., foot as in 12 inches vs. the foot of the bed),
2. mathematical meanings are more precise (e.g., product as the solution to a multiplication problem vs. the product of a company),

3. terms specific to mathematical contexts (e.g., polygon, parallelogram, imaginary number),
4. multiple meanings (e.g., side of a triangle vs. side of a cube),
5. discipline-specific technical meanings (e.g., cone as in the shape vs. cone as in what one eats),
6. homonyms with everyday words (e.g., pi vs. pie),
7. related but different words (e.g., circumference vs. perimeter),
8. specific challenges with translated words (e.g., mesa vs. table),
9. irregularities in spelling (e.g., obelus vs. obeli),
10. concepts may be verbalized in more than one way (e.g., 15 minutes after vs. quarter past),
11. students and teachers adopt informal terms instead of mathematical terms (e.g., diamond vs. rhombus) (Riccomini et al., 2015, p. 238).

PAUSE AND THINK

Students may be surprised about how 'young' some words in mathematics are. Investigate the meaning of the following words and find out their origin. You may need to use a dictionary or an online site that has etymological details:

- negative number
- vulgar fraction
- septendecillion
- googol
- algebra
- Cartesian plane.

For the senior students:

- chaos theory
- fractal
- cycloid
- Gaussian method of elimination
- kissing circles.

SHORT-ANSWER QUESTIONS

1. Do you know mathematical words or phrases in other languages? How could you incorporate this discussion into a lesson, project or homework activity?
2. Look up the word 'trapezium' in British and American dictionaries and compare them. You will see that it is not a unified definition. Are you surprised by this? Perhaps we are used to the idea

▶ ▶ that mathematics seems to be a very black-and-white subject, but it is good to suggest to students that this is not always the case. Consider everyday words, such as 'restaurant'. Does a restaurant have to be indoors, or must it provide table service? The answer to this question helps to characterise the 'restaurantness' of an establishment. Restaurant is a somewhat fuzzy concept. People are usually comfortable with this uncertainty. Therefore, for a word to be understood and used, it is not necessary to know its tightly defined meaning (Leung, 2005). As a teacher, you can explore the fuzziness of word meanings, even generalising and extending meaning from one instance to another. Learning vocabulary, particularly in terms of its associated concepts and linguistic properties, is an ongoing activity that can be fostered in the classroom.

Definitions

Definitions of mathematical concepts are presented differently at different school levels and play different roles in different mathematical practices. To illustrate, Morgan (2005) compared the language used in two textbooks: one for intermediate Year 10/11 and the other for higher Year 10/11. Read both below and note any differences. If you are familiar with functional grammar, you will notice differences in agency and modality, but if not, you should still see the different ways the textbook authors are talking to and engaging the student.

General Certificate in Secondary Education (GCSE) intermediate textbook

In investigation 15:1, you found that the ratio $\frac{\text{shortest side}}{\text{longest side}}$ i.e. $\frac{\text{opposite}}{\text{hypotenuse}}$ is the same
   for each of these triangles. This ratio is given a special name; it is called the sine of $40°$
   or sine $40°$.
The ratio $\frac{\text{adjacent}}{\text{hypotenuse}}$ is called the cosine $40°$. The ratio $\frac{\text{opposite}}{\text{adjacent}}$ is called tangent $40°$.
The abbreviations sin A, cos A and tan A are called trigonometrical ratios, or trig.
   ratios.

General Certificate in Secondary Education (GCSCE) higher textbook

The ratios sin θ and cos θ may be defined in relation to the lengths of the sides
   of a right-angled triangle.

$$\sin θ \text{ is defined as } \frac{\text{length of opposite side}}{\text{length of hypotenuse}}$$

$$\cos θ \text{ is defined as } \frac{\text{length of adjacent side}}{\text{length of hypotenuse}}$$

Since θ < 90, sin θ and cos θ defined in this way have meaning for angles less than $90°$.
We will now look at an alternative definition for sin θ and cos θ which has meaning for
angles of any size. (Example from Morgan, 2005)

As pointed out by Morgan (2005), while the first text reads like a set of unquestionable facts ('the ratio is ...'), the second provides ambiguity or choice ('the ratio cos may be defined ...'), which suggests the mathematics definitions can be negotiated.

More than words

Understanding nouns and adjectives are not the only difficulties of the language. The grammar used in mathematical text often renders it unreadable to novice learners (O'Halloran, 2015). The increase in text complexity as students move through high school is exacerbated by the vertical knowledge structure of mathematics which builds on previous concepts.

To understand mathematical language, students need to be able to understand the words and meanings (semantics) and the structure (syntax). Consider, for example, the importance of the word 'before' in the following two statements. Why is there a difference in the wording 'down to a stop' and 'eventually stopping'?:

- The car is moving with a constant speed **before** it slows down to a stop.
- The car starts from rest **before** moving and then eventually stopping.

Chapter 10 discusses some further issues around solving such word problems.

Table 2.1 summarises some of the possible language issues in mathematics. Try to think of similar examples in both the Junior and Senior Curriculum.

Table 2.1 Examples of language issues in mathematics

Issue	Example	Discussion
Subordinate clauses and text density (syntax and orthography)	Find the side length of a square whose diagonal is of length 10 cm.	Simplify this by breaking it into two sentences: A square has a diagonal of 10 cm. Find the side length.
Topic prominence (syntax)	Instead of: How long is this line? Use: Look at this line. How long is it?	Get students to pay attention to the topic of the sentence, again breaking it into two sentences.
Word order (syntax)	What number is two less than five? Find this square's side length.	The first example is difficult in English. Rewrite it: Consider the number 5. Pick a number two less than this. Perhaps change it to: Find the side length of this square.
Passive voice (syntax)	When 15 is added to a number the result is 21.	Change to an active statement: Add 15 to a number so you get 21.
Redundancy (syntax)	Ben is taller than George. Dave is shorter than George. Who is the tallest?	Compare this to: Ben is taller than George. Dave is shorter than George. Of the three, who is the tallest? This removes ambiguity.
Conditional statements (syntax)	How much string *would have* remained if 3 cm *had been* cut from it?	Simplify this to: I cut 3 cm from the string. What remains?
Morphological complexity (orthography)	Parallelogram Quadrilateral	Words in mathematical English tend to be complex. Assist students by breaking it down into morphemes and explaining each. Parallel//ogram Quadri//lateral

Morphological complexity: the proportion of the lexicon's total description length that is due to the description lengths of the affixes and signatures (Bane, 2008, p. 73). There is little morphological complexity in Chinese, but English is more complex and Hungarian more complex still.

Morpheme: the most elemental unit of meaning; it is the minimal linguistic sign in which there is an arbitrary union of a sound and a meaning that cannot be further analysed.

Table 2.1 (*cont.*)

Issue	Example	Discussion
Content words (semantics)	Diameter and radius are linked conceptually but not semantically.	This is a feature of English. Explain to students the origin of these words.
Connecting words (semantics)	If a number is greater than 10 or divisible by three, what could the number be?	A feature of mathematical English is this 'inclusive or'. This is different from 'I am going to the shop' or 'going to the movies'.
Script layout (in graphs and diagrams, and a combination of the two) (orthography)	Vertical and horizontal labels on a graph	Can a student pick a point on the graph and describe that point by using the labels?

Source: Adapted from Galligan, 2001.

SHORT-ANSWER QUESTIONS

1. Look at the statement at the introduction of this chapter: 'There are just four numbers, after unity, which are the sums of the cubes of their digits' (Parker, 2014). Turn this statement into something more symbolic and think about the time spent on the words to extract meaning. The phrase 'sums of the cubes of their digits' is a very dense statement (dense because all the words are necessary to extract meaning). The interesting part of the statement 'there are just four numbers' gets lost in the attempt to turn the 'word problem' into a mathematical one.

2. Which sentence correctly represents this algebraic equation?

$$2w - 9 = 43$$

   a. Nine less than twice a number is forty-three.
   b. Nine is forty-three more than twice a number.
   c. Nine is forty-three less than twice a number.
   d. Nine less than forty-three is twice a number

   Are there other ways to write the sentence? Could it be written in active rather than passive voice (i.e. *If you have twice a number*; or *double a number . . .*)?

Problem solving: a problem of language?

Word problems in mathematics are often cited as an area of difficulty for students. As a teacher, you should carefully consider the context of the problem itself, as many word problems are pseudo real world. If the problem is of this form, create an important teaching moment, and then let students know this is why you are asking them to solve the problem. For other problems (such as the two-way table in the short-answer questions at the end of this section), teachers should simplify the problem to allow for data collection.

Consider this classic problem, called the 'student–professor problem', that is also cited in Chapter 10 (Clement, Lochhead & Monk, 1981). Try it:

> There are six times as many students as professors. Write this in a mathematical sentence.

It is pseudo real world, but the importance of the problem is the syntax, which can contribute to high errors. Typically, students write $S = 6P$ instead of $P = 6S$. Researchers suggest that positioning 'information and the unknown in sentences may have an impact on the ease of processing the sentence' (MacGregor, 1993, cited in Galligan, 2001, p. 116), and this is particularly true for comparison problems in English, such as the example above. Many comparison problems in Chinese have a relational proposition that is explicitly shown. A Chinese multiplicative example is 'translated syntactically as *men are women six times*' (there are six times as many men as women) (Galligan, 2001, p. 116). Here the relational proposition is clear. In a similar additive comparison problem, *men compare women more five* (there are five more men than women). This again suggests 'a clear relational proposition, but there is a subtle change in syntax' (Galligan, 2001, p. 116). This creates 'a semantic difference and there is evidence to suggest a difference in the processing in Chinese' (Galligan, 1997; Lopez-Real, 1997, cited in Galligan, 2001, p. 116).

When students encounter a mathematical word problem, a teacher should highlight the necessity to read carefully, understand the words and sentences, and then solve the problem. In a well-cited study on students' errors in solving mathematical word problems (Ellerton & Clements, 1996), the major error highlighted was students' inability to read the problem and then translate it into correct mathematical form. The authors drew on research by Newman (1983) that challenged teachers to redefine what is 'basic' in school mathematics and highlighted errors in *reading, comprehension, transformation, process* and *encoding*. The questions you can ask the students include:

1. Please read the question to me (reading).
2. Tell me what the question is asking you to do (comprehension).
3. Tell me a method you can use to find an answer to this question (transformation).
4. Show me how you worked out the answer to the question. Explain to me what you are doing as you do it (process skills).
5. Now write down your answer to the question (encoding).

As a result of Newman's research, it has been found that many 'remedial' mathematics programs pay little attention to the first three questions above.

When students 'read' mathematics it is not just the words and symbols, but also diagrams. Graphs and charts, like word problems, are very dense (Lowrie, Diezmann & Logan, 2011). Kemp and Kissaine (2010) suggest that table reading and interpreting skills can be taught and learned, and it should not be assumed that students will be able to perform these tasks without instruction. For example, to understand the graph in Figure 2.1 from the Australian Institute of Health and Welfare (2020), students need to have an understanding of such things as the marks on the number lines, their relative positions, and the colour and orientation conventions. They also need to be able to extract the information related to both axes, the legend and the notes.

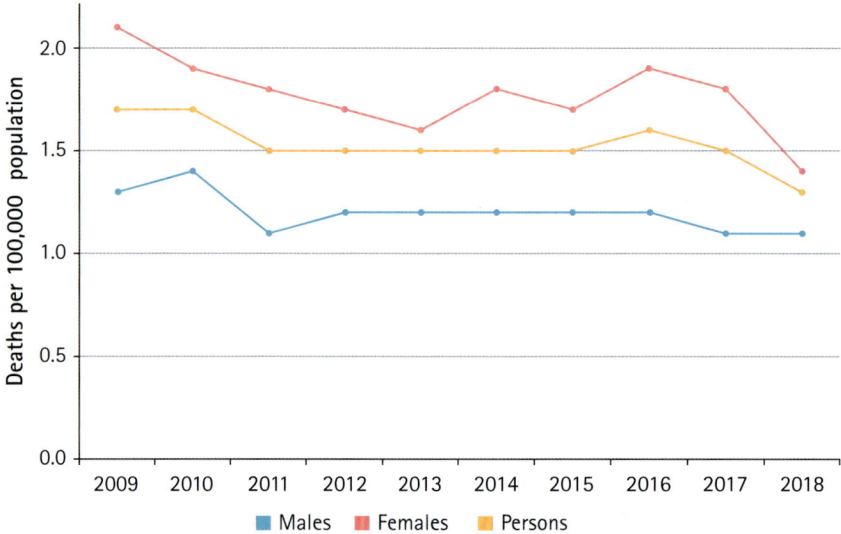

Figure 2.1 Age-standardised death rate due to asthma for all ages, by sex, 2009–2018

Notes:

1. Rates have been age-standardised to the 2001 Australian Standard Population as at 30 June 2001. For all ages, the age groups: 0–14, 15–24, 35–44, 45–54, 55–64, 65–74, 75–84, 85+. For 5–34, the age groups: 5–14, 15–24, 25–34.

2. Asthma classified according to the International Classification of Disease, 10th revision (ICD-10) codes J45 and J46.

3. Because attribution of death due to asthma is more certain among those aged 5–34, this age group is commonly used for examining time trends. In older people, other causes of death, in particular chronic obstructive pulmonary disease, cause difficulties in the attribution of causes of death.

4. Year refers to year of registration of death. Deaths registered in 2015 and earlier are based on the final version of cause of death data; deaths registered in 2016 are based on revised version; and deaths registered in 2017 and 2018 are based on preliminary version. Revised and preliminary versions are subject to further revision by the Australian Bureau of Statistics.

Source: Australian Institute of Health and Welfare, 2020.

PAUSE AND THINK

There are many more graphs like the one on Australia's health shown in Figure 2.1 that could be used for class discussion. Can you think of ways that graphs such as these could be used? And in what section of the curriculum would they be used?

SHORT-ANSWER QUESTIONS

1. The following table is based on a worksheet by MacGregor and Moore (1991, p. 134). Can you create your own worksheet similar to this for use in your classroom (at different levels)?

If we	multiply	$20	from	. . .	we get	. . .
	increase	4 kg	by			
	share		to			
	halve		into			
	reduce		between			
	divide					
	add					

2. Solve the following two problems. As you do so, consider the questions, where did you have to spend time thinking about the problem? How much of this time was related to the language structure of the question? How much time was spent on understanding the question and on solving the problem?

   a. The First Fleet carried enough food to keep its passengers alive for two years in Australia. The rations issued to sailors, marines and officers each week were:
   - Beef 4 pound
   - Hardtack 7 pound
   - Pork 2 pound
   - Cheese 12 ounces
   - Dried peas 2 pints
   - Butter 6 ounces
   - Oatmeal 3 pints
   - Vinegar pint

   The male convicts got one third less, while female convicts got two thirds of the male ration, or slightly less than half the naval standard. On paper, this was not a bad allowance. In practice it meant scurvy, and the meat was mostly bone and gristle.

   i. Calculate the weekly ration of each item for the male convicts.

   ii. Calculate the weekly ration of each item for the female convicts.

   iii. Show that the female ration is 'slightly less than half' the naval ration.

   (Example from Hughes (1986), *The Fatal Shore*)

   b. A survey was conducted asking students about their cereal preference, and the results are shown below. What is the probability that a randomly chosen male student preferred porridge?

	Wheaties	Choc Bubbles	Porridge
Female	104	54	18
Male	56	36	3

▶ ▶ Record yourself explaining to a fellow student (who pretends to be a school student) what you would do to solve one of the problems above. In particular, use the Newman's Error Analysis framework.

Second language learners

The mathematics register in English is of particular concern for teachers when there are second language learners in the classroom. This is because the cognitive effort required to process certain mathematical word problems is high in English (Clement, Lochhead & Monk, 1981; MacGregor, 1991). Research has therefore suggested the importance of learning and doing mathematics in students' native tongues rather than in a second language such as English (e.g. Nathan et al., 1993; Spanos et al., 1988), or even using two languages (Moschkovich, 2007). While most researchers agree that the native tongue allows for more efficient conceptual processing, there may be differences in the nature of that conceptual processing (Galligan, 1997, 2001). In addition, teachers should allow students to choose the language they prefer for carrying out arithmetic computation, either orally or in writing (Moschkovich, 2007).

As a teacher, it is important that you are aware of the difficulties students may have with reading mathematics and therefore ensure you make both written and oral communication as clear and simple as possible, but also appropriate to the level of the learner. A New Zealand study investigating the relationship between English language and mathematics learning for non-native speakers (Barton & Neville-Barton, 2005) found that for one senior high school, there was on average a 15 per cent disadvantage in the test because it was in English, and it appeared that it was the syntax more than the vocabulary that caused these problems. In many cases, teachers were unaware of this disadvantage. Furthermore, in many instances, the students were not aware of the extent of their difficulties (Barton & Neville-Barton, 2005). The main problems appeared to be with prepositions, word order, logical structures such as implication, conditionals and negation, and mathematics couched in everyday contexts. Chapter 6 discusses issues around teaching mathematics to English language learners.

If you have second language learners in your class, the UK's National Association for Language Development in the Curriculum (2011, p. 4) suggests 'to find ways to incorporate the use of students' home languages into mathematics lessons ... Supporting students to develop the mathematical aspects of their home language is likely to have a positive effect on their mathematical English. Mathematics teachers should also not assume that a student who appears to communicate effectively in "everyday" English will have the necessary proficiency to communicate at the same level in mathematics'.

Earlier in the chapter, we spoke of 11 categories of difficulties associated with learning the words of mathematics. If you have EAL students in your class, discuss the equivalent words in their language. Are there similar ambiguities? This is important and complex – how will you as a teacher ensure that the meaning is consistent in English and their first language?

Think of the meanings of the words unknown versus variable. $3x + 7 = 10$ is an equation with an unknown (i.e. we can find the value of x). $3x + 7 = y$ is an equation with two variables (i.e. the x and the y vary). Look up the meaning of these words and you will see that these statements could be contested. For example, in the *Collins Dictionary of Mathematics* it states that: 'An unknown is a variable, or the quantity that it represents, whose value is to be discovered by solving an equation' (Borowski & Borwein, 1989, p. 616). Goos, Stillman and Vale (2007) state that 'the notion of variable is fundamentally different from the concept of unknown'. Have you noticed students having trouble with these concepts? Ask your students: what does the symbol n in the expression $2n + 3$ stand for?

Language in the classroom

Consider the importance of language in helping students to learn and understand concepts; this may be through their dialogue with each other or by writing an explanation in words. Communication is an essential part of mathematics and mathematics education in the classroom (National Council of Teachers of Mathematics, 2000, p. 60).

Sullivan (2011) identifies six key principles of effective teaching of mathematics. Principle five, on structuring lessons, is to 'adopt pedagogies that foster communication and both individual and group responsibilities, use students' reports to the class as learning opportunities, with teacher summaries of key mathematical ideas' (p. 6). Sullivan argues that students can benefit from 'either giving or listening to explanations of strategies or results' (p. 37), and that this can best be facilitated by the teacher emphasising mathematical communication and justification. Allowing students to express understanding in their own words also helps them to demonstrate their learning, which thus supports their learning (Sammons, 2018). A key element of this style of teaching and learning is students having the opportunity to see the variability in responses given by other students and then confirming that this variability can indicate there are other underlying mathematical concepts involved (Watson & Sullivan, 2008). This type of communication can occur at any segment of a lesson, including the end.

Cheeseman (2003) argues that a lesson review must contain reflective activities that ask students to talk about the concept and their thinking. It is beyond the scope of this chapter to investigate in detail the communication strategies that can be used. However, to create a positive classroom culture, Sullivan (2011) recommends that communication needs to be actively encouraged and planned. As a teacher, you need to create a learning community by including tasks that ensure that all students participate in both group and whole-of-class discussions. Research suggests that students learn not only from teachers and the tasks, but from other students as well (Sammons, 2018).

An approach used in teaching primary students is also worth investigating. In primary settings, students and teachers move from everyday language to materials language to mathematical language and to symbolic language as they develop conceptual understanding (Larkin, Jamieson-Proctor & Finger, 2012). The first three are usually oral and the latter written. At the same time, when teaching the concept, a teacher moves from using familiar objects, substituted objects, photographs and diagrams, to non-word symbols. While this can be seen in young students' learning (Irons, 2014), it can also be seen in the secondary classroom. For example, in teaching the concept of function, teachers need to recognise 'the organising power of the concept of functions from middle school mathematics through more advanced topics in high school' and beyond (Leinhardt, Zaslavsky & Stein, 1990, p. 1). Teachers also need to see the potential to discuss the connections between graphical and algebraic representation of functions and how that understanding increases over time. Too often, functions appear in Year 11 as something alien or new, yet the concept may have been introduced in primary school using everyday language.

As students advance through school, however, teachers need to 'wean pupils off the use of informal everyday language and to privilege the use of formal technical vocabulary' (Leung, 2005, p. 127). One method of weaning is to reframe the way students talk in mathematically acceptable language. This provides teachers with the opportunity to enhance connections between language and conceptual understanding. O'Connor and Michaels (1996) use the term 'revoicing' to mean the repeating, rephrasing or expansion of student talk in order to clarify or highlight content, extend reasoning, include new ideas or move discussion in another direction. Walshaw and Anthony (2008, p. 119) suggest that in classrooms where revoicing is used, 'there is a greater tendency for students to provide the explanations … and for the teacher to repeat, expand, recast, or translate student explanations for the speaker and the rest of the class'.

Typical errors in language

When you speak in a classroom, it is easy to make small mistakes. For example, a preservice teacher recently reflected on a recording she did that explained the phrase 'expanding an expression'. In one short (less than five-minute screencast) she picked up on five errors in her explanation:

> … enjoy my 'warts and all', spur of the moment screencast I recorded. The main issue I pick up on my screencast is the misuse of terminology. Firstly, I use the term 'equation' where I should be using 'expression' and I think I said solve rather than expand. I also mention the mnemonic F.O.I.L which I don't think is quite right for this circumstance … Then I fall into the dreadful use of English with the word 'timesing' rather than multiplying. Finally, there is the tricky negative versus minus …

Below are five language 'bloopers' that teachers make in the everyday language of the classroom:

1. **Negative or minus**: Students will often say minus 10 when they mean negative 10. The first is an operation of subtraction; the second is a number. The expression $x - 10$, 'x minus 10', can be changed to an equivalent expression $x + {}^-10$, 'x plus negative 10'.
2. **Timesing and plussing**: These are not English words. Use correct words such as 'multiplying' and 'adding'.
3. **Mathematical tricks**: There are various mnemonics that are a memory aid, such as
   FOIL (first … last)
   BIMDAS (brackets … subtraction)
   SOHCAHTOA (sin … adjacent).

Make sure students do not use these mnemonics as an explanation of how mathematics works. The explanation of the expansion of the product of two binomials (i.e. $(x + a)(y + b) = xy + xb + ay + ab$ should include the phrase 'distributive law').

4. **Equation or expression**: $3x + 7$ is an expression. $3x + 7 = 10$ is an equation (it has an equals sign in it). $3x + 7$ is an expression of two terms (i.e. they are separated by a $+$ symbol), and the first term has two factors (x and 3). See Chapter 10 for more discussion on these two terms.

5. **Unknown versus variable**: $3x + 7 = 10$ is an equation with an unknown (i.e. we can find the value of x). $3x + 10 = y$ is an equation with two variables (i.e. the x and the y vary). See Chapter 10 for more discussion on these two terms.

Talking about errors

Have you tried to ask students to talk about errors they have made? Think back to when you did not understand some mathematics. Did you talk about those errors to anyone? Students often find it difficult to express themselves when they make errors. They may not have the language to talk about the issues clearly; they may have been humiliated in the past; or the atmosphere of the classroom may not be conducive to such discussions. However, a teacher should encourage students to change their mindset, explaining that their mind grows when they make errors and then go on to act on those errors. On the issue of growth mindset, you can read discussions by Boaler (2013) or a book by Ryan and Williams (2007) which has an appendix with common errors and a section on discussion prompt sheets.

Balance between talking and listening

An effective classroom teacher needs to have an eye on the 'mathematical horizon', using listening skills and attending to what students say, what they mean and what they do not say. In addition, while teachers need to communicate effectively, they also need to be careful about giving too much feedback, while also using an appropriate level of mathematical language (Maddern & Court, 1989). For example, Maddern and Court (1989) cite a teacher who had shown a keen desire for talk to occur in the classroom, but kept the actual mathematics talk to a minimum, and had created a positive learning environment. This meant there were limited opportunities to 'speak mathematics' and, on a number of occasions, no pauses were made for students' thinking time. Students were sometimes 'talked over'. Students often process differently to their teachers, and while some students may think more quickly, others need time – especially EAL students or students with learning difficulties.

Enabling opportunities for talking is especially important in high school where students are expected to use more mathematical and symbolic language. As mentioned earlier in this chapter, students need fluency in the mathematics register in order to be successful mathematics learners and they need to be taught explicitly to translate between everyday language and the mathematics registers. Marton and Tsui (2004, p. 532) suggest that 'the teacher who makes a difference ... is focused on shaping the development of novice mathematicians who speak the precise and generalizable language of mathematics'. This could take the form of argumentation where students take and defend a particular idea. While this process depends on a skilful teacher who can pre-empt mathematical conversations, it is a skill that 'provides a site for aligning students with each other and with the content of the academic work while simultaneously socializing them into particular ways of speaking and thinking' (O'Connor & Michaels, 1996, p. 65).

A study by Goos (2004) based on a senior secondary school class in Queensland shows how a teacher can initiate a culture of inquiry in the classroom. While the students tended to use informal language in discussions among themselves, the teacher insisted that students use mathematical language when explaining their solutions to the whole class. The teacher's approach can be reflected in an interview extract with this teacher:

> I do think it's important that they're able to communicate with other people and their peers. They will learn at least as much from each other as they will with me. . . . I often force them to say things because they need to be able to use the language because language itself carries very specific meanings; and unless they have the language to be able to, obviously communicate, but I think it also has something to do with their understanding as well. (Goos, 2004, p. 271)

The student's voice can also be heard in this same class. Here, two students (Dean and Adam) are exploring a problem. Dean comments:

> Adam helps me [. . .] see things in different ways. Because, like, if you have two people who think differently and you both work on the same problem you both see different areas of it, and so helps a lot more. More than having twice the brain, it's like having ten times the brain . . . (Goos, 2004, p. 278)

Questioning

Many books discuss the importance of good questioning in the classroom. Often such questions are categorised as closed (what is the value of x in this equation?) or open (how do you know the value of x is 10?). While open-ended questions are seen to promote active learning, there are situations where closed questions are useful.

Wiggins and McTighe (1998) suggest six types of questions connected to six facets of understanding. Table 2.2 shows these six facets with some examples.

Questioning may also differ depending on whether the questions are to the whole class, a small group or an individual student. Basic principles for whole class questioning include (Huetinck & Munshin, 2008):

- Do not ask a question unless you want a thoughtful response.
- Try not to embarrass a student if they cannot answer a question but ensure that all students have a chance to answer at least one question during the week.
- Ensure that you address the problem of a student saying, 'I don't know' or 'I have no idea', and if this type of answer persists, express your concern to the student (privately) that you want the student to increase their mathematical understanding.

Table 2.2 Facets of understanding and examples

Facet	Sample question
Understanding	We know that −3 times −5 is 15. But why is this so? How can you explain this to someone else?
Interpretation	What happens when we add a large number to a series of numbers where we want to find the mean? What happens to the median?
Application	Why would a real estate agent use the median instead of the mean to help sell a house?
Perspective	In what circumstances would the median be used instead of the mean?
Empathy	We know that −3 times −5 is 15. Why do you think people find this a difficult concept to understand and remember?
Self-knowledge	We know that −3 times −5 is 15. Do you remember when you found this difficult to understand and what was it that helped you realise this was true?

Source: Adapted from Wiggins & McTighe, 1998.

- Write higher-order thinking questions in your lesson plans.
- Ensure that you have 'wait time' (more than three seconds).

Also be aware of the series or sequence in which the questions are asked. Herbel-Eisenmann and Breyfogle (2005) suggest the use of a series of 'focusing' questions. These could include questions that could focus on translating, clarifying, interpreting, applying, synthesising or evaluating (reflecting Bloom's Taxonomy). Johnson (1982) suggests leaving a question unanswered at the end of the lesson but ensuring that you come back to it later. He also suggests to not overuse the questions: 'Do you have any questions?' and 'Does everyone understand?'

If you have access to mini whiteboards, you can gauge students' level of understanding without embarrassing individual students. A more high-tech version of this is learner–response devices (using clickers or mobile phones) that can provide teachers with instant feedback. Using a collaborative classroom management software program where all students have a tablet wirelessly connected to a teacher's learning management system can be used by the teacher to individually gauge a student's performance and ask students to answer questions, displaying the results on a screen without the student standing in front of the class. You could also utilise your Learning Management System where interactive questions can be posed using software such as Camtasia® or Panopto® and answered anonymously.

SHORT-ANSWER QUESTION

The questions below can be improved. Write alternatives to elicit deeper understanding:

- What is Pythagoras' theorem?
- What is the derivative of $y = x^2 + 2x + 4$?
- Would you say x is 10?
- Does everyone understand?
- Has anyone worked out the answer to question 4?

IN PRACTICE

Potential mathematical language difficulties for students

In developing lesson plans (discussed further in Chapter 9), consider the potential mathematics language difficulties of your students. The following is a work sample developed by Thomas et al. (2015) for a typical algebra lesson, showing both the learning objectives and part of the learning sequence. Once you have identified the learning goals, then you need to:

- examine the mathematical language and skill demands that are involved (including listening and reading as well as speaking and writing)
- identify the language barriers. Students may be asked to recognise, comprehend and utilise mathematical verbal phrases (e.g. increased by, more than, difference, multiplied by, quotient, divided into) in order to interpret algebraic expressions (e.g. $q + 5$, $\$29.95 \times T + \$5$).
- examine the solutions to these barriers (e.g. peer pair or small group learning to help generate verbal ideas, use cue cards to remind students of order of operations).

Writing Expressions Lesson Plan

Content Standard
Represent and analyze mathematical situations and structures using algebraic symbols
- Write equivalent forms of equations, inequalities, and systems of equations and solve them with fluency
- Understand the meaning of equivalent forms of expressions, equations, inequalities, and relations – mentally or with paper and pencil in simple cases and using technology in all cases
- Use symbolic algebra to represent and explain mathematical relationships (NCTM Principles & Standards, p. 395)

Learning Objectives
Students will be able to use verbal phrases to translate real world situations into algebraic expressions.

Prior Knowledge
Evaluate algebraic expressions involving exponents
Utilize the order of operations
Use all of the basic operations with whole numbers, fractions, and decimals
Recognize and explain the importance and use of variables
Describe algebraic expressions as a combination of variables, numbers, and operations that describe a real world pattern or situation

Future Knowledge
Write equations and inequalities
Represent functions using rules, tables, and graphs

Assessments
In their daily homework assignment the students will be asked to
- Translate ten verbal phrases into expressions (i.e., 5 more than a number q = q + 5)
- Write expressions for seven situations (i.e., # of pages of a 25 page chapter left to read if you've read r pages = 25 − r)
- Solve six word problems by using verbal phrases to write expressions (i.e., Tickets to the concert cost $29.95 each. There is a $5 surcharge for ordering on-line. Write an expression for the cost (in dollars) of ordering tickets. Then find the total cost if you ordered 6 tickets. = $29.95 x T + $5; $29.95 x 6 + $5 = $184.70)

Figure 2.2 Sample Lesson Plan – Writing Expression
Source: Thomas et al., 2015.

Instruction

Time	Instruction (Teacher/Student)
10 min.	**Warm-up/Bell Ringer** Students are asked to note the order of operations in sequence and then show the steps to evaluate several expressions. a) $30 \times 2 \div 2^2 - 5$ b) $15 - 9 + 3$ c) $16 - 2^3 + 4$ d) $50 - (3^2 + 1)$ Student should record in their math journals the following: Grouping Symbols: (); []; fraction bar Exponents Multiply & Divide – left to right Add & Subtract – left to right a) $30 \cdot 2 \div 2^2 - 5$ b) $15 - 9 + 3$ $30 \cdot 2 \div 4 - 5$ $6 + 3$ $60 \div 4 - 5$ $\boxed{9}$ $15 - 5$ $\boxed{10}$ c) $16 - 2^3 + 4$ d) $50 - (3^2 + 1)$ $16 - 8 + 4$ $50 - (9 + 1)$ $8 + 4$ $50 - 10$ $\boxed{12}$ $\boxed{40}$
20-25 min	**Key Concept Introduced** Introduce the idea of a **verbal phrase** – words and numbers used to indicate mathematical operations For example: the sum of 4 and a number n Some key verbal phrases that indicate operations include: **Addition**: sum, plus, total, increased by, more than - *order is switched for this one* **Subtraction**: difference, less than, minus, decreased by **Multiplication**: times, product, multiplied by, of **Division**: quotient, divided by, divided into It is critical to remember that order makes a difference when working with subtraction and division. This means that "the difference of a number n and 5" is written n-5 NOT 5-n, and "the quotient of a number s and 5" is written $\frac{s}{5}$ NOT $\frac{5}{s}$.

Figure 2.2 (*cont.*)

PAUSE AND THINK

Based on the example above, write a lesson plan that takes into consideration the language issues in a particular section of the curriculum or reconsider ones you have written to ensure you have accounted for such language difficulties.

Record yourself explaining to a student how to solve these problems: ▶▶

▶ ▶

Audio & video recordings

a. Solve $12x - 10 = 6x + 32$.

b. A club has 86 members with 14 more female members than male members. How many males and females are members of the club?

The online resources for this text include recordings of student teachers explaining how to solve the problem above. Compare your explanation to that of the other student teachers.

Conclusion

Language plays a pivotal role in the teaching and learning of mathematics. As a teacher, you need to be fluent in the everyday language of the classroom and the particular language of mathematics. In addition, you need to recognise the important role language plays in the ease with which students understand mathematical concepts.

As a teacher, you need to be aware of the impact that mathematical language has on learning by considering that:

- words used in mathematics may have a different meaning in everyday language
- 'definitions' in mathematics may have different meanings depending on the level of the student
- syntactic and semantic structure of word problems can make interpretation of a word problem much more difficult
- second language learners have particular difficulties in the mathematics classroom, but that they may also bring teachable moments through rich discussion on the mathematics register in other languages.

A key to effective teaching of mathematics is the appropriate use of language in the classroom, both oral and written. A teacher should encourage meaningful, directed and fluent student talk at the appropriate level and at the appropriate times. This involves modelling the correct language, encouraging speaking and framing questions to promote active learning.

BRINGING IT TOGETHER

1. In geometry, the words 'diameter' and 'radius' are closely related, but the words themselves do not have any semantic link. In Chinese, diameter is 直径 [zhíjìng], literally meaning 'straight path', and radius is 半径 [bànjìng], meaning 'half path'. How would you structure a section of a lesson around a discussion on mathematical word meaning?

2. What is the difference between a square metre and a metre squared? Think of the structure of the English language. We write m^2, so in reading from left to right the 'metre' is first encountered, but we say square metres.

Download: Technical communication

3. Solve a set of simultaneous equations as accurately as you can using the conventions of formal mathematics (refer to the Technical Communication resource in the online resources for this text).

$$y = x - 5 \qquad \text{(A1)}$$

$$y = \frac{1}{2}x - 1 \qquad \text{(A2)}$$

Record yourself solving these equations as if you were teaching students and check that your language is correct. Listen particularly to the words and phrases you use. For example:

- Did you use the word 'simply' at any stage or 'just do'? (Remember, students are learning this, so it is not simple.)
- Did you use the correct terms such as 'equations' and 'expressions' in the right context?
- Did you use the terms 'unknown' and 'variable' in the right way?
- Did you put '=' signs under '=' signs?
- Did you explain from one line to the next what to do and write it down, e.g. 'substitute $x = 2$ into (A1)'?
- Did you have the final statement in proving your solution is correct as something like '$10 = 10$'?

As a teacher, you should decide on the level of formality of mathematics language. You should not correct students every time they make such language errors, but as a teacher, you should set an example. If you make errors as above, try to correct yourself, and then explain your error.

4. Go back to the 11 categories of difficulties associated with learning the words of mathematics. Think of a lesson you have recently taught or are about to teach. Was there a teaching moment in that lesson that could be used to highlight the language issue in mathematics?

5. Reflect on a mathematics lesson you are going to teach. Can you estimate how much time you plan on talking versus your students talking? Consider how much time you will allocate for each of your students to have a chance to speak mathematics (either to the class or in small groups). Listen to the recordings of student teachers explaining key concepts and solving problems available on the website for this text in order to help you generate your lesson plan.

REFERENCES

Further resources

Adams, T.L., Thangata, F. & King, C. (2005). 'Weigh' to go! Exploring mathematical language. *Mathematics Teaching in the Middle School*, 10(9), 444–8.

Australian Curriculum, Assessment and Reporting Authority (ACARA) (2018). *Australian Curriculum*. Sydney: ACARA. Retrieved from https://www.australiancurriculum.edu.au/

Australian Institute for Teaching and School Leadership (AITSL) (2011). *Australian Professional Standards for Teachers*. Melbourne: AITSL.

Australian Institute of Health and Welfare (AIHW) (2020). *Asthma*. Cat. no. ACM 33. Canberra: AIHW. Retrieved from https://www.aihw.gov.au/reports/chronic-respiratory-conditions/asthma/contents/deaths

Bane, M. (2008). Quantifying and measuring morphological complexity. In C.B. Chang & H.J. Haynie (eds), *Proceedings of the 26th West Coast Conference on Formal Linguistics*. Somerville, MA: Cascadilla Press.

Barton, B., Fairhall, U. & Trinick, T. (1998). Tikanga Reo Ttitai: Issues in the development of a Maori mathematics register. *For the Learning of Mathematics*, 18(1), 3–9. https://doi: 10.1080/09500780508668668

Barton, B. & Neville-Barton, P. (2005). *The relationship between English language and mathematics learning for non-native speakers*. Wellington: Unitec. Retrieved from http://www.tlri.org.nz/sites/default/files/projects/9211_summaryreport.pdf

Boaler, J. (2013). Ability and mathematics: The mindset revolution that is reshaping education. *FORUM*, 55(1), 143–52.

Borowski, E.J. & Borwein, J.M. (1989). *Collins dictionary of mathematics* (2nd edn). London: Harper Collins.

Cheeseman, J. (2003). Orchestrating the end of mathematics lessons. In B. Clarke, A. Bishop, H. Forgasz & W.T. Seah (eds), *Making mathematicians* (17–26). Brunswick: Mathematical Association of Victoria.

Clement, J., Lochhead, J. & Monk, G. (1981). Translational difficulties in learning mathematics. *American Mathematical Monthly*, 88(4), 286–90. https://doi: 10.2307/2320560

Council of Australian Governments (2008). *National Numeracy Review Report, May 2008*. Canberra: Commonwealth of Australia. Retrieved from http://hdl.voced.edu.au/10707/143183

Ellerton, N.F. & Clements, M.A. (1996). Newman error analysis: A comparative study involving Year 7 students in Malaysia and Australia. In P.C. Clarkson (ed.), *Technology and mathematics education. Proceedings of the 19th Annual Conference of the Mathematics Education Research Group of Australasia (MERGA)* (vol. 1, 186–93). Melbourne: MERGA.

Fuson, K. & Kwon, Y. (1991). Chinese based regular and European irregular systems of number words: Disadvantages for English speaking children. In K. Durkin & B. Shire (eds), *Language in mathematical education: Research and practice* (211–26). Milton Keys: Open University Press.

Galligan, L. (1997). Differences in problem processing: A comparison of English and Chinese mathematical word problems. In A. Begg (ed.), *People in mathematics education. Proceedings of the 20th Annual Conference of the Mathematics Education Research Group of Australasia (MERGA)* (vol. 1, 177–83). Rotorua: MERGA.

Galligan, L. (2001). Possible effects of English – Chinese language differences on processing of mathematical text: A review. *Mathematics Education Research Journal*, 13(2), 112–32. https://doi: 10.1007/BF03217102

Galligan, L. & Hobohm, C. (2013). Investigating inking devices to support learning in mathematics. In V. Steinle, L. Ball & C. Bardini (eds), *Mathematics education: Yesterday, today and tomorrow. Proceedings of the 36th Annual Conference of the Mathematics Education Research Group of Australasia (MERGA)* (322–9). Melbourne: MERGA.

Goos, M. (2004). Learning mathematics in a classroom community of inquiry. *Journal for Research in Mathematics Education*, 35(4), 258–91. https://doi:10.2307/30034810

Goos, M., Stillman, G. & Vale, C. (2007). *Teaching secondary school mathematics research and practice for the 21st century*. Crows Nest: Allen & Unwin.

Halliday, M.A.K. (1978). *Language as the social semiotic: The social interpretation of language and meaning*. Baltimore, MD: University Park Press.

Herbel-Eisenmann, B.A. & Breyfogle, M.L. (2005). Questioning our patterns of questioning. *Mathematics Teaching in the Middle School*, 10(9), 484–9.

Huetinck, L. & Munshin, S.N. (2008). *Teaching mathematics for the 21st century: Methods and activities for grades 6–12* (3rd edn). Hoboken, NJ: Pearson.

Hughes, R. (1986). *The fatal shore*. New York: Alfred A. Knopf Inc.

Irons, R. (2014). Language is the core for the concept of addition. *Educating Young Children: Learning and Teaching in the Early Childhood Years*, 20(1), 38–41.

Johnson, D.R. (1982). *Every minute counts, making your math class work*. Palo Alto, CA: Dale Seymour Publications.

Jourdain, L. & Sharma, S. (2016) Language challenges in mathematics education: A literature review. *Waikato Journal of Education*, 21(2), 43–56.

Kemp, M. & Kissane, B. (2010). A five step framework for interpreting tables and graphs in their contexts. In C. Reading (ed.), *Data and context in statistics education: Towards an evidence-based society. Proceedings of the Eighth International Conference on Teaching Statistics* (ICOTS8, July 2010) (vol. 8), Ljubljana, Slovenia. Voorburg: International Statistical Institute.

Larkin, K., Jamieson-Proctor, R. & Finger, G. (2012). TPACK and pre-service teacher mathematics education: Defining a signature pedagogy for mathematics education using ICT and based on the metaphor 'mathematics is a language'. *Computers in the Schools*, 29(1–2), 207–26.

Leinhardt, G., Zaslavsky, O. & Stein, M.K. (1990). Functions, graphs, and graphing: Tasks, learning, and teaching. *Review of Educational Research*, 60(1), 1–64. https://doi:10.3102/00346543060001001

Leung, C. (2005). Mathematical vocabulary: Fixers of knowledge or points of exploration? *Language and Education*, 19(2), 127–35. https://doi: 10.1080/09500780508668668

Lopez-Real, F. (1997). Effect of the different syntactic structures of English and Chinese in simple algebra problems. In A. Begg (ed.), *People in mathematics education. Proceedings of the 20th Annual Conference of the Mathematics Education Research Group of Australasia* (vol. 2, 317–23). Rotorua.

Lowrie, T., Diezmann, C.M. & Logan, T. (2011). Understanding graphicacy: Students' making sense of graphics in mathematics assessment tasks. *International Journal for Mathematics Teaching and Learning*, 1–32.

MacGregor, M. (1991). *Making sense of algebra: Cognitive processes influencing comprehension*. Waurn Ponds: Deakin University Press.

MacGregor, M. & Moore, R. (1991). *Teaching mathematics in the multilingual classroom: A resource for teachers and teacher educators*. Melbourne: Institute of Education, University of Melbourne.

Maddern, S. & Court, R. (1989). *Improving mathematics practice and classroom teaching*. Nottingham: Shell Centre for Mathematics Education.

Marton, F. & Tsui, A. (2004). *Classroom discourse and the space of learning*. Mahwah, NJ: Lawrence Erlbaum.

Morgan, C. (2005). Word, definitions and concepts in discourses of mathematics, teaching and learning. *Language and Education*, 19(2), 102–16. https://doi: 10.1080/09500780508668666

Moschkovich, J. (2007). Using two languages when learning mathematics. *Educational Studies in Mathematics*, 64(2), 121–44. https://doi: 10.1007/s10649-005-9005-1

Nathan, G., Trinick, T., Tobin, E. & Barton, B. (1993). Tahi, Rua, Toru, Wha: Mathematical counts in Maori renaissance. In M. Stephens, A. Waywood, D. Clarke & J. Izard (eds), *Communicating mathematically: Perspectives from classroom practice and current research*. Camberwell: Australian Council for Educational Research.

National Association for Language Development in the Curriculum (2011). *Some issues concerning EAL in the mathematics classroom*. Retrieved from https://www.naldic.org.uk/Resources/NALDIC/Initial%20Teacher%20Education/Documents/Maths1.pdf

National Council of Teachers of Mathematics (NCTM) (2000). *Principles and standards for school mathematics*. Reston, VA: NCTM. Retrieved from http://www.ms.uky.edu/~lee/ma310sp09/Standards_for_School_Mathematics_Communication.pdf

Newman, M.A. (1983). *Strategies for diagnosis and remediation*. Sydney: Harcourt, Brace Jovanovich.

O'Connor, M.C. & Michaels, S. (1996). Shifting participant frameworks: Orchestrating thinking practices in group discussion. In D. Hicks (ed.), *Discourse, learning and schooling* (63–103). New York: Cambridge University Press.

O'Halloran, K.L. (2015). The language of learning mathematics: A multimodal perspective. *The Journal of Mathematical Behaviour*, 40(Part A), 63–74. https://doi: 10.1016/j.jmathb.2014.09.002

Parker, M. (2014). *Things to make and do in the fourth dimension*. New York: Farrar, Straus and Giroux.

Pimm, D. (1991). Communicating mathematically. In K. Durkin & B. Shire (eds), *Language in mathematical education: Research and practice* (17–23). Milton Keys: Open University Press.

Riccomini, P.J., Smith, G.W., Hughes, E.M. & Fries, K.M. (2015). The language of mathematics: The importance of teaching and learning mathematical vocabulary. *Reading & Writing Quarterly*, 31(3), 235–52. https://doi: 10.1080/10573569.2015.1030995

Ryan, J. & Williams, B. (2007). *Children's mathematics 4–15: Learning from errors and misconceptions*. Maidenhead: McGraw-Hill Education.

Sammons, L. (2018). *Teaching students to communicate mathematically*. Alexandria, VA: Association for Supervision and Curriculum Development.

Spanos, G., Rhodes, N., Dale, T. & Crandall, J.A. (1988). Linguistic features of mathematical problem solving: Insights and applications. In R. Cocking & J. Mestre (eds), *Linguistic and cultural influences on learning mathematics*. Mahwah, NJ: Erlbaum.

Sullivan, P. (2011). *Teaching mathematics: Using research-informed strategies*. Canberra: Australian Council for Educational Research. Retrieved from http://research.acer.edu.au/aer/13

Thomas, C.N., Van Garderen, D., Scheuermann, A. & Lee, E.J. (2015). Applying a universal design for learning framework to mediate the language demands of mathematics. *Reading & Writing Quarterly*, 31(3), 207–34. https://doi: 10.1080/10573569.2015.1030988

Walshaw, M. & Anthony, G. (2008). The teacher's role in classroom discourse: A review of recent research into mathematics classrooms. *Review of Educational Research*, 78(3), 516–51. https://doi:10.3102/0034654308320292

Watson, A. & Sullivan, P. (2008). Teachers learning about tasks and lessons. In D. Tirosh & T. Wood (eds), *Tools and resources in mathematics teacher education* (109–35). Rotterdam: Sense Publishers.

Wiggins, G. & McTighe, J. (1998). *Understanding by design*. Alexandria, VA: Association for Supervision and Curriculum Development.

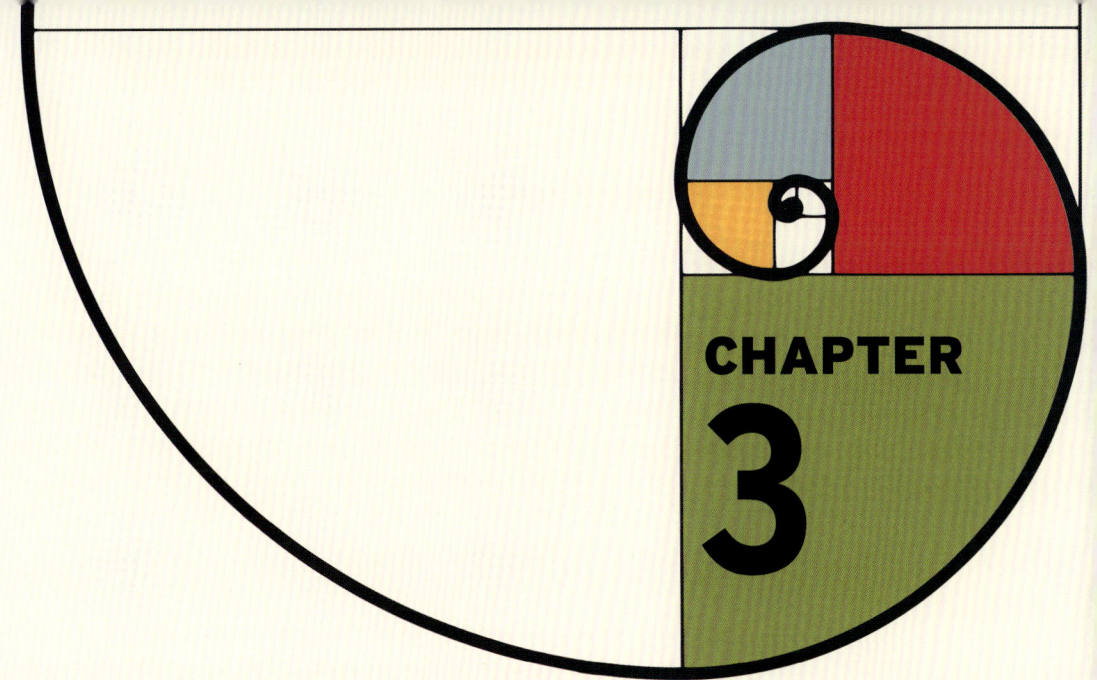

Making mathematical connections

Gregory Hine

LEARNING OBJECTIVES

The learning objectives of this chapter are directly linked to the Australian Professional Standards for Teachers (APST) (Australian Institute for Teaching and School Leadership [AITSL], 2011). After studying this chapter, you should be able to:

- articulate the importance of helping students make mathematical connections in their learning (APST 1.2)
- define the term 'mathematical connections' in a secondary context (APST 2.1)
- understand teachers' practices in promoting mathematical connections (APST 2.1)
- outline key features, instructional approaches and challenges associated with STEM education (APST 2.1)
- outline key instructional approaches and tasks that can help students connect with mathematics (APST 1.2, 2.1).

Introduction

This chapter commences with an examination of the importance of making mathematical connections in the secondary classroom. Next, various conceptualisations of the term 'mathematical connections' will be presented in an attempt to define the term itself. Then, the work of practitioners and researchers is offered in the way of useful and appropriate instructional practices and guidelines. Science, technology, engineering and mathematics (STEM) education will also be articulated as a way of making mathematical connections clear and relevant to secondary students. Finally, a variety of activities are included for teachers to help students make mathematical connections both within the discipline and to real-world settings.

The importance of making mathematical connections

The Australian Curriculum, Assessment and Reporting Authority (ACARA) recognises the importance of making mathematical connections. Specifically, ACARA (2020a) articulates this importance in its curriculum *Aims*, which are to ensure that students:

- are confident, creative users and communicators of mathematics, able to investigate, represent and interpret situations in their personal and work lives and as active citizens
- develop an increasingly sophisticated understanding of mathematical concepts and fluency with processes, and are able to pose and solve problems and reason in Number and Algebra, Measurement and Geometry, and Statistics and Probability
- recognise connections between the areas of mathematics and other disciplines and appreciate mathematics as an accessible and enjoyable discipline to study.

The importance of helping students of mathematics make connections both within the discipline and to the real world is a claim that receives consistent international reinforcement (Boaler, 2002; Gainsburg, 2008; Sullivan, 2011). Although learners might make connections spontaneously, teachers cannot 'assume that the connection will be made without some intervention' (Weinberg, 2001, p. 26). Implied within this statement is the exhortation for teachers to take an active role in making mathematical connections clear to students. For instance, Ma (1999) contended that to help students make connections, teachers must understand mathematics as an interrelated web of ideas, and possess the appropriate pedagogical content knowledge to know which strategies and activities best facilitate student learning. The National Council of Teachers of Mathematics (NCTM) asserts that teachers must be instrumental in developing a deep conceptual understanding of mathematics within students and assist in departing from a perception of mathematics as a 'set of isolated facts and procedures' (NCTM, 2009, p. 3). Moreover, the Australian Curriculum: Scope and Sequence documents can be viewed as a conceptual 'road map' of content which can be used strategically by educators in planning units of work replete with meaningful connections.

Pedagogical content knowledge: the teacher's interpretations and transformations of content knowledge in the context of facilitating student learning.

The literature base underscores the importance of making mathematical connections across mathematical concepts (Businskas, 2008) and to the real world (Boaler, 1997; Gainsburg, 2008). Leikin and Levav-Waynberg (2007, p. 350) assert that 'connecting mathematical ideas means linking new ideas to related ones and solving challenging mathematical tasks by seeking familiar concepts and procedures that may help in new situations'. As discussed in Chapter 1, Skemp (1987) discussed two types of mathematical understanding critical for progress in mathematics learning: relational and instrumental. While instrumental understanding implies habitual learning, or acquiring 'rules without reasons', it 'does not promote thinking about how familiar concepts and procedures can help in new situations' (Leikin & Levav-Waynberg, 2007, p. 350). By contrast, relational understanding requires the learner to make connections between different mathematical concepts and to make progress by linking these concepts to prior knowledge (Skemp, 1987). Mathematical connections are thought to benefit student learning (Gainsburg, 2008) by enhancing students' capacity to remember, appreciate and use mathematics (Businskas, 2008) and to develop conceptual understanding (Anthony & Walshaw, 2009a). Additionally, the use of real-world connections can motivate mathematics learning (National Academy of Sciences, 2003) and help students apply mathematics to real problems – particularly those in the workplace (Gainsburg, 2008; National Research Council, 1998). To make connections in the secondary classroom, authors argue that teachers must constructively use students' prior knowledge (Hattie, 2009; Hattie et al., 2016; Swan, 2005) so that a 'deeper and more lasting understanding' can occur (NCTM, 2000, p. 64). After discerning what students can and cannot do, it is recommended that teachers build connections from previous lessons, experiences and thinking (Anthony & Walshaw, 2009b; Clarke & Clarke, 2004).

Instrumental understanding: the acquisition of rules without the promotion of thinking about how familiar concepts and procedures can help in new situations (Leikin & Levav-Waynberg, 2007).

Relational understanding: learning to make connections between mathematical concepts and advancing understanding by linking those concepts to prior knowledge (Leikin & Levav-Waynberg, 2007).

PAUSE AND THINK

Think back to a time when a mathematical concept made perfect sense to you. What was the concept being taught, and how was it taught by the teacher? How will you aim for more of this 'connective' thinking in your own mathematics lessons?

Mathematics connections in Australia: status quo

In addition to ACARA, various documents point to the importance of mathematical connections. For instance, the Australian Education Report (Sullivan, 2011) drew upon research findings and other sets of recommendations for teaching actions, developing six principles to guide teaching practice. Of these six principles, *Principle 2: Making connections* is elaborated for teachers as contextualising students' mathematical and experiential knowledge by crafting stories to 'establish a rationale for the learning' (Sullivan, 2011, p. 26).

A Trends in International Mathematics and Science Study (TIMSS) found Australian students continued to rank in the top third of all participating countries. Interestingly,

48 per cent of Singaporean Year 8 students demonstrated performance at the *Advanced International Benchmark*, while only nine per cent of Australian Year 8 students reached this same benchmark (Thomson et al., 2012). In a previous study, the *Advanced International Benchmark* was attained by 44 per cent of Singaporean students and only 7 per cent of Australian students (Thompson & Fleming, 2004). The Advanced International Benchmark is included in Table 3.1.

Table 3.1 TIMMS Advanced International Benchmark Descriptor

Advanced International Benchmark	Students can organise and draw conclusions from information, make generalisations and solve non-routine problems. Students can solve a variety of fraction, proportion and percent problems and justify their conclusions. Students can express generalisations algebraically and model situations. They can solve a variety of problems involving equations, formulas and functions. Students can reason with geometric figures to solve problems. Students can reason with data from several sources or unfamiliar representations to solve multi-step problems.

Source: Thomson et al., 2012, p. 21.

In an analysis of the Singaporean mathematics curriculum, Kaur (2001, p. 141) noted an emphasis on 'the development of mathematical concepts and skills, and the ability to apply them to solve problems'. During lessons, she observed mathematics teachers placing emphasis on students solving non-routine problems. In an Australian context, research conducted into teaching approaches for Year 8 students suggested that few complex problem-solving opportunities are being provided (Hollingsworth, 2003). Furthermore, and consistent with Skemp's notion of instrumental understanding, Stacey (2003, p. 119) appraised the average Australian mathematics lesson as one exhibiting 'a syndrome of shallow teaching, where students are asked to follow procedures without reasons'. Stigler and Hiebert (1999) proposed that one factor influencing the lack of adoption of problem-solving strategies concerns primarily the teachers' knowledge and beliefs about mathematics teaching and learning.

Analysts of the 2012 Programme for International Student Assessment (PISA) report (De Bortoli & Macaskill, 2014) highlight that along with England and the United States, Australian students demonstrated excellent performance in problem solving. According to De Bortoli and Macaskill (2014), this may suggest that 'in these countries, top performers in mathematics have access to – and take advantage of – the kinds of learning opportunities that are also useful for improving their problem-solving skills' (p. 46). Additionally, when compared to other countries 'Australian students are comparatively stronger on both the *Exploring and Understanding* and the *Representing and Formulating* processes, and are relatively weaker on the *Planning and Executing* process' (p. 46). Despite this acclaim, there has been a significant decline in the performance of Australian students who are classified as top performers, average performers and low performers. In other words, there have been significant declines for Australian students at the 10th, 25th, 75th and 90th percentiles between PISA 2003 and PISA 2012 (Thomson et al., 2012, p. 44). Furthermore, from 2003 (when mathematical literacy was first assessed as a major domain) to PISA 2018, Thomson et al. (2019, p. 112) have noted that '. . . Australia's average score has declined by 33 points (more than one year of schooling)'. Following the earlier comparison between Singaporean and Australian students, the proportion of Singaporean students (40%) who achieved Proficiency Level 5 or 6 far exceeded that of Australian students (15%). These Proficiency Levels have been included in Table 3.2.

Table 3.2 PISA Proficiency Levels 5 and 6

Level 6	[Students] conceptualise, generalise and use information. They are capable of advanced mathematical thinking and reasoning; have a mastery of symbolic and formal mathematical operations and relationships; and can formulate and precisely communicate their findings, interpretations and arguments.
Level 5	[Students] develop and work with models for complex situations; select, compare and evaluate appropriate problem-solving strategies for dealing with complex problems; work strategically using broad, well-developed thinking and reasoning skills; and reflect on their work and formulate and communicate their interpretations and reasoning.

Source: Thomson, de Bortoli & Buckley, 2013, p. 13.

Given the evidence that Year 8 students experience little complex problem solving in mathematics classes, as well as the results from PISA and TIMSS, there is considerable room for change (Anderson, 2005). To this point, Stacey (2003, p. 122) recommends that 'there needs to be a greater emphasis on explicit mathematical reasoning, deduction, connections and higher-order thinking in lessons'. Drawing on the work of scholars within this section, one can posit that Australian students require access to learning activities that intentionally promote mathematical connections.

SHORT-ANSWER QUESTIONS

1. ACARA emphasises the importance of mathematical connections in the *Aims* of the Australian Curriculum. Restate these aims in your words and argue why each aim is important for you as a mathematics teacher.
2. Richard Skemp posited two types of mathematical understanding critical for progress in mathematics learning: *relational* and *instrumental*. Define each type of understanding and provide an example of each.
3. In your own words, define the term *pedagogical content knowledge*. In your definition, what key words did you use?
4. Data from a TIMSS reflect that 48 per cent of Singaporean Year 8 students demonstrated performance at the *Advanced Benchmark*, while only 9 per cent of Australian Year 8 students reached this same benchmark. What explanations could account for this discrepancy in achievement?

What are mathematical connections?

So, what is meant by the term 'mathematics connection'? If we commence with a general definition for the term 'connection' as 'a relationship between two things or ideas'

(Cambridge Dictionary, 2020), further questions must be asked in a mathematical context. For instance, Businskas (2008, p. 7) proposed that a connection is a feature of the content matter (e.g. a relationship between ideas), a relationship that is constructed by the learner (e.g. a mental construction in the learner's mind) or a 'process that is part of the activity of doing mathematics'. In the United States, the NCTM outlines six principles which are 'statements reflecting basic precepts that are fundamental to a high-quality mathematics education' (2015, p. 2). The second principle is Curriculum, which is articulated as:

> In a coherent curriculum, mathematical ideas are linked to and build on one another so that students' understanding and knowledge deepen and their ability to apply mathematics expands. An effective mathematics curriculum focuses on important mathematics that will prepare students for continued study and for solving problems in a variety of school, home, and work settings. A well-articulated curriculum challenges students to learn increasingly more sophisticated mathematical ideas as they continue their studies.

This principle focuses explicitly on both those connections which link mathematical ideas together, as well as the appreciation and application of mathematical concepts to contexts outside the classroom.

Connections: feature, relationship or process?

Much of the literature presents the notion of mathematical connections as a process (Boaler, 2002; Evitts, 2004; Ma, 1999). For instance, Ma (1999) contended that having a profound understanding of mathematics required an ability to connect ideas within a topic and to central concepts of the discipline. In a similar vein, Boaler (2002, p. 11) highlighted that 'the act of observing relationships and drawing connections, whether between different functional representations or mathematical areas, is a key aspect of mathematical work, in itself, and should not only be thought of as a route to other knowledge'.

CONNECTION   Thomas Evitts: making connections through problem solving

Thomas Evitts was interested in how preservice teachers used problem-solving activities as a process for making mathematical connections. He observed that preservice teachers were engaged in making a variety of connections, which were categorised as: modelling, representational, structural, procedure-concept, or between strands of mathematics.

- Modelling: Attempting to find some aspect of prior mathematical knowledge that could be used to portray some real-world component of the problem in a mathematical way.
- Representational: Using two or more representations to talk about the same mathematical idea.
- Structural: Discussing and using similarities found between a real or mathematical component of the problem and another real or mathematical situation.

- Procedure-Concept: Describing or using procedures via a rule or formula-based approach. Indicating a conceptual basis for utilising a procedure.
- Between Strands of Mathematics: 'Crossing over' from one strand of mathematics to another when analysing the problem, and using references to other areas of mathematics (Evitts, 2004, p. 56).

Questions

1. Look at the list of Evitts' categories and select several worded problems from a secondary mathematics text. After solving each problem, write a few sentences to summarise the process you used to solve the problem. Which of Evitts' categories did your approaches best correspond to?
2. Explain how this knowledge of your problem-solving approaches could be used in the secondary classroom.

Coxford (1995) conceptualised connections as features, or very broad ideas linking different topics in mathematics. In this conceptualisation, he identified three categories of mathematical connections – unifying themes, mathematical processes and mathematical connectors. First, *unifying themes* draw attention to the connected nature of mathematics. For instance, the theme of change may connect algebra, polynomials, differential calculus and geometry. Second, key *mathematical processes* include representation, application, problem solving and reasoning. To illustrate, lower secondary students (Years 7–10) should develop competency in moving fluidly through the concrete–representational–abstract notions of fractions, decimals and percentages. Coxford (1995, p. 7) maintained that these connections 'are vital if students are to make sense out of later operations on numbers'. Third, *connectors* are mathematical ideas (e.g. graph, algorithm, variable, proportion, function) which arise in relation to studying a wide spectrum of topics. According to Coxford, these ideas 'permit the student to see the use of one idea in many different and, perhaps, seemingly unrelated situations' (1995, p. 10). While this section has dealt with the notion of connections as features and processes, one should be careful not to dismiss the extent to which mathematical connections exist in the minds of learners. On this point, Hodgson (1995, pp. 14–15) commented that 'if students are unable to establish connections, then the connections cannot be used in problem situations regardless of whether they exist or not'. In other words, the province of mathematical efficacy remains the enterprise of the learner, and the importance of teachers in making connections known explicitly to learners is underscored.

Enrico Fermi

CONNECTION

Enrico Fermi was an Italian physicist who made important discoveries in nuclear physics and quantum mechanics. In 1938, Fermi received the Nobel Prize for physics. Fermi was well known for his ability to make good estimates in situations where little information was known. A 'Fermi Question' asks for a quick estimate of a quantity that seems difficult to determine precisely. Fermi questions:

- are posed with limited information given
- require that students ask many more questions

- demand communication
- utilise estimation
- emphasise process rather than 'the' answer.

Download: Navajo Math
Fermi Questions

Before answering the questions below, access the Navajo Math Fermi Questions resource (https://navajomath .math.ksu.edu/wp-content/uploads/2015/03/fermi_questions_handouts_and_lesson_plan.pdf), also available in the online resources.

Questions

1. Look at the list of Fermi questions (pp. 5–6) and select one problem. How could you devise a lesson plan where this selected problem would feature? What strand, sub-strand and content descriptor(s) from the Australian Curriculum: Mathematics would you align this lesson with?
2. Explain how the Fermi Approach 'fits into' your teaching and learning philosophy.

Teachers' conceptualisation of connections

Teachers view mathematical connections as a function of instructional practice (Thompson, 1992), and a component of their mathematical content knowledge (Gainsburg, 2008). Businskas (2008, pp. 150–1) found that secondary teachers perceived mathematics as an interconnected web of concepts, and such connections comprised mathematical knowledge that became a strategic and integral element of their teaching. Another important finding was that some teachers expressed that emphasising mathematical connections within lessons would be time-consuming and could detract from their broader teaching responsibilities (e.g. covering the curriculum, preparing students for external assessments). Fennema and Franke (1992) concluded that when teacher content knowledge has been defined in a manner consistent with the nature of mathematics (or when a conceptual organisation of knowledge was considered) there existed a positive relationship between teacher content knowledge and their instruction. Building on this idea, Ball and Bass (2000, p. 89) viewed this relationship as the foundation for developing students' understanding, or as either a 'pedagogically useful understanding' or pedagogical content knowledge. Concerning the latter term, Shulman (1986, p. 9) asserted that 'to think properly about content knowledge requires going beyond knowledge of the facts or concepts of a domain. It requires understanding of the structures of the subject'.

PAUSE AND THINK

What do you consider to be the structures of mathematics that mathematics teachers need to understand?

Constraints

Teachers' attempts at engaging students in real-world problem solving can be constrained by many factors. One factor includes meeting requirements for standardised testing and externally mandated curricula, although some teachers believe real-world connections can prepare students for tests (Gainsburg, 2008). Sullivan (2011) notes that teachers experience students who avoid risk taking and do not persist when challenges arise. After discerning the avoidance strategies used by students, teachers can also become complicit in adopting avoidance strategies of their own (Sullivan, Tobias & McDonough, 2006). For instance, teachers sometimes modify tasks at the planning stage if they expect that students will not engage with the tasks without considerable assistance (Tzur, 2008; Sullivan, 2011). Other teachers reduce the potential demand of the task to avoid the challenge of dealing with students who have given up (Desforges & Cockburn, 1987; Sullivan, 2011) or those who do not respond to activites as anticipated (Charalambous, 2008). Gainsburg (2008) found that although 'some teachers valued tasks that require critical thinking or promote literacy development, more feared that complex, ill-structured or language-intensive tasks would overwhelm students' (p. 219). As such, the key finding supported the notion that teachers were concerned more about over-challenging their students than under-challenging them.

SHORT-ANSWER QUESTIONS

1. In your own words, define the term 'mathematical connections'. In your definition, what key words did you use?
2. There is a considerable amount of literature which describes mathematical connections as a 'feature', 'relationship' or a 'process'. As a preservice teacher, which one of these terms fits most comfortably with your current view of mathematical connections?
3. List and describe some of the commonly espoused views of teachers regarding mathematical connections.
4. What are some constraints teachers face in engaging students in real-world problem-solving activities?

Teachers' practices in promoting mathematical connections

The literature is replete with practical ideas for teachers to make mathematical connections (Gainsburg, 2008; Little, 2019; Siemon, Banks & Prasad, 2018; Sullivan, 2011). Commonly, such ideas include careful scaffolding of learning (Stephens, 2009) and using problem solving as an instructional strategy (Anderson, 2009). Given the implementation of an Australian Curriculum, teachers have many opportunities to teach problem solving and use real-life problems as a focus of learning in mathematics lessons. Drawing upon earlier comments from Stacey (2003), there are concerns about the extent to which Australian students solve problems other than those of low procedural complexity. While some teachers may use non-routine and problem-centred tasks in lessons (Anderson & Bobis, 2005), research suggests that many do not (Anderson, Sullivan & White, 2004). Instead, there remains a preference for teachers to rely on presenting lessons focused on the types of questions found in examinations and in textbooks (Anderson, 2009; Doorman et al., 2007;

Kaur & Yeap, 2009; Vincent & Stacey, 2008). In this section, we look at some of what the current literature offers in a way of general instructional guidelines for making connections, the role of scaffolding within mathematical learning and various problem-solving strategies teachers can use.

PAUSE AND THINK

How do you plan to scaffold student learning? Which approach(es) will you predominantly use?

General guidelines for engaging students in mathematics

Sullivan (2011) listed broad recommendations for teachers wishing to engage students in learning mathematics. These recommendations include:

- examining the Australian mathematics curriculum and its 'big ideas' in order to 'inform long-term and daily planning' (2011, p. 61)
- in compulsory years, maintaining an emphasis on teaching and assessing skills in numeracy and practical mathematics
- identifying opportunities to encourage students' engagement, including connecting learning to prior knowledge and the purpose of the learning, and decision making
- making mathematics and numeracy a meaningful learning experience, through lessons filled with engaging and varied tasks
- identifying where common content knowledge is required for certain tasks, and strategising how this knowledge can be embedded into classroom teaching
- exploring relevant pedagogies for classes where extended learning may be required, and to also further support students who may have difficulty in understanding current specialised mathematics content knowledge.

In addition to these recommendations, finding relevance for what teachers do in their instructional lessons, Sullivan (2011) contends that these same items are important mathematics education components for the professional development of prospective primary and secondary teachers.

IN PRACTICE

Making connections: instructional practices

In a summary of mathematics education literature, Gainsburg (2008, p. 200) posited a range of specific instructional practices to assist teachers in creating connections for students:

- simple analogies (e.g. relating negative numbers to subzero temperatures)
- classic 'word problems' (e.g. 'Two trains leave the same station . . .')
- the analysis of real data (e.g. finding the mean and median heights of classmates)

- discussions of mathematics in society (e.g. media misuses of statistics to sway public opinion)
- 'hands-on' representations of mathematics concepts (e.g. models of regular solids, dice)
- mathematically modelling real phenomena (e.g. writing a formula to express temperature as an approximate function of the day of the year).

Big Ideas in mathematics education

SCENARIO

The notion of Big Ideas in mathematics education has existed for some time now, but how familiar are you with this terminology? Queensland College of Teachers (2015) explores Big Ideas and their importance to mathematics education in its Research Digest. In this article, two prevailing views about Big Ideas are unpacked, and five scenarios are described to provide mathematics teachers with insights about how to use Big Ideas in their classrooms.

Before answering the questions below, access the Queensland College of Teachers Research Digest, No. 11.

Questions

1. How would you describe the term *Big Ideas in mathematics education* and what these ideas seek to achieve?
2. From the five scenarios presented, which do you feel you would be able to use on your next practicum experience? Why?

Scaffolding

Bearing in mind that scaffolding (Wood, Bruner & Ross, 1976) enables workers to operate at their current level with assistance – and to make progress in their work – we look at how this term can be applied to an educational context. As introduced in Chapter 1, the work of Vygotsky (1978) suggested that the zone of proximal development describes the region that exists between what students can do independently and that which can only be completed with the assistance of others. For students to learn, Vygotsky believed that activities which 'bridged' or 'scaffolded' this zone were required – as opposed to tasks pitched at students' current level of learning. Mathematics teachers who possess high-level knowledge work with students who clearly have more basic forms of that knowledge (Stephens, 2009). To understand and build on what the students already know, Fennema and Romberg (1999) suggested that teachers use an approach of 'cognitively guided instruction'. Such an approach places a greater demand on teachers' mathematical knowledge for teaching, as students' responses and strategies can steer the lesson in various directions. According to Stephens (2009, p. 30), the teacher's role is then 'to draw together those different directions with a clear focus on enhancing students' understanding'.

Zone of proximal development: the region that exists between what students can do independently and that which can only be completed with the assistance of others.

In addition, Stephens (2009) identified a range of interaction patterns or scaffolding practices undertaken by teachers during mathematics lessons. Although these patterns and practices were drawn from a primary context, these instructional practices can support secondary teachers to make knowledgeable choices to meet the learning needs of all students in the most appropriate manner possible (Stephens, 2009). The 12 scaffolding practices that contribute to improved student learning outcomes are listed and described in Table 3.3. According to Stephens (2009, p. 30) 'when teachers used these scaffolding practices it had an effect on their own perceptions of what to teach and how to help students learn – that is, on their MPCK'. Consequently, there was a significant shift in what teachers perceived to be associated with effective mathematics teaching during the project (Stephens, 2009).

Table 3.3 Scaffolding practice

Scaffolding practice	Application of practice
1. **Excavating**: Drawing out, digging, uncovering what is known, making it transparent	Teacher systematically questions to find out what students know or to make the known explicit. Teacher explores children's understanding in a systematic way.
2. **Modelling**: Demonstrating, directing, instructing, showing, telling, funnelling, naming, labelling, explaining	Teacher shows students what to do and/or how to do it. Teacher instructs, explains, demonstrates, tells, offers behaviour for imitation.
3. **Collaborating**: Acting as an accomplice, co-learner/problem-solver, co-conspirator, negotiator	Teacher works interactively with students in-the-moment on a task to jointly achieve a solution. Teacher contributes ideas, tries things out, responds to suggestions of others, invites comments/opinions on what she/he is doing, accepts critique.
4. **Guiding**: Cuing, promoting, hinting, navigating, shepherding, encouraging, nudging	Teacher observes, listens, monitors students as they work, asks questions designed to help them see connections, and/or articulate generalisations.
5. **Convince me**: Seeking explanation, justification, evidence, proving	Teacher actively seeks evidence, encourages students to be more specific. Teacher may act as he/she doesn't understand what students are saying, encourages students to explain, to provide/obtain data.
6. **Noticing**: Highlighting, drawing attention to, valuing, pointing to	Teacher draws students' attention to particular features without telling students what to see/notice (i.e. by careful questioning, rephrasing or gestures), encourages students to question their sensory experience.
7. **Focusing**: Coaching, tutoring, mentoring, flagging, redirecting, re-voicing, filtering	Teacher focuses on a specific gap (i.e. a concept, skill or strategy) that students need to progress. Teacher maintains a joint collective focus and provides an opportunity for students to bridge the gap themselves.
8. **Probing**: Clarifying, monitoring, checking	Teacher evaluates students' understanding using a specific question/task designed to elicit a range of strategies, presses for clarification, identifies possible areas of need.
9. **Orienting**: Setting the scene, contextualising, reminding, alerting, recalling	Teacher sets the scene, poses a problem, establishes a context, invokes relevant prior knowledge and experience, provides a rationale (not necessarily at the beginning of the lesson, but at the beginning of a new task/idea).
10. **Reflecting/reviewing**: Sharing, reflecting, recounting, summarising, capturing, reinforcing, reflecting, rehearsing	Teacher orchestrates a recount of what was learnt, a sharing of ideas and strategies. This typically occurs during whole class share time at the end of a lesson where learning is made explicit, key strategies are articulated, valued and recorded.

Table 3.3 (*cont.*)

Scaffolding practice	Application of practice
11. **Extending**: Challenging, spring boarding, linking, connecting	Teacher sets significant challenge, uses open-ended questions to explore extent of children's understanding, facilitate generalisations, provide a context for further learning.
12. **Apprenticing**: Inviting peer assistance, peer teaching, peer mentoring	Teacher provides opportunities for more learned peers to operate in a student-as-teacher capacity, endorses student/student interaction.

Source: Stephens, 2009, p. 31.

SHORT-ANSWER QUESTIONS

1. As a preservice secondary mathematics teacher, what is your position on using real-world situations to teach problem-solving strategies?
2. Define the term *mathematical knowledge for teaching* in your own words. In your definition, what key words did you use?
3. Look at the scaffolding practices for making mathematical connections recommended by Stephens (2009). Which of these practices have you already tried while on practicum experience? Estimate the amount of instructional time and the amount of learning time for these practices (make sure these percentages sum to 100%).

Science, technology, engineering and mathematics (STEM) education

Science, technology, engineering and mathematics (STEM) education has maintained significant interest among educators, politicians and industry personnel worldwide for decades (Fan & Ritz, 2014). Tytler et al. (2008) state that 'The Australian Academy of Sciences (2006) has identified high-end mathematical skills as crucial to Australia's success in a wide range of fields, including: mining and resources, manufacturing and trade, biotechnology, statistics and finance, and environmental risk assessment' (p. 10). Around the same time as this identification, Tytler and Symington (2006) underscored the value of having a scientifically informed and oriented public to support research and development in emerging technological and science-driven areas. At a school level, both mathematics and science are critically important subjects in preparing students for future roles interacting with STEM or working as STEM professionals – as well as engaging them both conceptually and aesthetically (Zembylas, 2005). Despite this importance, science and mathematics are viewed differently by educators and curriculum designers. To illustrate, mathematics is considered to be much more sequential and structured than science (Siskin, 1994), whereby students must master one concept before commencing the next. However, science is perceived as a more topic-based subject (and less strictly sequential) with progression through topics governed less via conceptual mastery. Nevertheless, Tytler et al. (2008, p. 116) note that for both subjects 'ultimately the interest resides in an appreciation of the conceptual explanations and structures of the discipline and their power in making sense of the world'.

There remains debate over the best method to deliver STEM education. For instance, Bybee (2010) advocates an *integrated approach* based on contemporary STEM issues. An integrated approach to STEM education 'removes the walls placed between each of the STEM content areas and teach[es] them as one subject' (Breiner et al., 2012; Morrison & Bartlett, 2009; Roberts & Cantu, 2012, p. 113). The integrated approach begins with an age-appropriate challenge or problem that engages students. As students explore options and gain an understanding of the problem, they must access the respective STEM disciplines and apply knowledge and skills to the problem. The *silo approach* to STEM education is characterised by a teacher-driven classroom where each individual STEM subject is taught in isolation (Dugger, 2010). The concentrated study of each individual subject allows students to learn course content at greater depth, with little opportunity to 'learn by doing' (Morrison, 2006). A third approach is the *embedded approach*, which requires students to acquire content knowledge through an exploration of 'real-world situations and problem-solving techniques within social, cultural and functional contexts' (Chen, 2001; Roberts & Cantu, 2012, p. 112). According to Roberts and Cantu (2012), an embedded approach aids learning as it reinforces and complements materials students learn in other classes. To illustrate, a technology education teacher uses embedding to emphasise technological content (as it would be done via the silo approach) and this maintains the integrity of the subject matter. In contrast to the silo approach, however, embedding promotes learning through a variety of contexts (Rossouw, Hacker & de Vries, 2010).

PAUSE AND THINK

Thus far in this section a number of instructional approaches have been outlined for STEM. Which of these approaches do you feel is a good 'fit' for your teaching and learning philosophy, and why?

Features of STEM programs

Within the literature, successful STEM programs have a common set of features. Typically – and from a mathematical perspective – STEM programs engage students in mathematical problem-solving situations that stimulate intellectual activity (Tytler et al., 2008), and offer opportunities for students to develop deep, connected understandings of mathematics that will enable them to solve unfamiliar, non-routine problems (Kilpatrick, Swafford & Findell, 2001). Following an evaluation of STEM programs within secondary schools, Brody (2006) classified key features of STEM programs as: (a) exposure to strong content knowledge in mathematics and science based on academic instruction and hands-on demonstration, (b) an appreciation of the utility of STEM subjects in the workplace, (c) access to role models working in STEM fields and (d) collaboration with peers who share interests in STEM. In the United States, Sanders (2009) notes that the 'flavour' of present-day, integrative STEM education efforts resembles several accreditation standards that were developed from past engineering education reform efforts.

To advance STEM education within schools, a fundamental clarification of STEM literacy is required (Tytler et al., 2008). These authors also contended that translating a working definition of STEM literacy into school programs and instructional practices requires a way of organising education so that the respective disciplines can be integrated and instructional materials designed, developed and implemented (Tytler et al., 2008). One author suggested that such a definition of STEM literacy refers to how STEM knowledge: is acquired, used and applied; helps to drive STEM disciplines as forms of human endeavour; assists in shaping the material, intellectual and cultural world; and actively contributes to humankind's capacity to become well-informed and constructive citizens (Bybee, 2010, p. 31).

George Polya
CONNECTION

George Polya was a Hungarian mathematician who immigrated to the United States in 1940. Polya has been nicknamed 'the father of problem solving' and his main contributions to mathematics education came through his work on teaching others how to solve problems. In 1945, he published the book *How To Solve It*, which has sold over one million copies and has been translated into 17 languages. This book outlines Polya's four basic principles of problem solving:

1. Preparation (Understand the problem)
2. Thinking Time (Devise a plan)
3. Insight (Carry out the plan)
4. Verification (Look back).

Before answering the questions below, access the Polya Problem Solving Worksheet (http://web.mnstate .edu/peil/M110/Worksheet/PolyaProblemSolve.pdf).

Questions

1. From the resource indicated above, select one of the worded questions. Devise a lesson plan where this question will feature in the main body of the lesson. What content and problem-solving strategies do you feel will be essential scaffolding for your students to answer the question?
2. In a secondary school context, under what circumstances do you feel Polya's approach to solving problems will be a successful way to teach students mathematics?

STEM in action

The work of Tytler et al. (2008) reviewed extant literature concerning supports of and barriers to STEM education in Australia during the primary–secondary transition. In particular, this work highlighted various enrichment initiatives where students participated in contemporary STEM occupations as part of their school STEM curriculum. Such enrichment and enhancement initiatives can be:

> embedded in curriculum materials or achieved by links being made to STEM professionals through excursions or incursions, web-based explorations of new developments, curriculum modules designed to embed this sort of material, or school-community linked units of work. (Tytler et al., 2008, p. 126)

According to Stagg (2007, p. 12), 'direct contact between students and people working in scientific jobs tends to be identified by the students themselves as the most effective way to learn about careers'. Other researchers note the enthusiastic reactions STEM-based initiatives generate within students (Cripps Clark, 2006), as well as school teachers deriving significant professional learning from working with scientists. One project highlighted by Tytler et al. (2008) is the Australian School Innovation in Science, Technology and Mathematics (ASISTM) project. Through strategic partnerships between clusters of schools, scientific and industrial organisations, universities and government organisations, this project developed innovative curriculum projects for students. One of these projects is entitled 'Bloodstain Pattern Analysis', run through the University of Western Australia. In this project, Year 10 students participate in a range of activities that require them to make observations, collect, analyse and interpret data before forming a conclusion. For the project – which aims to engage students in real-life forensic science – teacher and student support is made available through learning modules, background information and chapter-based resources. In a study of the ASISTM project, Tytler et al. (2008) designed an innovation framework to interpret the experiences provided. Specifically, these scholars concluded two key benefits were that the technology projects offered schools access to expensive and contemporary technologies, and a number of projects focused specifically on alerting students to STEM career opportunities in their local regions.

CONNECTION — 'Teachers' perceptions of STEM integration and education: a systematic literature review'

Authors Kelly Margot and Todd Kettler are deeply interested in understanding 'teachers' beliefs and perceptions related to STEM talent development' (Margot & Kettler, 2019, p. 1). They feel that 'teachers, as important people within a student's talent development, hold prior views and experiences that will influence their STEM instruction' (Margot & Kettler, 2019, p. 1). Through thematic analysis of empirical journal articles published within scholarly journals between 2000 and 2016 (in English), Margot and Kettler attempt to understand what is known about teachers' perceptions of STEM education.

Access Margot and Kettler's article before answering the questions below.

Questions

1. Through their analysis, Margot and Kettler determined the overall values, beliefs and levels of confidence of teachers with regard to STEM education. What precisely did these authors determine, and how well do your values, beliefs and confidence level correspond with their findings?

2. According to the authors, what are some of the missing components of STEM education? How do you think you are able to contribute to STEM education overall?

Challenges for STEM education

Advancing STEM education presents several significant challenges for educators (Bybee, 2010; Hoachlander & Yanofsky, 2011). As mentioned earlier, there is no consensus on what is the best approach for STEM instruction. For instance, when considering the benefits of an integrated STEM approach, one must weigh up the associated shortfalls; for example, teaching through integrative approaches requires considerable pedagogical training and a mastery of all four disciplines. Equally, one should be reminded of the benefits of other approaches – such as the interactive approach, which develops interaction between STEM subjects (Williams, 2011). According to Williams (2011, p. 32), this interactive approach can be achieved 'by fostering cross-curricular links in a context where the integrity of each subject remains respected'. A second challenge involves actively planning for technology and engineering learning opportunities in school programs (Bybee, 2010). While 'scaling up' these courses in schools may appear to be concomitant with a 'silo approach', the appropriate inclusion of technology and engineering in science and mathematics education appears a reasonable way to meet this challenge (Bybee, 2010). Third, while Australia implements a national mathematics curriculum from Kindergarten through to Year 12, there is no attention directly focused on STEM education by way of syllabus documents (e.g. scope and sequence), general capabilities or proficiency standards. Instead, it has been left up to educators and curriculum designers to develop units of STEM work carefully and creatively to prepare students for future roles in STEM-related careers.

SHORT-ANSWER QUESTIONS

1. List and briefly describe the three common approaches to STEM education.
2. Explain how schools are able to plan for and deliver a successful STEM education program.
3. According to current literature, what are three challenges for STEM education? How might these challenges be overcome?

Putting activities into practice

In the final section of this chapter, it seems appropriate to examine some activities that help students make mathematical connections. By their nature, these activities are problem based and therefore require a problem-solving approach. According to Anderson (2009, p. 342), problem solving is 'recognised as an important life skill involving a range of processes including analysing, interpreting, reasoning, predicting, evaluating and reflecting'. Developing students who are competent problem solvers is a complex task that requires a range of knowledge, skills and dispositions (Stacey, 2005). Moreover, and consistent with Figure 3.1, Stacey (2005) contends that students require good communication skills, the ability to work cooperatively, as well as a teacher who possesses the appropriate personal attributes for organising and directing their efforts.

The general capabilities of the Australian Curriculum outline the importance of problem solving in learning mathematics. Clearly, problem solving is an important component of

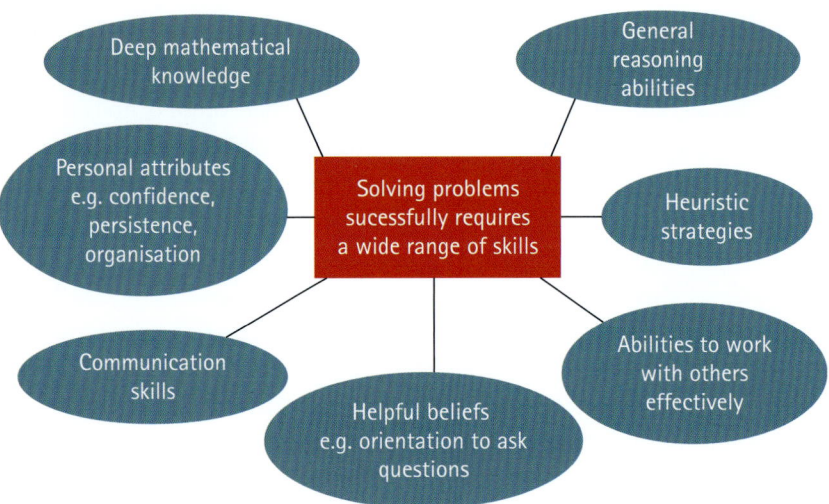

Figure 3.1 Factors contributing to successful problem solving
Source: Stacey, 2005, p. 342.

doing mathematics, and consequently students should be given regular opportunities to solve complex, real-world problems. For instance, the Literacy and Numeracy capabilities are included in Table 3.4.

Table 3.4 General capabilities (Literacy and Numeracy)

Literacy	Students use literacy to understand and interpret word problems and instructions that contain the particular language features of mathematics. They use literacy to pose and answer questions, engage in mathematical problem solving, and to discuss, produce and explain solutions.
Numeracy	It is important that the mathematics curriculum provides the opportunity to apply mathematical understanding and skills in context, both in other learning areas and in real world contexts.

Source: ACARA, 2020b.

PAUSE AND THINK

In your experience as a learner, what were some lessons or activities that you regard as useful for making mathematical connections? List the key features of these lessons and recall how the teacher facilitated the learning process. How do you plan to help your students make these connections in your classroom?

The famine problem

One question that can promote mathematical connections in the secondary classroom is the famine problem (Stillman & Galbraith, 1998). The question is presented as such:

Every day during the month of July, Relief Aid Abroad trucked supplies of food into the famine stricken areas of Nacirema. On the first day, 1000 tonnes were shifted; on the second day 1100 tonnes were shifted; on the third day, 1200 tonnes were shifted and so on until a maximum amount was reached. The supply of food then declined by 100 tonnes per day until the end of the month. If the total food supplied for the month was 59 300 tonnes, on which day of the month was the maximum amount trucked out? (pp. 159–60)

Stillman and Galbraith (1998) suggest that a student's solution to the problem contains the following essential aspects in some form:

- Obtain $T_n = 900 + 100n$ for amount trucked on days $n = 1, 2, 3 \ldots m$ where $T_m = 900 + 100m$ gives maximum on day m.
- Obtain $T_{m+r} = 900 + 100(m - r)$ for amount trucked on day $n = m + r$ so $T_{31} = 200m - 2200$ for amount trucked on day $n = m + r = 31$.
- Obtain $S_m = m(950 + 50m)$ for total trucked from $n = 1$ to m and $S_{31-m} = (31 - m)(150m - 700)$ for total trucked from $n = m + 1$ to 31.
- Note that $S_m + S_{31} - m = 59\,300$ for total trucked during month leading to $m^2 - 63m + 810 = 0$ with solutions $m = 18, 45$.
- Note that $m < 31$ (days in July) so $m = 18$ (maximum amount is trucked on July 18).

Looking at the verbal nature of the problem it is necessary that students:

- recognise the arithmetic progression (AP) pattern in the data
- form T_n and S_n expressions in terms of correct variables identified from the context
- interpret key words such as:
  – 'maximum' T_n stops increasing when $n = m$
  – 'decline per day' common difference is negative in a second AP
  – 'total food' addition of sums of two different APs (1998, p. 160).

Looking at the real-world aspect of the context, it is necessary that students additionally import a numerical datum (31), not explicitly provided, into the solution process:

- 31 days in July to give the link $m + r = 31$ between the APs
- 31 days in July means that $m < 31$ so $m = 45$ is rejected (1998, p. 160).

Stillman and Galbraith (1998) intentionally designed this problem 'to entail a large complex dataset for initial perception, representation and analysis'. In looking at the nature of the problem, it was expected that students would make mathematical connections with a number of variables and relationships in generating a solution based on abstract reasoning. For this to occur, students would need to augment their working memory with an external representation (e.g. pen and paper).

The Newman Procedure for Analysing Errors on Written Tasks

IN PRACTICE

Did you ever struggle with a written mathematical task (worded problem)? The Newman Procedure (Newman, 1977; 1983) is based on the research of M. Anne Newman, who sought to determine how

▶ ▶

and why students made errors on worded problems. The Newman Procedure can be described as a set of questions to assist mathematics educators in pinpointing at what hierarchical stage students make errors. These stages are summarised below, together with a possible prompt a teacher could use in the classroom:

1. Reading – Read the question to me
2. Comprehension – Tell me what the question is asking you to do
3. Transformation – Describe what method you could use to find an answer to the question
4. Process Skills – Show me how you worked out the answer, and explain each step to me while you do it
5. Encoding – Write down the answer to the question.

The Licorice Factory

Lovitt and Williams have developed The Licorice Factory, an activity for middle school students that makes connections between integers and properties of integers (e.g. factors, multiples) (Lovitt & Williams, 2015). To commence this activity, teachers can introduce the story to their class:

> Once upon a time in Lolly Land, there was a grand Licorice Factory. In Lolly Land, customers order their licorice in any lengths from 1 to 100. For example, if a customer wants some 36-length licorice, a Factory Employee goes to the 36 machine, feeds in the unit length pieces, and the machine stretches it out to length 36. In the Factory there is room to walk between the machines, and soon you will be given a Floor Plan.
>
> All was going well until one day someone ordered a Number 6 length licorice and . . . they found the Number 6 machine was broken! The boss was very worried until one of the workers figured out how to make Number 6s without that machine. What do you think she proposed that they should do? She said: We could feed the unit pieces into Number 2 machine and get length 2 licorice and then feed those into the Number 3 machine and they will be stretched three times to become length 6.
>
> So, they tried it and it worked! They didn't really need the Number 6 machine. The boss gave a year's supply of licorice to the worker who thought up this idea . . . Then a week later the Number 10 machine broke . . . When the boss realised that more than one machine was not needed, he offered a lifetime supply of licorice to the worker who could tell him all the machines he could shut down. (Lovitt & Williams, 2015)

Although this is the complete situation, teachers may find that it is useful to stop the class at various points throughout the story to model the 'stretching' of licorice (both with real licorice, and with unifix cubes), and to allow students to try it for themselves. Producing large, laminated cards with numbers 1–36 and placing these on the classroom floor can also assist in creating the factory floor where students can walk around and stand on numbers (machines). At this stage, distribute the Factory Floor Plan to the students (this also works nicely if they work in partners with one large, paper-based Floor Plan). Their task is to 'cross 'off' as many machines as possible, but at the same time, they must still have access to machines that can produce licorice to any length from 1–100.

Some key questions to ask the students here are:

- What strategy are you using to cross off machine numbers?
- How does this activity require knowledge of numbers?
- Are there any patterns in numbers that make the process of crossing off easier?
- Are you able to justify *every* decision to cross off the machines on your Floor Plan?

Students may recall and use some number facts from their prior learning, e.g. *prime numbers*, *factors*, *multiples* and *composite numbers*. Encourage them to use these terms when asserting a position to 'cross off' or 'not cross off' a machine.

After creating a Master List of Licorice Machines that will remain working after as many as possible have been shut down, conclude as a class that the machines still working are the *prime numbers* from 1–100. Using Number 24 Machine as an example, demonstrate on the board how 24 can be written as $2^3 \times 3$. This can be done by introducing the 'Factor tree' approach to break the machines down into their prime factors (use 24 and 36) – see Figure 3.2.

Ask students to choose 10 Machines that have been shut down, and then to write each of the numbers as a product of its prime factors.

To conclude, ask the students what they have learned as a result of the activity. This activity emphasises a shift from additive thinking to multiplicative thinking, introduces the ideas of prime numbers and composite numbers, factors, multiples and number relationships. To extend the task, a teacher can draw students' attention to divisibility tests (e.g. the sieve of Eratosthenes), number properties (e.g. amicable numbers, deficient and abundant numbers), and various famous number puzzles (Riemann hypothesis, Goldbach conjecture).

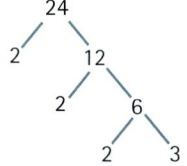

Figure 3.2 Writing 24 as a product of its prime factors, i.e. $2 \times 2 \times 2 \times 3 = 2^3 \times 3$

The weighted dice

This activity commences with the teacher posing a question concerned with simple probability: Can we alter the outcome of rolling a die if one of the sides has been altered or 'loaded'? As a corollary, students are asked if they think an odd number is twice as likely to occur with a loaded die. To make loaded dice, Kincaid (2015) suggests that the materials needed are: regular playing dice, small fishing sinkers and a $\frac{1}{8}$ inch drill bit. With these materials, drill a hole from the '4' side to the '3' side along the corner where the '5' and '6' sides meet. After pressing sinkers into the holes created by drilling, the result will be a loaded die (and the mass has been increased by approximately 1.8 g, or 39 per cent of the original mass). In Figure 3.3 an original die is compared with a loaded die.

Figure 3.3 An original die and a loaded die
Source: Kincaid, 2015, p. 18.

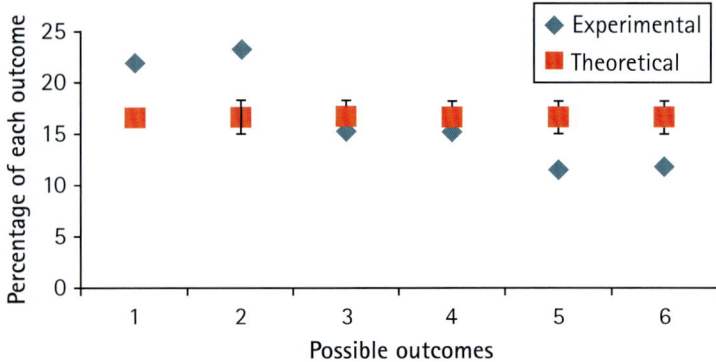

Figure 3.4 Results: single die rolls
Source: Kincaid, 2015, p. 20.

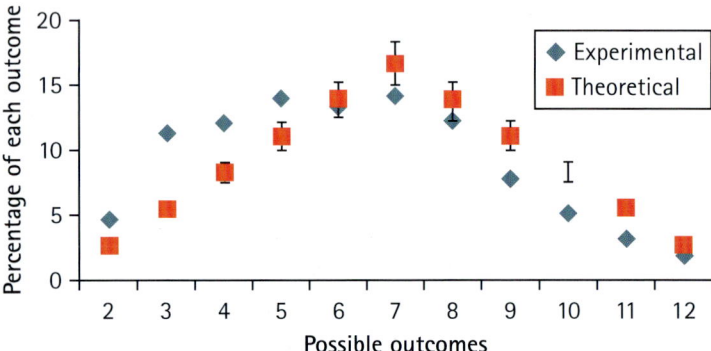

Figure 3.5 Results: two dice sum rolls
Source: Kincaid, 2015, p. 21.

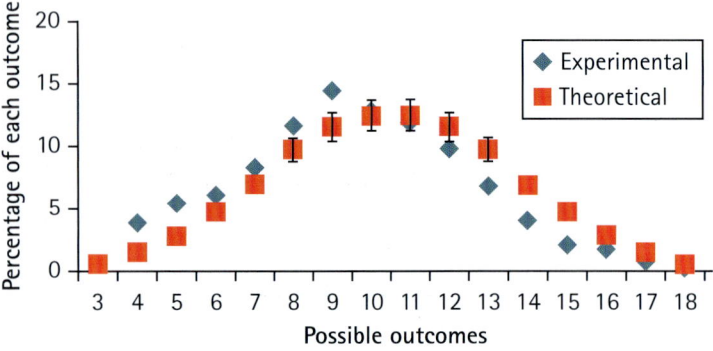

Figure 3.6 Results: three dice sum rolls
Source: Kincaid, 2015, p. 21.

After creating a class set of loaded dice, Kincaid engaged students in a dice rolling activity for single die rolls ($n = 1440$), two dice sum rolls ($n = 1439$) and three dice sum rolls ($n = 1120$). The results for each roll are displayed in Figures 3.4–3.6.

As can be noted from Figures 3.4–3.6, Kincaid was able to engage his students in an activity where theoretical and experimental probabilities could be compared. To illustrate, from the single die roll activity the theoretical probability of rolling a 1, 2, 3, 4, 5 or 6 is $\frac{1}{6}$ or $16.\dot{6}\%$. Using a loaded die, the probability of rolling a 1 or a 2 has increased by approximately 10%. Obtaining a 3 or 4 has approximately the same probability, while $P(5)$ and $P(6)$ are approximately 10% lower than their respective theoretical probabilities.

With the two dice rolls, Kincaid asked his class to determine the probability of obtaining a 9 with regular dice. This calculation can be expressed as:

$$P(9) = P(3) \times P(6) + P(4) \times P(5) + P(6) \times P(3)$$
$$= \frac{1}{6} \times \frac{1}{6} + \frac{1}{6} \times \frac{1}{6} + \frac{1}{6} \times \frac{1}{6} + \frac{1}{6} \times \frac{1}{6}$$
$$\cong 0.11 \text{ or } 11\%$$

Comparing this theoretical probability with the experimental probability, students saw that $P(9)$ with loaded dice decreased from the 'expected' to approximately 7.783%. In addition to providing a fun and interesting real-world application of probability theory, this activity can be used to make mathematical connections between Australian Curriculum strands Number and Algebra and Statistics and Probability. Moreover, these connections can be made by extending the understanding of predictability of single or multi-dice sums, and through an examination of data with calculations of fractions, decimals and percentages.

SHORT-ANSWER QUESTIONS

1. Re-read *the famine problem* in this section. Look at the Australian Curriculum and select the student year level this problem should be taught to.

2. Re-read *The Licorice Factory* in this section. Look at the Australian Curriculum and select the student year level this problem should be taught to.

3. Re-read *the weighted dice* in this section. Look at the Australian Curriculum and select the student year level this problem should be taught to.

Conclusion

Mathematical connections benefit student learning by enhancing students' capacity to remember, appreciate and use mathematics and to develop conceptual understanding. Additionally, the use of real-world connections can motivate mathematics learning and help 'students apply mathematics to real problems – particularly those in the workplace' (Tirosh, 2008).

To make mathematical connections, teachers can use simple analogies and classic 'word problems', use and analyse real data, discuss the use of mathematics in society and mathematically model real phenomena.

Science, technology, engineering and mathematics (STEM) education initiatives are identified typically by the following key features: (a) exposure to strong content knowledge in mathematics and science based on academic instruction and hands-on demonstration, (b) an appreciation for the utility of STEM subjects in the workplace, (c) access to role models working in STEM fields and (d) collaboration with peers who share interests in STEM.

When planning to use contextual word problems, teachers should consider some general guidelines (e.g. simple analogies, classic word problems, real data analysis, mathematical discussions, hands-on representation of concepts, modelling real phenomena) as well as the roles that scaffolding and problem solving will play in students' learning.

BRINGING IT TOGETHER

1. Why should teachers help students to make mathematical connections in their learning?

2. How do you define the term *mathematical connections* in a secondary context?

3. Outline some instructional practices teachers can use in helping students make mathematical connections.

4. What are some key features, instructional approaches and challenges associated with STEM education?

5. During an instructional unit on circles, a student in your Year 10 class asks you how certain formulas were derived (e.g. $C = 2\pi r$ and $A = \pi r^2$). How do you respond so that there is an opportunity for strong mathematical connections to be taught to the class from the Number and Algebra strand to that of Measurement and Geometry?

6. Re-read *the weighted dice* problem and develop a lesson plan for this problem to be taught in a STEM context. You might like to commence by selecting an instructional approach first before developing the plan.

REFERENCES

Further resources

Anderson, J. (2005). Implementing problem solving in mathematics classrooms: What do teachers want? In P. Clarkson, A. Downton, D. Gronn, M. Horne, A. McDonough, R. Pierce & A. Roche (eds), *Building connections: Theory, research and practice. Proceedings of the 28th Annual Conference of the Mathematical Education Research Group of Australasia (MERGA)*. Melbourne: MERGA.

Anderson, J. (2009). *Mathematics curriculum development and the role of problem solving. Australian Curriculum Studies Association National Biennial Conference*. Canberra: Australian Curriculum Studies Association.

Anderson, J.A. & Bobis, J. (2005). Reform-oriented teaching practices: A survey of primary school teachers. In H.L. Chick & J.L. Vincent (eds), *Proceedings of the 29th Conference of the International Group for the Psychology of Mathematics Education (PME)* (vol. 2, 65–72). Melbourne: PME.

Anderson, J., Sullivan, P. & White, P. (2004). The influences of perceived constraints on teachers' problem-solving beliefs and practices. In I. Putt, R. Faragher & M. McLean (eds), *Mathematics education for the third millennium: Towards 2010. Proceedings of the 27th Annual Conference of the Mathematics Education Research Group of Australasia (MERGA)* (vol. 1, 39–46). Sydney: MERGA.

Anthony, G. & Walshaw, M. (2009a). Characteristics of effective teaching of mathematics: A view from the west. *Journal of Mathematics Education*, 2(2), 147–64.

Anthony, G. & Walshaw, M. (2009b). *Effective pedagogy in mathematics*. Educational Series 19. Brussels: International Academy of Education; Geneva: International Bureau of Education.

Australian Academy of Sciences (2006). *Mathematics and statistics: Critical skills for Australia's future*. The national strategic review of mathematical sciences research in Australia. Melbourne.

Australian Curriculum, Assessment and Reporting Authority (ACARA) (2020a). *Aims*. Sydney: ACARA. Retrieved from https://www.australiancurriculum.edu.au/f-10-curriculum/mathematics/aims/

Australian Curriculum, Assessment and Reporting Authority (ACARA) (2020b). *General capabilities*. Sydney: ACARA. Retrieved from http://www.australiancurriculum.edu.au/mathematics/general-capabilities

Australian Institute for Teaching and School Leadership (AITSL) (2011). *Australian Professional Standards for Teachers*. Melbourne: AITSL.

Ball, D.L. & Bass, H. (2000). Interweaving content and pedagogy in teaching and learning to teach: Knowing and using mathematics. In J. Boaler (ed.), *Multiple perspectives on the teaching and learning of mathematics* (83–104). Westport, CT: Ablex.

Boaler, J. (1997). *Experiencing school mathematics: Teaching styles, sex, and setting*. Philadelphia, PA: Open University Press.

Boaler, J. (2002). Exploring the nature of mathematical activity: Using theory, research and 'working hypotheses' to broaden conceptions of mathematics knowing. *Educational Studies in Mathematics*, 51(1), 3–21.

Breiner, J.M., Harkness, S.S., Johnson, C.C. & Koehler, C.M. (2012). What is STEM? A discussion about conceptions of STEM in education and partnerships. *School Science and Mathematics*, 112(1). https://doi: 10.1111/j.1949-8594.2011.00109.x

Brody, L. (2006). *Measuring the effectiveness of STEM talent initiatives for middle and high school students.* Paper presented at the annual meeting of the National Academies Center for Education. Washington, DC.

Businskas, A.M. (2008). *Conversations about connections: Secondary mathematics teachers conceptualise and contend with mathematical connections* (Unpublished doctoral dissertation). Simon Fraser University, Canada.

Bybee, R.W. (2010). Advancing STEM education: A 2020 vision. *Technology and Engineering Teacher*, 70(1), 30–5.

Cambridge Dictionary (2020). *Connections.* Retrieved from https://dictionary.cambridge.org/dictionary/english/connection?q=connections

Charalambous, C.Y. (2008). Mathematical knowledge for teaching and the unfolding of tasks in mathematics lessons: Integrating two lines of research. In O. Figuras, J.L. Cortina, S. Alatorre, T. Rojano & A. Sepulveda (eds), *Proceedings of the 32nd Annual Conference of the International Group for the Psychology of Mathematics Education (PME)* (vol. 2, 281–8). Morelia: PME.

Chen, M. (2001). A potential limitation of embedded teaching for formal learning. In J. Moore & K. Stenning (eds), *Proceedings of the Twenty-Third Annual Conference of the Cognitive Science Society* (194–9). Edinburgh: Lawrence Erlbaum Associates, Inc.

Clarke, D.M. & Clarke, B.A. (2004). Mathematics teaching in grades K–2: Painting a picture of challenging, supportive, and effective classrooms. In R.N. Rubenstein & G.W. Bright (eds), *Perspectives on the teaching of mathematics* (66th Yearbook of the National Council of Teachers of Mathematics (NCTM), 67–81). Reston, VA: NCTM.

Coxford, A.F. (1995). The case for connections. In P.A. House & A.F. Coxford (eds), *Connecting mathematics across the curriculum.* Reston, VA: National Council of Teachers of Mathematics.

Cripps Clark, J. (2006). *The role of practical activities in primary school science.* Melbourne: Deakin University.

De Bortoli, L. & Macaskill, G. (2014). *Thinking it through: Australian students' skills in creative problem solving.* Melbourne: Australian Council for Educational Research.

Desforges, C. & Cockburn, A. (1987). *Understanding the mathematics teacher: A study of practice in first schools.* London: The Palmer Press.

Doorman, M., Drijvers, P., Dekker, T., Van den Heuvel-Panhuizen, M., de Lange, J. & Vijers, M. (2007). Problem solving as a challenge for mathematics education in The Netherlands. *ZDM Mathematics Education*, 39(5/6), 405–18.

Dugger, W. (2010). *Evolution of STEM in the U.S. 6th Biennial International Conference on Technology Education Research.* Retrieved from http://citeseerx.ist.psu.edu/viewdoc/summary?doi=10.1.1.476.5804

Evitts, T.A. (2004). *Investigating the mathematical connections that pre-service teachers use and develop while solving problems from reform curricula* (Unpublished doctoral dissertation). Pennsylvania State University College of Education, United States.

Fan, S. & Ritz, J. (2014). *International views of STEM education.* In PATT-28 Research into Technological and Engineering Literacy Core Connections (7–14). Orlando: International Technology and Engineering Educators Association. Retrieved from http://www.iteea.org/Conference/PATT/PATT28/Fan%20Ritz.pdf

Fennema, E. & Franke, M. (1992). Teachers' knowledge and its impact. In D. Grouws (ed.), *Handbook of research on mathematics teaching and learning* (147–64). New York: Macmillan.

Fennema, E. & Romberg, T.A. (1999). *Mathematics classrooms that promote understanding.* Mahwah, NJ: Lawrence Erlbaum.

Gainsburg, J. (2008). Real-world connections in secondary mathematics teaching. *Journal of Mathematics Teacher Education*, 11, 199–219.

Hattie, J. (2009). *Visible learning: A synthesis of over 800 meta-analyses relating to achievement.* New York: Routledge.

Hattie, J., Fisher, D., Frey, N., Gojak, L.M., Moore, S.D. & Mellman, W. (2016). *Visible learning grades K–12: What works best to optimize student learning.* Thousand Oaks, CA: SAGE.

Hoachlander, G. & Yanofsky, D. (2011). Making STEM real: By infusing core academics with rigorous real-world work, linked learning pathways prepare students for both college and career. *Educational Leadership*, 68(3), 60–5.

Hodgson, T.R. (1995). Connections as problem-solving tools. In P.A. House & A.F. Coxford (eds), *Connecting mathematics across the curriculum.* Reston, VA: National Council of Teachers of Mathematics.

Hollingsworth, H. (2003). The TIMSS 1999 video study and its relevance to Australian mathematics education research, innovation, networking and opportunities. In L. Bragg, C. Campbell, G. Herbert & J. Mousley (eds), *Mathematics education research: Innovation, networking, opportunity. Proceedings of the 26th Annual Conference of the Mathematics Education Research Group of Australasia (MERGA)* (7–16). Sydney: MERGA.

Kaur, B. (2001). TIMSS & TIMSS-R – Performance of grade eight Singaporean students. In C. Vale, J. Horwood & J. Roumeliotis (eds), *2001 A Mathematical Odyssey. Proceedings of the 38th Annual Conference of the Mathematical Association (MA) of Victoria* (132–44). Brunswick: MA.

Kaur, B. & Yeap, B.H. (2009). Mathematical problem solving in Singapore schools. In B. Kaur, B.H. Yeap & M. Kapur (eds), *Mathematical problem solving: Yearbook 2009* (3–13). Singapore: Association of Mathematics Education and World Scientific.

Kilpatrick, J., Swafford, J. & Findell, B. (2001). *Adding it up: Helping children learn mathematics.* Washington, DC: National Academy Press.

Kincaid, R. (2015). Weighted dice: A study in applications of probability. *Ohio Journal of School Mathematics*, 72, 18–22.

Leikin, R. & Levav-Waynberg, A. (2007). Exploring mathematics teacher knowledge to explain the gap between theory-based recommendations and school practice in the use of connecting tasks. *Educational Studies in Mathematics*, 66(3), 349–71.

Little, K. (2019). *Connecting mathematics with science to enhance student achievement: A position paper.* Retrieved from https://merga.net.au/common/Uploaded%20files/Annual%20Conference%20Proceedings/2019%20Annual%20Conference%20Proceedings/RP_Little.pdf

Lovitt, C. & Williams, D. (2015). *The licorice factory.* Retrieved from http://www.maths300.com

Ma, L. (1999). *Knowing and teaching elementary mathematics.* Mahwah, NJ: Erlbaum.

Margot, K.C. & Kettler, T. (2019). Teachers' perceptions of STEM integration and education: A systematic literature review. *International Journal of STEM Education*, 6(2). https://doi.org/10.1186/s40594-018-0151-2

Morrison, J. (2006). *TIES STEM education monograph series: Attributes of STEM education.* Baltimore, MD: TIES, (2): 5.

Morrison, J. & Bartlett, B. (2009). *STEM as a curriculum: An experimental approach.* Retrieved from http://www.labaids.com/docs/stem/EdWeekArticleSTEM.pdf

National Academy of Sciences (2003). *Engaging schools: Fostering high school students' motivation to learn.* Washington, DC: National Academy Press.

National Council of Teachers of Mathematics (2000). *Principles and standards for school mathematics.* Reston, VA: National Council of Teachers of Mathematics.

National Council of Teachers of Mathematics (2009). *Guiding principles for mathematics curriculum and assessment.* Retrieved from http://old.nctm.org/uploadedFiles/Math_Standards/NCTM%20Guiding%20Principles%206209.pdf

National Council of Teachers of Mathematics (2015). *Executive summary: Principles and standards for teaching mathematics.* Retrieved from http://www.nctm.org/uploadedFiles/Standards_and_Positions/PSSM_ExecutiveSummary.pdf

National Research Council (1998). *High school mathematics at work: Essays and examples for the education of all students*. Washington, DC: National Academy Press.

Newman, M.A. (1977). An analysis of sixth-grade pupils' errors on written mathematical tasks. *Victorian Institute for Educational Research Bulletin*, 39, 31–43.

Newman, M.A. (1983). *Strategies for diagnosis and remediation*. Sydney: Harcourt, Brace Jovanovich.

Queensland College of Teachers (2015). *Research Digest No. 11*, August 2015. Retrieved from https://cdn.qct.edu.au/pdf/Research%20Periodicals/QCTResearchDigest2015_11.pdf

Roberts, A. & Cantu, D. (2012). *Applying STEM instructional strategies to design and technology curriculum*. Norfolk, VA: Old Dominion University.

Rossouw, A., Hacker, M. & de Vries, M. (2010). Concepts and contexts in engineering and technology education: An international and interdisciplinary Delphi study. *International Journal of Technology and Design Education*, 21(4), 409–24.

Sanders, M. (2009). STEM, STEM education, STEMmania. *The Technology Teacher*, 68(4), 20–6.

Shulman, L.S. (1986). Those who understand: Knowledge growth in teaching. *Educational Researcher*, 15(2), 4–14.

Siemon, D., Banks, N. & Prasad, S. (2018). Multiplicative thinking: A necessary STEM foundation. In T. Barkatsas, N. Carr & G. Cooper (eds), *STEM education: An emerging field of inquiry* (74–100). https://doi.org/10.1163/9789004391413

Siskin, L.S. (1994). *Realms of knowledge: Academic departments in secondary schools*. London: The Falmer Press.

Skemp, R. (1987). *The psychology of learning mathematics*. Hillsdale, NJ: Lawrence Erlbaum Associates.

Stacey, K. (2003). The need to increase attention to mathematical reasoning. In H. Hollingsworth, J. Lokan & B. McCrae, *Teaching mathematics in Australia: Results from the TIMSS 1999 video study* (119–22). Camberwell: Australian Council of Educational Research.

Stacey, K. (2005). The place of problem solving in contemporary mathematics curriculum documents. *Journal of Mathematical Behaviour*, 24, 341–50.

Stagg, P. (2007). *Careers from science*. An investigation for the Science Education Forum: Centre for Education and Industry.

Stephens, M. (2009). *Numeracy in practice: Teaching, learning and using mathematics*. Melbourne: Department of Education and Early Childhood Development.

Stigler, J.W. & Hiebert, J. (1999). *The teaching gap: Best ideas from the world's teachers for improving education in the classroom*. New York: Free Press.

Stillman, G. & Galbraith, P.L. (1998). Applying mathematics with real world connections: Metacognitive characteristics of secondary students. *Educational Studies in Mathematics*, 36(2), 157–95.

Sullivan, P. (2011). *Teaching mathematics: Using research-informed strategies*. Melbourne: Australian Council for Educational Research.

Sullivan, P., Tobias, S. & McDonough, A. (2006). Perhaps the decision of some students not to engage in learning mathematics in school is deliberate. *Educational Studies in Mathematics*, 62(1), 81–99.

Swan, M. (2005). *Improving learning in mathematics: Challenges and strategies*. Sheffield: Department of Education and Skills Standards Unity.

Thompson, A.G. (1992). Teachers' beliefs and conceptions: A synthesis of the research. In D.A. Grouws (ed.), *NCTM handbook of research on mathematics teaching and learning* (127–46). New York: Macmillan.

Thompson, S. & Fleming, N. (2004). *Summing it up: Mathematics achievement in Australian schools in TIMSS 2002*. Melbourne: Australian Council for Educational Research.

Thomson, S., de Bortoli, L. & Buckley, S. (2013). *PISA 2012: How Australia measures up*. Melbourne: Australian Council for Educational Research.

Thomson, S., de Bortoli, L., Underwood, C. & Schmid, M. (2019). *PISA 2018: Reporting Australia's results (Volume 1: Student performance)*. Melbourne: Australian Council for Educational Research.

Thomson, S., Hillman, K., Wernet, N., Schmid, M., Buckley, S. & Munene, A. (2012). *Highlights from TIMSS & PIRLS 2011 from Australia's perspective*. Melbourne: Australian Council for Educational Research.

Tirosh, D. (2008). The tension between the general and the specific in an international mathematics teacher education. *Journal of Mathematics Teacher Education*, 11, 165–9.

Tytler R. (2007). Reimagining science education: Engaging students in science for Australia's future. *Australian Educational Review*, 51. Retrieved from http://research.acer.edu.au/aer/3

Tytler, R., Osborne, J., Williams, G., Tytler, K. & Clark, J. C. (2008). *Opening up pathways: Engagement in STEM across the primary-secondary school transition*. Retrieved from http://nla.gov.au/nla.arc-88047

Tytler, R. & Symington, D. (2006). Science in school and society: Teaching science. *The Journal of the Australian Science Teachers Association*, 52(3), 10–15.

Tzur, R. (2008). A researcher perplexity: Why do mathematical tasks undergo metamorphosis in teacher hands? In O. Figuras, J.L. Cortina, S. Alatorre, T. Rojano & A. Sepulveda (eds), *Proceedings of the 32nd Annual Conference of the International Group for the Psychology of Mathematics Education (PME)* (vol. 1, 139–47). Morelia: PME.

Vincent, J. & Stacey, K. (2008). Do mathematics textbooks cultivate shallow teaching? Applying the TIMSS video study criteria to Australian eighth-grade mathematics textbooks. *Mathematics Education Research Journal*, 20(1), 82–107.

Vygotsky, L.S. (1978). *Mind in society: The development of higher psychological processes*. Cambridge, MA: Harvard University Press.

Weinberg, S.L. (2001). *Is there a connection between fractions and division? Students' inconsistent responses*. Paper presented at the Annual Meeting of the American Educational Research Association, April 10–14, 2001. Seattle, WA.

Williams, J. (2011). STEM education: Proceed with caution. *Design and Technology Education: An International Journal,* 16(1). Retrieved from https://ojs.lboro.ac.uk/DATE/article/view/1590

Wood, D.J., Bruner, J.S. & Ross, G. (1976). The role of tutoring in problem solving. *Journal of Child Psychiatry and Psychology*, 17(2), 89–100.

Zembylas, M. (2005). Three perspectives on linking the cognitive and the emotional in science learning: Conceptual change, socio-constructivism and poststructualism. *Studies in Science Education*, 41, 91–116.

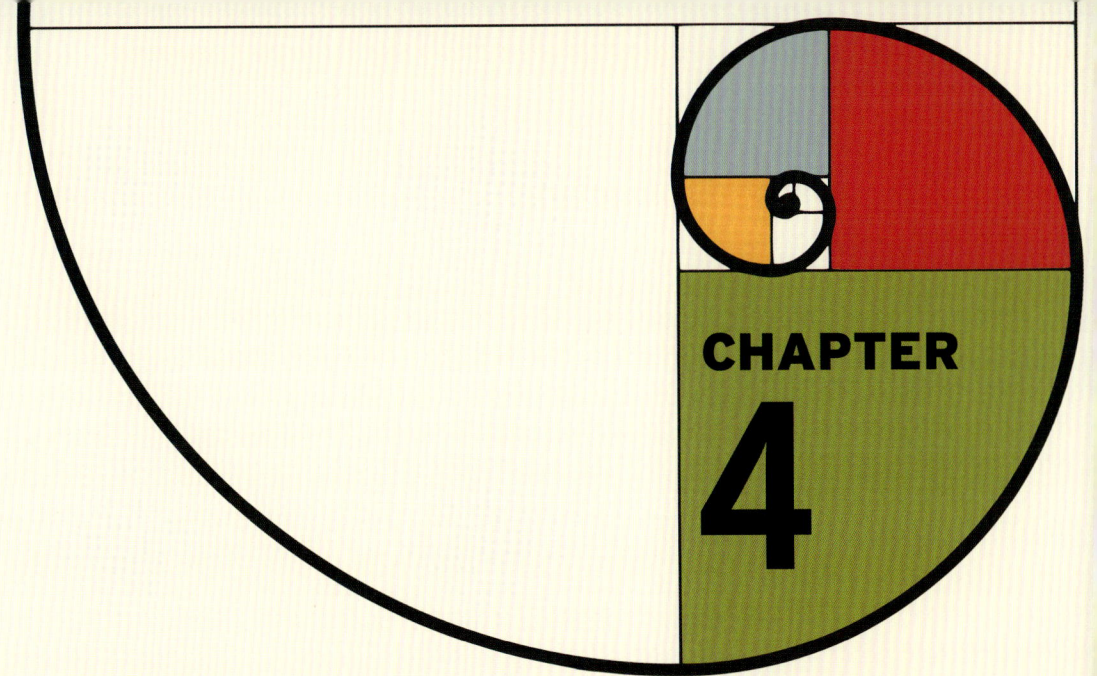

Using technology in mathematics education

Bruce White

LEARNING OBJECTIVES

The learning objectives of this chapter are directly linked to the Australian Professional Standards for Teachers (APST) (Australian Institute for Teaching and School Leadership [AITSL], 2011). After studying this chapter, you should be able to:

- articulate the evidence that supports why it is important to incorporate technology into the teaching and learning of mathematics (APST 4.5)
- describe your strengths and areas of development in terms of the TPACK model (APST 2.6, 4.5)
- be able to choose the appropriate technology for the teaching and learning of mathematics and have a range of strategies to implement these technologies (APST 2.6, 3.3, 4.5)
- find and adapt resources from the internet and know how to use them in class and online (APST 3.4).

Introduction

This chapter will look at the general use of technology for mathematics teaching and learning, starting with an examination of research to mount a case for the importance of using technology to teach mathematics. The Technological Pedagogical Content Knowledge (TPACK) framework (Koehler & Mishra, 2009), which identifies the knowledge teachers need to teach effectively with technology, will be outlined and will give you one way of looking at your readiness to implement technology into your teaching. This will be followed by an overview of the key technologies used in mathematics education, and finally a brief discussion of how technology is being used for ongoing teacher professional learning.

One of the features of the use of technology in teaching and learning is that it is constantly evolving and so this chapter will look at ways of using the affordances of technology rather than specific devices and programs. As such, it is important for you as a preservice teacher to spend time learning and being aware of the current technological tools. Learn how to use them and explore their use in mathematics education. It is neither possible nor necessary in a single chapter to give detailed instructions on how to use all of the wide variety of devices available and so the focus will be on how to use an example of the device with particular functionality in a variety of ways for teaching. An example of this is devices that allow you to sketch and write mathematics, such as tablet computers and interactive whiteboards; these can be used to give visual representations that are important for the learning of mathematics and to solve mathematical problems.

Why use technology in the teaching and learning of mathematics?

There are several terms that are used when describing the use of technology in mathematics education. Information and communication technology (ICT), learning technology, digital technology, information technology, educational technology and e-learning are some of the more common ones. Often these terms are used interchangeably. However, for the purposes of this chapter, technology will be used to describe digital devices such as calculators, laptops, tablets, interactive whiteboards, online communications such as email, social media and blended learning. This of course excludes analogue technologies such as rulers, set squares and even slide rules, which were an analogue version of a computer. In essence, all forms of digital technology that can be used in mathematics education.

The use of technology in the teaching and learning of mathematics is important for several reasons. Research evidence (Cheung & Slavin, 2013; Drijvers, 2018; Slavin, Lake & Groff, 2009; Rakes et al., 2010) concludes that there are modest gains in student achievement when technology is used. For example, a meta-analysis by Cheung and Slavin (2013) looked at 74 studies carried out in K–12 schools and found small but significant increases in student achievement when technology-based programs or applications (apps) were used to

---

Technology: term used in this chapter to encompass the range of software, devices and online materials that can be used in teaching and learning.

Information and communication technology (ICT): an alternative term used in some of the references for the software, devices and online materials that can be used in teaching and learning.

Blended learning: combining face-to-face with virtual learning.

Application (app): a software program that can be downloaded to a mobile device.

support the learning of mathematics in K–12 classrooms. Interestingly, they found the effect size was greater for primary students than for secondary students. An unexpected finding was that with newer studies, the effect size decreased. Other studies indicate a better attitude towards mathematics (Attard et al., 2020; Ellington, 2003; Heid, 2018) and improvement in students' problem-solving skills in mathematics (Ellington, 2003; Goos, 2010) because of technology use in the teaching and learning of mathematics. Several studies have looked at the use of calculators (Burrill et al., 2002; Drijvers et al., 2016; Ellington, 2003) and again these have indicated small but positive gains in student outcomes, improved problem-solving skills and a better understanding of mathematical concepts. Students expect teachers to use technology. Ken Clements, in the introduction to the *Third International Handbook of Mathematics Education*, noted 'the world of mathematics education is changing very rapidly, and [that] technology is a major factor influencing the directions of change' (Clements et al., 2013, p. viii). As such, it is critical for you as preservice teachers to be aware of the impact of technology. Technology also enables you as a teacher, and your students, to engage with mathematics in different ways such as modelling, simulations, dynamic geometry and large statistical data sets, which because they are often set in a real context have complex data that are difficult to analyse without the use of technology. Overlaying a function plot onto a picture of a bridge, for example, can be accomplished relatively easily using technology but is very time consuming without it, which may mean that the mathematics can be lost in the process of doing the task.

What is also clear from the research (Hoyles, 2010) is that simply having the technology at your disposal is not enough. Rather, it is the *way* it is used in the classroom that is crucial to its successful implementation and many teachers struggle with this (Attard et al., 2020; Zbiek & Hollebrands, 2008). Attard et al. connected the way technology is used to task design, indicating that developments in task design could signal a move away from drill and practice apps.

What the policy and curriculum documents require

Technology use has been embedded in mathematics curriculum documents (Australian Curriculum, Assessment and Reporting Authority [ACARA], 2020) as well as the APST (AITSL, 2011; National Council of Teachers of Mathematics [NCTM], 2014). The Australian Curriculum: Mathematics (ACARA, 2020), for example, has the following statement in the Rationale referring to the use of digital technologies: 'Mathematical ideas have evolved across all cultures over thousands of years, and are constantly developing. Digital technologies are facilitating this expansion of ideas and providing access to new tools for continuing mathematical exploration and invention' (ACARA, 2020).

Historically, mathematicians have been using a range of devices for mathematical exploration, from the Sumerians using clay tablets to keep track of business transactions, the abacus, logarithms and the slide rule through to modern computers and calculators. Mathematics has long been at the cutting edge of using technology.

Digital technologies are also embedded in several content strands of the Australian Curriculum: Mathematics (ACARA, 2020). For example, 'Carry out the four operations with rational numbers and integers, using efficient mental and written strategies and appropriate digital technologies (ACMNA183)' and 'Solve problems involving profit and loss, with and without digital technologies (ACMNA189)'. These are typical of the descriptors used and, as indicated by Goos (2010), at times the use of ICT seems to be somewhat tacked on and not always incorporated in a meaningful way. Also, the technology is used 'to facilitate the traditional content and skills rather than affect the knowledge and possible learning that can occur where the use of technology becomes central' (Atweh & Goos, 2011, p. 226). These statements do not identify any specific technologies, which highlights that the use of technology in mathematics is not a source of study, but rather a tool for solving mathematical problems and teaching and learning mathematics.

The APST identifies three focus areas (2.6, 3.4 and 4.5) that explicitly address the importance of teachers using and integrating ICT effectively into their teaching (AITSL, 2011).

CONNECTION Australian Association of Mathematics Teachers

The Australian Association of Mathematics Teachers (AAMT) is a federation of mathematics teaching associations in each state and territory. As such, it is a voice for mathematics teaching and learning in Australian schools. The AAMT website (https://aamt.edu.au/) has a range of resources for teachers and students including position papers on contemporary issues in the teaching, learning and assessing of mathematics. The following preamble from the AAMT position statement on the use of digital technologies acknowledges that technology changes rapidly and that while it can be useful to support learning, the use of technology does raise issues for curriculum teaching and assessment in mathematics:

> Digital learning supports and can provide feedback on personalised learning of students in school mathematics at all levels. The responsible use of relevant technologies by students is a significant contribution to enhancing the skills of all members of Australian society, with information and communication technologies as human-centred means of enhancing our personal and working lives. Access to technologies in school mathematics raises important issues for curriculum design, teaching, and assessment of learning, and for the capacity of schools to provide and support the use of such technologies. Social networking tools and online learning in virtual environments are important parts of teaching and learning in mathematics. Classrooms are beyond school walls and this impacts on pedagogies and student learning styles. The web platforms offer opportunities for collaboration, and the proliferation of resources on the internet requires students to make critical judgements on the accuracy of information. (AAMT, 2014)

Questions

1. The position paper was written in 2014. Are there any more contemporary technologies that you are aware of that would change what is written?
2. How important is it to have a national position on topics in mathematics education?

The National Council of Teachers of Mathematics (NCTM) position statement has a strong emphasis on the use of technology to support students' understanding of mathematics:

> It is essential that teachers and students have regular access to technologies that support and advance mathematical sense making, reasoning, problem solving, and communication. Effective teachers optimize the potential of technology to develop students' understanding, stimulate their interest, and increase their proficiency in mathematics. When teachers use technology strategically, they can provide greater access to mathematics for all students. (NCTM, 2011)

Technological Pedagogical Content Knowledge

Pedagogical Content Knowledge (PCK) (Shulman, 1986) was the model developed by Shulman which connected the Content Knowledge (CK) (i.e. knowledge of the material being taught) with the Pedagogical Knowledge (PK) (i.e. the knowledge about how to teach). The intersection is PCK, which would be how to teach mathematics. There have been several models developed from Shulman's initial work. Ball, Hill and Bass (2005) refined it to look at the Mathematical Knowledge for Teaching (MKT). While the **TPACK** framework (Koehler & Mishra, 2009) is an extension to include the use of technology, Mishra (Thompson & Mishra, 2008) also used the description of Total PACKage to describe TPACK to indicate that the technology must be integrated into the teaching practice. TPACK adds to the Shulman PCK model by looking at what technology skills teachers need to effectively embed technology into their teaching in a similar vein to PCK. It identifies an additional domain of Technological Knowledge (TK) and three intersections (see Figure 4.1):

TPACK: Technological Pedagogical Content Knowledge, a model that can be used to represent the interaction of technology in Shulman's PCK model.

1. The intersection of TK with PK to identify Technological Pedagogical Knowledge (TPK) – how to use technology to teach in general terms.
2. The intersection of TK with CK to describe Technological Content Knowledge (TCK) – using technology to solve mathematical problems.
3. The intersection of all three domains to describe TPACK for the discipline of mathematics is how to teach mathematics with technology.

An Australian project looked at the readiness of Australian preservice teachers to use technology in their teaching and found that generally there was a range of confidence levels in how, where and when to use technology (Finger et al., 2015). This was consistent with the data for teaching mathematics (Figure 4.2). The survey data also showed that generally preservice teachers believed that technology was useful to support student learning and again this was supported by the data for its perceived usefulness in teaching mathematics (Figure 4.3).

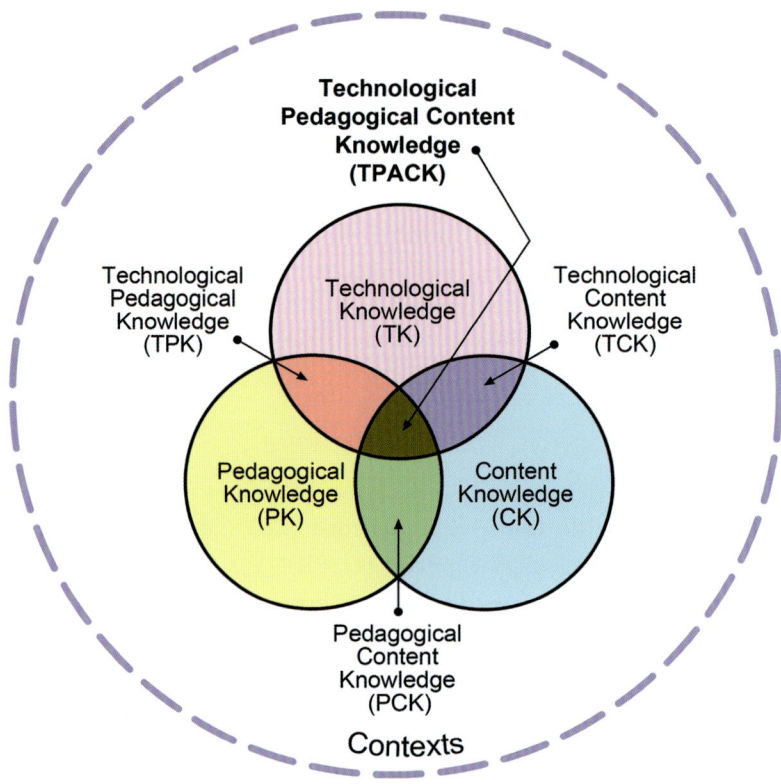

Figure 4.1 The Technological Pedagogical Content Knowledge (TPACK) model
Source: tpack.org.

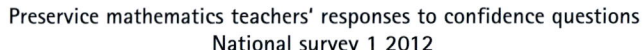

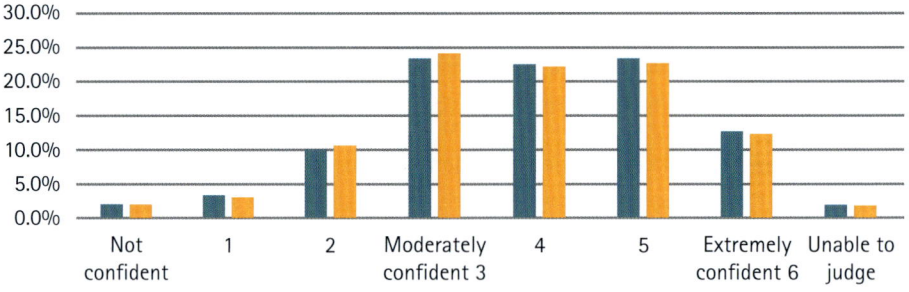

Figure 4.2 Graph showing the student responses from the Teaching Teachers for the Future project National survey 1, confidence in designing and implementing tasks that use technology to teach mathematics
Source: data from Finger et al., 2015.

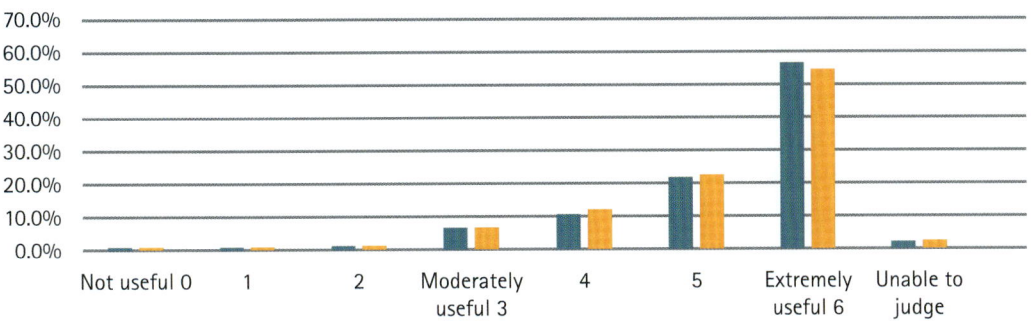

design learning sequences, lesson plans and assessment that use ICT to develop students' mathematics knowledge, attitudes and skills, n = 6331

implement meaningful use of ICT by students in achieving mathematics curriculum goals, n = 6240

Figure 4.3 Graph showing the student responses from the Teaching Teachers for the Future project National survey 1, perceived usefulness in designing and implementing tasks that use technology to teach mathematics
Source: data from Finger et al., 2015.

PAUSE AND THINK

Take some time to think about your strengths and weaknesses as they relate to the TPACK model, and what you need to do to develop further.

SHORT-ANSWER QUESTIONS

1. Research one of the early devices, such as an abacus or a slide rule, that have been used in mathematics. Would a historical perspective of technology support your students' understanding of mathematics?

2. Compare and contrast the AAMT and the NCTM positions. Do they have a different focus? How does this compare with the implementation of technology in the Australian Curriculum: Mathematics?

Mathematics software

There is a wide range of computer software available on different operating systems (Windows, Mac OS, Android, etc.) which are useful in mathematics education. There are also different types of online and offline devices available (computer, laptop, tablet, phone, etc.). The software falls into two groups: software directly related to the teaching of mathematics and software that is used for general purpose across a wide range of areas (e.g. communications tools and web browsers). A range of programs for mathematics teaching have been in use for a number of years and have had systematic reviews done on them, so it is possible to analyse these reviews when making decisions on their use within the classroom or school. However, when using new apps and software programs, it is important that as a teacher you evaluate their suitability and safety within your classroom. There are a range of approaches that are used to evaluate software; for example, Ladel and Kortenkamp (2016) developed an approach based on Artifact-Centric Activity Theory which uses activity theory as a basis for the evaluation, and Ginsburg, Jamalian and Creighan (2013) proposed that cognitive psychology could be used to inform the design and evaluation of early mathematics software. It is important to ensure that the evaluation tool you use and the pedagogical approach you will be using relates to the type of software. There are many online tools available, so always check the underlying educational theory that was used to construct the tool and ensure it relates to the way in which you intend to use the software.

Activity theory: originally developed by the Russian psychologists Leontiev and Vygotsky, the focus is on 'activity as the purposeful, transformative and developing interaction between subject and object' (Ladel & Kortenkamp, 2016 p. 25).

Mathematics specific software

There are programs that have been designed as tools to solve problems using mathematics, including Excel©, Maple©, Mathematica©, MATLAB© and SPSS©, which can also be used to support mathematics teaching in secondary schools. There are also programs that have been designed as tools to support the teaching of mathematics, including Desmos, GeoGebra, TinkerPlots, Fathom, Nspire, Autograph and Geometers Sketchpad. The diversity is too great to list all the available programs in this chapter and the list is also constantly changing. Therefore, the focus here will be on types of programs available and while some programs will be named, this is for illustrative purposes only and you will need to keep yourself abreast of the current offerings. One of the questions that you will need to ask is the amount of time required to learn to use the program versus learning with the tool (Pierce et al., 2011). A program such as a spreadsheet is multipurpose and used for a variety of types of mathematics problems, but while these may be used in other subjects (science, geography, etc.), they may take some time for a student to learn to use effectively. In contrast, a single-purpose app for a tablet device may only do one thing but may take very little time to learn how to use. Students may also confuse learning the program with learning mathematics. This investment of time to learn to use a program effectively is one that must be considered when choosing which program you will use.

Spreadsheets

One of the most used tools in mathematics is the electronic spreadsheet. Spreadsheet programs have been commonly used since the original VisiCalc was first released in 1979, and although the functionality and power of the spreadsheets has increased dramatically, the idea of using cells to hold text, numbers or a formula has remained the same. The computational power of the spreadsheet and its ability to recalculate if one or more cells are changed has led to their being very popular in a range of areas, including the finance world, and as such they are very popular when teaching financial mathematics. They are also able to perform most if not all of the statistical calculations required for secondary school mathematics (depending on the package used), and the ability to do 'what if' type calculations means that they are also very useful for modelling. There is some evidence to suggest that the use of spreadsheets assists in the transition from arithmetic to algebra (Filloy, Rojano & Rubio, 2001) because of the use of cell references.

As a teaching tool, a spreadsheet can be used to set up compound interest problems, for example, that show the rate of growth in a simple yet effective manner. Another example is using the spreadsheet to look at the classic birthday problem (Neal et al., 2014) – see Figure 4.4.

A	B	C	D	E
Number of people	Number of available days	Col B value/365	Cumulative product of entries in C =p(no matches)	Chance of birthday matches (=1 . Col D valve)
1	365	1.0000	1.0000	0%
2	364	0.9973	0.9973	0.27%
3	363	0.9945	0.9918	0.82%
4	362	0.9918	0.9836	1.64%
5	361	0.9890	0.9729	2.71%
6	360	0.9863	0.9595	4.05%
7	359	0.9836	0.9438	5.62%
8	358	0.9808	0.9257	7.43%
9	357	0.9781	0.9054	9.46%
10	356	0.9753	0.8831	11.69%
11	355	0.9726	0.8589	14.11%
12	354	0.9699	0.8330	16.70%
13	353	0.9671	0.8056	19.44%
14	352	0.9644	0.7769	22.31%
15	351	0.9616	0.7471	25.29%
16	350	0.9589	0.7164	28.36%
17	349	0.9562	0.6850	31.50%
18	348	0.9534	0.6531	34.69%
19	347	0.9507	0.6209	37.91%
20	346	0.9479	0.5886	41.14%
21	345	0.9452	0.5563	44.37%
22	344	0.9425	0.5243	47.57%
23	343	0.9397	0.4927	50.73%
24	342	0.9370	0.4617	53.83%
25	341	0.9342	0.4313	56.87%
26	340	0.9315	0.4018	59.82%
27	339	0.9288	0.3731	62.69%
28	338	0.9260	0.3455	65.45%
29	337	0.9233	0.3190	68.10%
30	336	0.9205	0.2937	70.63%
34	332	0.9096	0.2047	79.53%
35	331	0.9068	0.1856	81.44%
36	330	0.9041	0.1678	83.22%
37	329	0.9014	0.1513	84.87%
38	328	0.8986	0.1359	86.41%
39	327	0.8959	0.1218	87.82%
40	326	0.8932	0.1088	89.12%
41	325	0.8904	0.0968	90.32%
48		0.8712	0.0394	96.06%
49	317	0.8685	0.0342	96.58%
50	316	0.8658	0.0296	97.04%
51	315	0.8630	0.0256	97.44%
52			0.0220	97.80%
57	309	0.8466	0.0099	99.01%
58	308	0.8438	0.0083	99.17%
59	307	0.8411	0.0070	99.30%
60	306	0.8384	0.0059	99.41%
61	305	0.8356	0.0049	99.51%
69	297	0.8137	0.0010	99.90%
70	296	0.8110	0.0008	99.92%
71	295	0.8082	0.0007	99.93%
79	287	0.7863	0.0001	99.99%
80	286	0.7836	0.0001	99.99%
81	285	0.7808	0.0001	99.99%
88	278	0.7616	0.0000	100.00%
89	277	0.7589	0.0000	100.00%
90	276	0.7562	0.0000	100.00%
91	275	0.7534	0.0000	100.00%
92	274	0.7507	0.0000	100.00%

Figure 4.4 The probability that there will be at least one shared birthday in a group for different-sized groups of people
Source: Neal et al., 2014.

Dynamic geometry software

The ability to use dynamic geometry software (DGS) opens up a range of possibilities for the study of geometry, allowing for simple constructions, looking at the properties of shapes and developing conjectures using the features of the DGS software, in particular the ability to measure and drag objects. The use of this software has led to improvements in understanding (Chan & Leung, 2014; Guven, Baki & Cekmez, 2012; Jiang, White & Rosenwasser, 2011), increased engagement and increased problem-solving skills (Sinclair, 2003). There is a range of software available that is relatively easy to use and combines well with devices that have a touch interface, such as an interactive whiteboard (IWB) or tablet computer.

One example of use would be the demonstration of the Pythagorean equality; that is, ask students to construct a right-angled triangle followed by the measurement of sides. Substitution into the Pythagorean equality allows a demonstration of the relationship between the sides, which in turn allows students to form conjectures. The dynamic nature

Dynamic geometry software (DGS): software that allows the user to create and manipulate geometrical objects.

of the software, and its ability to measure, supports students to transition from exploration (i.e. conjecture) to the construction of geometric proof (Sinclair & Robutti, 2013). Exploring real-world examples, for example, examining how a scissor lift works, is also made easier because of some of the features of the DGS (Pierce & Stacey, 2010).

CONNECTION | From example to proof

An example of the power of DGS is the simple construction of a triangle, followed by the measurement of the internal angles and a simple sum to total the value of the angles. Dragging any of the points shows how the value of the individual angles may change but the total remains at 180. Battista (2007) hypothesised that students notice this invariance, which he called *transformational-saliency hypothesis*. Students cannot help but notice the invariance.

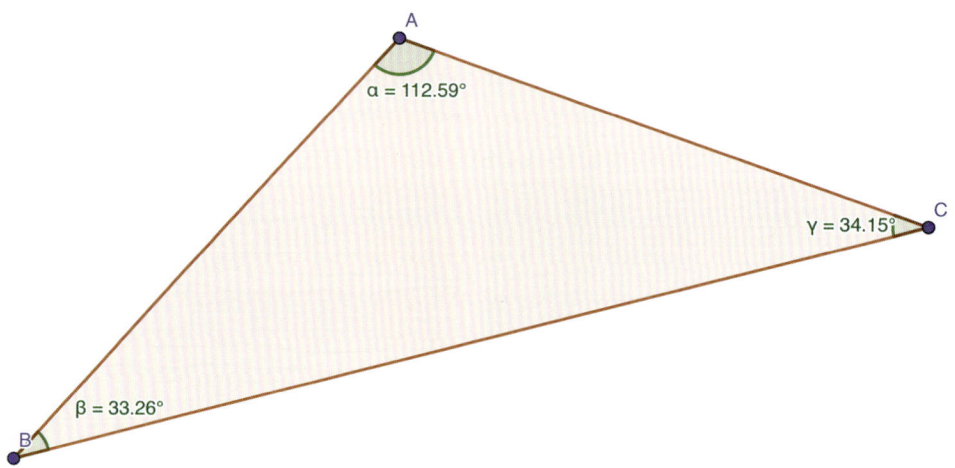

Figure 4.5 Screenshot of triangle constructed with GeoGebra

Try this for yourself.

- First construct a triangle using any of the DGS programs available to you.
- Use the tools within the program to measure the internal angles.
- Then sum the internal angles and note that they sum to 180°.
- Drag one of the vertices to change the internal angles of the triangle and note the sum. This shows the invariance of the sum, from which you can make a conjecture.
- Are you able to prove this easily?

Questions

1. Do you think that the invariance is obvious?
2. Would your students be able to recognise the difference between invariance, conjecture and proof?

Some implementations of the dynamic geometry software allow you to export the constructions to the internet so you can create your own interactive learning objects.

One example of this is GeoGebra tube (tube.geogebra.org). This website has a range of materials that have been uploaded by others and allows you to upload your materials also.

Statistics software

The advent of specialised statistics software increases the potential for student understanding of this branch of mathematics. There are a number of specialised statistics packages available, including SPSS (www.spss.com), SAS (www.sas.com), R (www.r-project.org/) and Minitab (www.minitab.com). However, these are not commonly used in schools. As previously discussed, spreadsheets also provide the facility to do statistical analyses, as do the graphic and CAS calculators that will be discussed in the next section.

CAS: Computer Algebra Systems, software or a calculator which will solve algebraic equations, and often includes graphing capabilities.

Statistics software has meant that students are now able to work with much larger and more realistic data sets and to experiment with the data and model various scenarios. This means that students' understanding of statistics can be further developed. The issue with many of these specialised programs is that they have many more features than are required in schools. This can be confusing for students and you are paying for the features that you would not use.

There are statistics packages such as TinkerPlots (www.tinkerplots.com) and Fathom (fathom.concord.org) which are aimed specifically at the education market as teaching tools. Biehler et al. (2013) identified the following four requirements for statistics software to make it more useful for teachers:

- Students can practise graphical and numerical data analysis by developing an exploratory working style.
- Students can construct models for random experiments and use computer simulation to study them.
- Students can participate in 'research in statistics': that is, they participate in constructing, analysing and comparing statistical methods.
- Students can use, modify and create 'embedded' microworlds in the software for exploring statistical concepts.

Software such as Autograph, TinkerPlots and Fathom implement these through features such as drag-and-drop graphing, dragging points, linked multiple representations and simulations. These dynamic features enable students to explore data, develop connections between chance and data, explore different representations of data (Figure 4.6) and support the transition to formal hypothesis testing (Biehler et al., 2013). In Figure 4.6, which is a screen capture using the TinkerPlots software, the data have been graphically represented in two formats and the mean and median of the data are indicated on the horizontal axis. This multiple representation allows students the opportunity to see the relationship between the box plot and the actual data as well as the two measures of central tendency.

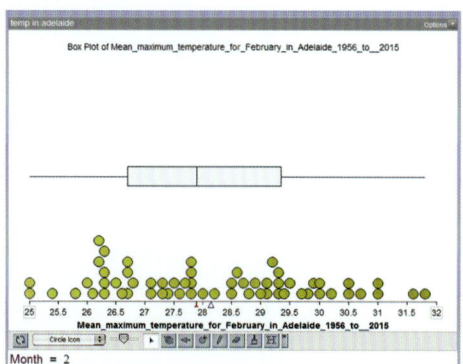

Figure 4.6 Screenshot of TinkerPlots graph: two representations of mean maximum temperatures for February

Access to statistics software and other technologies that allow students to do statistical calculations (such as spreadsheets) means that larger data sets can be used than if the students were doing the calculations manually. It also allows for individual data sets for assessment purposes. Students can create their own data sets in a range of ways, including using dataloggers and online surveys using tools such as SurveyMonkey. Secondary data sources are also readily available from sites such as the Bureau of Meteorology (www.bom .gov.au). While the CensusAtSchool project is no longer collecting data, the data sets that are available are excellent.

Computer algebra systems (CAS) software

CAS software is becoming increasingly available to schools. Computer algebra systems used to be only available through programs such as Mathematica, Maple or MATLAB, which were not commonly used in secondary schools. However, the CAS is now not only available in software designed for schools (e.g. TI-Nspire) but also on CAS calculators and handheld devices that are commonly used by schools. CAS software performs symbolic manipulation on algebraic objects, including the solution of algebraic equations, factoring, expanding and finding derivatives. The power of CAS is that it performs these operations accurately and quickly. It means that students and teachers can use CAS as a powerful functional tool to solve mathematics problems with confidence, and this also offers significant pedagogical opportunities because it allows teachers to set tasks that require high levels of mathematical thinking (Pierce & Bardini, 2015). CAS will be further discussed in the CAS calculator section of this chapter.

Multipurpose software

Increasingly, programs are available that offer connections with one or more of the software groups previously identified. These programs offer interconnections between spreadsheet, DGS, CAS, statistics and graphing packages, which increases the opportunity for teachers to highlight the connections between mathematical concepts. The connection between an equation, a spreadsheet of values and its graphical form is just one example of the power of these connections. These combinations support students to have multiple representations of a mathematics problem, which is generally accepted as being important. However, implementation does need careful planning in order to reduce cognitive load, maintain motivation and have clear learning outcomes (Pierce et al., 2011).

SCENARIO | **Using software to build mathematics connections**

A preservice teacher in their final placement chose to develop a unit of work incorporating the use of GeoGebra for visual representations when solving equations. His expectation was that this would improve students' mathematical reasoning skills. He used the software for explicit teaching of the concept in a mathematics class. GeoGebra allowed him to create detailed visual representations and examples. Because the software is dynamic, interactive and intuitive, these representations were clear and engaging for students.

Further, by making comparisons between examples and observing multiple representations with the software, the preservice teacher facilitated the connections of mathematical relationships. The students solved a variety of equations algebraically and used the software to check their results. The students developed a portfolio of activities that was used for assessment, including homework activities. Being able to check solutions using the software should have helped with the homework activities but the preservice teacher found that this was not always the case. The preservice teacher found that the homework activities were at times problematic with the students getting stuck and requiring help.

The preservice teacher found that by getting the students to use the software, their connection between the algebraic and graphic form of the equations was stronger. He also suspected that the use of the software had helped develop their mathematical reasoning skills. The preservice teacher found a range of resources that could be used with the software and would try these next time in order to make the unit more engaging for the students.

PAUSE AND THINK

What features would your ideal mathematics software have?

SHORT-ANSWER QUESTIONS

1. Access one of the DGS packages (there are some that have evaluation downloads and some that are free for personal use). Use the construction tools to create an object(s) to support your teaching of the Australian Curriculum: Mathematics content descriptor 'Use similarity to investigate the constancy of the sine, cosine and tangent ratios for a given angle in right-angled triangles (ACMMG223)' (ACARA, 2020). How does this support the learning of content described?

2. Using either Fathom (evaluation copy and support materials from fathom.concord.org) or TinkerPlots (preview copy support materials from www.tinkerplots.com/get), create an activity that will show students how outliers impact on boxplots. What questions would you need to ask to scaffold the students to make the connections?

3. Using a spreadsheet, recreate the birthday problem (see Figure 4.4) and then explain how this would support, or otherwise, a student's understanding of probability.

4. Using a multipurpose tool, represent $y = x^2$ graphically and as a table of values, and then represent $y = x^2 + 2$ and $y = x^2 - 2$ graphically and as a table of values. Does having the multiple representation help in your conceptual development? How do you think this differs from getting the students to calculate the table of values and then plot by hand?

Devices used for teaching, learning and doing mathematics

Mathematics teachers use a range of physical materials to support their teaching. Instruments such as rulers, protractors and compasses, and manipulatives such as blocks, dice and spinners are used to support students to develop an understanding of mathematical concepts. There is also a range of technological devices that can be used to both solve mathematical problems and support the learning of mathematics.

Calculators: graphics and CAS

Calculators are one of the more prevalent technologies used in secondary school mathematics teaching. Graphics/graphing calculators (GC) and CAS calculators, graphing calculators with an integrated computer algebra system (CAS), are the two main types of calculators used in schools. The choice is often determined by which ones are approved for use in the senior years' assessments. In jurisdictions where GCs are the only ones approved and the assessments are written with the expectation that students will have access to a GC from an approved list, schools use GCs and very few CAS calculators. However, in jurisdictions where CAS calculators can be used for assessment, schools will generally adopt them because of the advantages they have over the GC.

Research into the use of both GC and CAS calculators (Cheung & Slavin, 2013; Ellington, 2003) has indicated small but positive gains in student outcomes, improved problem-solving skills and better understanding of mathematical concepts. There are also gains recorded in problem-solving ability and increased engagement with mathematics. One of the perennial issues that is raised against the use of calculators, whether scientific, GC or CAS, is the concern that students will lose the ability to do mathematics 'by hand' and that their mental arithmetic skills will be reduced. This loss of skill can be used as a reason not to implement technology into teaching despite evidence that suggests that this is not the case and that these skills may improve. Pierce and Bardini (2015) indicated that according to the students they surveyed, many teachers were frequently not using CAS in the classroom. The students had the skills to use CAS calculators, as they were using them much more often than the teachers, therefore other factors such as loss of 'by hand' skills and insufficient time must be impacting on teachers' choices.

Calculators can be used both as a functional tool for solving mathematical problems and as a pedagogical tool. Graphics calculators are particularly useful for checking difficult calculations, solving statistics problems, graphical solutions to algebraic problems and matrix algebra. The calculator removes the restriction of using 'nice' whole numbers so that manual calculations are easy. It also means that larger data sets can be used, more meaningful problems can be set by the teacher, and more interesting investigations can be attempted by students. Another example would be that when investigating outliers in a statistics topic, the calculator allows the teacher to manipulate the data set and quickly see the impact on mean, median and mode, and standard deviation. Figure 4.7 illustrates a typical output from a GC. The speed of the GC also means that students and teachers can use a variety of graphical representations of data, see how they change when the data changes and determine which is the best representation. This functionality means that more

Graphics/graphing calculator (GC): calculator which has the functions of a scientific calculator but also allows the user to create and analyse statistical and algebraic graphs.

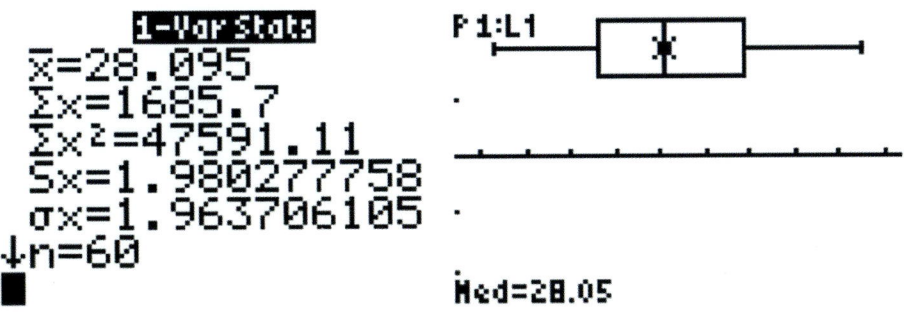

Figure 4.7 GC output of statistics problem

time can be spent in class on investigating and discussing results, rather than manual calculations.

As a teaching tool, graphics calculators allow students to manipulate variables and look at the outcome very quickly. An example of this is to graph $y = x^2$ then $y = x^2 + 1$, $y = x^2 + 2y = x^2 - 1$, etc. and to look at how the graph changes (Figure 4.8).

CAS calculators, in addition to the features mentioned, also allow for the manipulation of algebraic functions, including solving equations, differentiating functions, factoring and expanding.

The use of CAS calculators extends multiple representations and, as indicated in Figure 4.9, allows for images of real objects to be modelled and analysed.

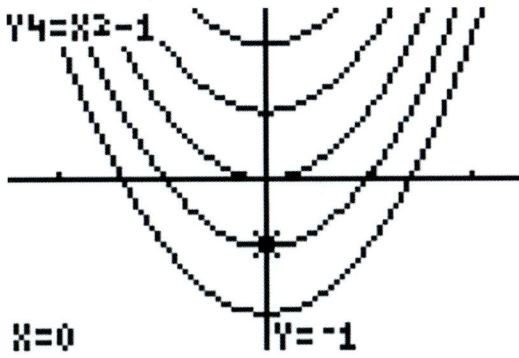

Figure 4.8 Screenshot from GC showing graphs of quadratics

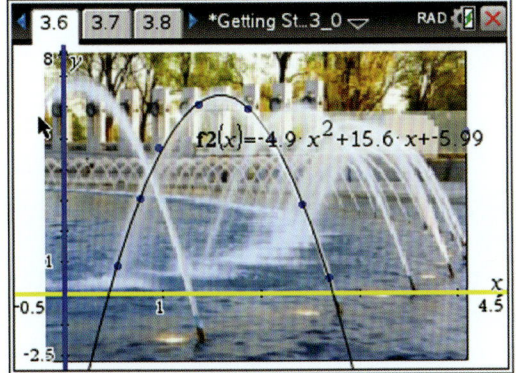
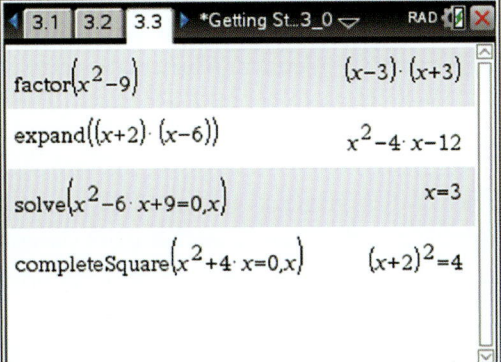

Figure 4.9 Screenshots from CAS calculator showing function plot and sample algebraic solutions

Touch and visual interfaces

The use of visual representations in mathematics is very important (David & Tomaz, 2012) and there are a number of technologies that can support their use in mathematics. Computer keyboards are not set up for mathematics and as such it has been a difficult subject for which to develop materials for teaching and to communicate with electronically.

Mathematical symbols until recently required complex markup languages or keystroke combinations to allow them to be incorporated into electronic communications. Specialised software such as Equation Editor and MathType have made this much easier in documents; however, emailing a mathematics function can require some ingenuity to compensate for the lack of symbols and layout that is possible. Even writing a simple algebraic fraction can be fraught with difficulty without the use of specialised software. The increasing popularity of tablets and interactive whiteboards has meant that the mathematics can be drawn, an image captured or in some cases converted to an editable object, and the image can be shared and worked on in a collaborative environment.

Interactive whiteboards

The interactive whiteboard (IWB) is a multipurpose tool that allows the user to interact on the whiteboard screen as if it were a large touch screen. The ability to interact through the IWB with a range of media, for example, video, images and animations, has seen interactive whiteboards being described as a technology hub (Miller & Glover, 2010). This highlights the role of the IWB as a multimodal media tool which, when used appropriately, has the potential to offer rich learning experiences to students (White, Barnes & Lawson, 2012).

The defining feature of an IWB is its touch screen interactivity, which enables the user to physically interact with the program running on the computer. The user can move objects around the screen, and create and manipulate objects with software such as GeoGebra (Betcher & Lee, 2009). Another feature that is very useful in the mathematics classroom is the ability to record what happens on the IWB and save this as a movie file. This can then be made available to students to view after the class or, in the case of the flipped classroom (discussed later in the chapter), before the class. This feature is very useful for revision purposes for students in senior years as they are often assessed on material long after they studied it, usually in the form of an end-of-year exam. This combination of IWB and online availability has shown improved motivation of students in mathematics classrooms (Heemskerk, Kuiper & Meijer, 2014).

Tablets

Tablet or slate devices, such as the iPad, and smartphones are becoming household items in Australia and their use in mathematics classrooms is increasing. They combine the interface of the touch screen with the personal features and portability that calculators offer and the flexibility, power and interconnectivity of computers in one device. This combination has significant potential as a teaching tool, although, for reasons similar to why CAS calculators have not become the device of choice (i.e. what is allowed in the high-stakes assessment), these devices will be restricted in their use in senior years. However, students often have access to tablet devices, either their own or belonging to their schools. They do offer some significant advantages over the GC and CAS calculators, with screen size, internet connectivity, inbuilt camera and touch interface being just a few. Consequently, some of the manufacturers have moved their CAS and GC capabilities to the tablet devices (e.g. TI-Nspire, GeoGebra). There is now software available that allows students to write mathematical equations using a pen, and the software will then not only convert the handwriting but will also solve the problem being written. This natural interface on a portable device adds another dimension to what is possible in mathematics classrooms.

PAUSE AND THINK

If the school you were teaching in asked you to put forward a case to introduce CAS in Year 7 to Year 9 mathematics, what would be your top five arguments?

SHORT-ANSWER QUESTIONS

1. Research the current rules in your state with regard to the use of graphics/CAS calculators in senior years' assessments. Has this impacted on how often and in what ways these calculators are used?
2. Design a task that could be used to teach 'Solve linear equations using algebraic and graphical techniques. Verify solutions by substitution (ACMNA194)' using either a GC or CAS calculator.
3. Examine the Year 8 and Year 9 algebra strand in the Australian Curriculum: Mathematics. How much of this would become trivial if students had their own CAS calculators?

Mathematics online

The internet has made a wide range of resources available to teachers. Resources such as the NRICH website, Scootle, Maths300, Cambridge HOTmaths and the National Library of Virtual Manipulatives provide activities and teacher resources that can be used as they are or adapted to suit a teaching situation. Some websites allow students to ask questions of experts to get help with their homework and access practice activities for a range of mathematical skills. Another form of resource comes from the Wolfram|Alpha website that allows students to solve simple mathematical problems by typing the question into what it calls a computational knowledge engine. Its stated aim is: 'Wolfram|Alpha's long-term goal is to make all systematic knowledge immediately computable and accessible to everyone.' (Wolfram|Alpha, 2020).

This technology forces teachers to examine the mathematics taught, and also opens up options for teachers and students about what is possible. Students can use the website to check their computations and research areas of interest beyond the school curriculum more easily. It does, of course, open up the possibility of students taking shortcuts and not mastering the concepts, similar to the arguments used against the use of calculators, and so requires teachers to look at alternative strategies to ensure that students achieve an understanding of the mathematics required.

Virtual learning environments and blended learning

Virtual learning environments is a term that is used to describe the range of technologies that are used to deliver learning via the web. The terms Course Management Systems or Learning Management Systems are also used to describe these technologies. They usually include

Virtual learning: the use of online materials to facilitate student learning.

features for collaboration, communication, content delivery and assignment processing. These technologies are being increasingly used in schools to support the face-to-face delivery of classes. This combination of face-to-face and online learning is usually called blended learning. Blended learning is the 'blend' of physical and virtual environments, with the virtual environment being used to supplement the physical classroom environment. The balance and manner of implementation varies to suit particular circumstances (International Society for Technology in Education, 2008). While there is little research into blended learning in secondary schools (Drysdale et al., 2013), it has been suggested that students appreciate the increased opportunities to learn outside the normal constraints of the classroom, with learning occurring across different mediums and at various times (White & Geer, 2010). Also, the suggested benefits of e-learning, such as time efficiency and location convenience, supplement the face-to-face advantages of one-to-one personal understanding and motivation (Brown, 2003; Fazal & Bryant, 2019). It takes more time to structure and incorporate the online materials into the blended structure, but some of this time may be recovered in future years (Ahmad, Shafie & Janier, 2008) as teachers take advantage of the blended environment.

There are many forms of blended learning. The flipped classroom is one where the instruction is completed online by the students prior to the face-to-face class. Often the instruction is in the form of a video presentation. The advantages given for this approach are that flipping the classroom allows for better use of the face-to-face class time for discussion group work. Some schools that have tried this approach in mathematics teaching (Fulton, 2012) give reasons including that it allows students to move at their own pace, that teachers can customise and update the curriculum and provide it to students 24/7, that students have access to multiple teachers' expertise and that teachers can make better use of the face-to-face classroom time. They also indicated that teachers could learn from each other's material and so it was good professional learning for them, and that students liked this form of learning.

Essential to the success of blended learning is the availability and use of high-quality online materials. There is a range of materials that are accessible for teachers to use when constructing the virtual environment. Online video tutorials are available from a number of sources, with one of the better known being the Khan Academy website (www.khanacademy .org). There are also numerous YouTube channels dedicated to learning mathematics – an example of this is one by Eddie Woo (www.youtube.com/channel/UCq0EGvLTyy-LLT1oUSO_0FQ). There is a range of virtual manipulatives available on websites such as the National Library of Virtual Manipulatives, which is a library of interactive, web-based virtual manipulatives for mathematics teaching and learning (nlvm.usu.edu/en/nav/vlibrary .html) and Scootle (www.scootle.edu.au), which is 'a national digital learning repository that provides Australian teachers with access to more than 20 000 digital learning items, provided by a wide array of contributors and aligned to core areas of the Australian Curriculum' (Scootle, 2021). Publishers have online materials to supplement textbooks, for example, the Cambridge HOTmaths materials. These resources can be used in a number of ways: they can be embedded into a virtual environment to supplement the face-to-face teaching, used as one-off activities or as part of a flipped classroom. They offer alternative explanations of concepts and also the ability to interact with simulation and animations to help conceptual development. The resources do not replace the teacher, and teachers must carefully look at how they fit within the whole student experience. There are websites that offer a form of programmed instruction where the students do a diagnostic test online and are then presented with learning activities that match their results. Textbooks are often packaged with online materials that match the approach used in the textbook and supplementary materials that are not able to be put into print.

Teacher professional learning

Technology also has a role to play in teacher professional learning. There is a variety of websites that are specifically designed for mathematics teachers. The Teaching Teachers for the Future project website (www.ttf.edu.au) has materials that look at how to integrate technology into teaching, with video footage of experienced teachers and preservice teachers trialling the examples on the website. Another example is the Top Drawer materials on the AAMT website (topdrawer.aamt.edu.au), which has teaching advice and activities for a range of mathematical concepts.

PAUSE AND THINK

1. Create your own video explanation of a concept (using a format similar to those on the Khan Academy website or the Eddie Woo YouTube channel) and use this to evaluate your explanation by trying it out on some students or your peers.
2. With the development of tools like Wolfram|Alpha, what do you think is the future of school mathematics?

SHORT-ANSWER QUESTIONS

1. View one of the instructional videos online for mathematics and critically appraise the video. How useful would it be in a flipped learning environment?
2. With all of the instructional materials that are available online, what is the role of the mathematics teacher in this virtual school environment?

Conclusion

Using technology improves students' understanding of mathematics, increases engagement and improves problem-solving skills. Students and teachers can engage with richer and more complex tasks because of the speed and accuracy of technology. While it may not be central to the Australian Curriculum: Mathematics, it is embedded in parts and its value is recognised. All students come to the classroom with a varying level of technological skill.

While this needs to be developed, it is the connection with pedagogy and the mathematics content to develop TPACK that is key. It is essential that preservice teachers familiarise themselves with a range of software and devices and evaluate these with respect to their pedagogical approach and classroom context. Doing this will enable teachers to make appropriate choices, ensure they are aware of a constantly changing field and keep up with current trends. It is important to know where to find quality resources and also how to adapt them and implement them into a teaching situation. The richness of the available materials is matched by their vast volume and so it is essential that teachers become discerning users of these materials.

BRINGING IT TOGETHER

1. What has been the most useful technological tool you have used when solving problems of a mathematical nature and why?
2. What mathematical content do you believe is most effectively taught using technology and why? Which is the least effective and why?
3. What evidence is there that using technology increases students' engagement with mathematics?
4. It has been proposed that the final exam for students doing mathematics in their last year of schooling be conducted online. As a teacher of mathematics, you have been asked to select two web-based programs that could be used for the online exam.
    a. What would you select and why?
    b. What is an example of a question that could be set in an exam and how does your question take advantage of the features of the software?
    c. Would this be a better assessment of student mathematical understanding than the current paper-based exam?

REFERENCES

Further resources

Ahmad, W.F.B.W., Shafie, A.B. & Janier, J.B. (2008). *Students' perceptions towards blended learning in teaching and learning mathematics: Application of integration. Proceedings of 13th Asian Technology Conference in Mathematic (ATCM08)*. Suan Sunanda Rajabhat, University Bangkok, Thailand.

Attard, C., Calder, N., Holmes, K., Larkin, K. & Trenholm, S. (2020). Teaching and learning mathematics with digital technologies. In J. Way, C. Attard, J. Anderson, J. Bobis, H. McMaster & K. Cartwright (eds), *Research in Mathematics Education in Australasia 2016–2019* (319–47). Singapore: Springer Singapore.

Atweh, B. & Goos, M. (2011). The Australian mathematics curriculum: A move forward or back to the future? *Australian Journal of Education*, 55(3), 214–28.

Australian Association of Mathematics Teachers (AAMT) (2014). *AAMT position paper on digital learning in school mathematics*. Retrieved from https://primarystandards.aamt.edu.au/content/download/32750/463134/file/Digital_learning.pdf

Australian Curriculum, Assessment and Reporting Authority (ACARA) (2020). *Australian Curriculum: Mathematics* (2nd edn). Sydney: ACARA.

Australian Institute for Teaching and School Leadership (AITSL) (2011). *Australian Professional Standards for Teachers*. Melbourne: AITSL.

Ball, D.L., Hill, H.C. & Bass, H. (2005). Knowing mathematics for teaching: Who knows mathematics well enough to teach third grade, and how can we decide? *American Educator*, 29(1), 14–17, 20–2, 43–6.

Battista, M.T. (2007). The development of geometric and spatial thinking. In F.K. Lester (ed.), *Second handbook of research on mathematics teaching and learning* (843–908). Charlotte, NC: Information Age.

Betcher, C. & Lee, M. (2009). *The interactive whiteboard revolution: Teaching with IWBs*. Camberwell: ACER Press.

Biehler, R., Ben-Zvi, D., Bakker, A. & Makar, K. (2013). Technology for enhancing statistical reasoning at the school level. In M.A. (Ken) Clements, A.J. Bishop, C. Keitel, J. Kilpatrick & F.K.S. Leung (eds), *Third international handbook of mathematics education* (643–89). New York: Springer.

Brown, R. (2003). Blending learning: Rich experiences from a rich picture. *Training and Development in Australia*, 30(3), 14–17.

Burrill, G., Allison, J., Breaux, G., Kastberg, S., Leatham, K. & Sanchez, W. (2002). *Handheld graphing technology in secondary mathematics: Research findings and implications for classroom practice*. Dallas, TX: Texas Instruments.

Chan, K.K. & Leung, S.W. (2014). Dynamic geometry software improves mathematical achievement: Systematic review and meta-analysis. *Journal of Educational Computing Research*, 51(3), 311–25.

Cheung, A.C. & Slavin, R.E. (2013). The effectiveness of educational technology applications for enhancing mathematics achievement in K–12 classrooms: A meta-analysis. *Educational Research Review*, 9, 88–113.

Clements, M.K., Bishop, A., Keitel-Kreidt, C., Kilpatrick, J. & Leung, F.K.S. (eds). (2013). *Third international handbook of mathematics education* (vol. 27). Springer Science & Business Media.

David, M.M. & Tomaz, V.S. (2012). The role of visual representations for structuring classroom mathematical activity. *Educational Studies in Mathematics*, 80(3), 413–31.

Drijvers, P. (2018). Empirical evidence for benefit? Reviewing quantitative research on the use of digital tools in mathematics education. In L. Ball, P. Drijvers, S. Ladel, H.-S. Siller, M. Tabach & C. Vale (eds), *Uses of technology in primary and secondary mathematics education* (161–75). Cham: Springer.

Drijvers, P., Ball, L., Barzel, B., Heid, M.K., Cao, Y. & Maschietto, M. (2016). *Uses of technology in lower secondary mathematics education: A concise topical survey*. ICME-13 Topical Surveys. Cham: Springer. https://doi.org/10.1007/978-3-319-33666-4_1

Drysdale, J.S., Graham, C.R., Spring, K.J. & Halverson, L.R. (2013). An analysis of research trends in dissertations and theses studying blended learning. *The Internet and Higher Education*, 17, 90–100.

Ellington, A. (2003). A meta-analysis of the effects of calculators on students' achievement and attitude levels in precollege mathematics classes. *Journal for Research in Mathematics Education*, 34, 433–63.

Fazal, M. & Bryant, M. (2019). Blended learning in middle school math: The question of effectiveness. *Journal of Online Learning Research*, 5(1), 49–64.

Filloy, E., Rojano, T. & Rubio, G. (2001). Propositions concerning the resolution of arithmetical-algebraic problems. In R. Sutherland, T. Rojano, A. Bell & R. Lins (eds), *Perspectives on school algebra* (155–76). Dordrecht: Kluwer Academic.

Finger, G., Romeo, G., Lloyd, M., Heck, D., Sweeney, T., Albion, P. & Jamieson-Proctor, R. (2015). Developing graduate TPACK capabilities in initial teacher education programs: Insights from the teaching teachers for the future project. *The Asia-Pacific Education Researcher*, 1–9.

Fulton, K. (2012). The flipped classroom: Transforming education at Byron High School: A Minnesota High School with severe budget constraints enlisted YouTube in its successful effort to boost math competency scores. *The Journal (Technological Horizons in Education)*, 39(3), 18.

Ginsburg, H.P., Jamalian, A. & Creighan, S. (2013). Cognitive guidelines for the design and evaluation of early mathematics software: The example of MathemAntics. In J. Sowder, L. Sowder & S. Nickerson (eds), *Reconceptualizing early mathematics learning* (83–120). Dordrecht: Springer.

Goos, M. (2010). *Using technology to support effective mathematics teaching and learning: What counts?* Retrieved from http://research.acer.edu.au/cgi/viewcontent.cgi?article=1067&context=research_conference

Guven, B., Baki, A. & Cekmez, E. (2012). Using dynamic geometry software to develop problem solving skills. *Mathematics and Computer Education*, 46(1), 6.

Heemskerk, I., Kuiper, E. & Meijer, J. (2014). Interactive whiteboard and virtual learning environment combined: Effects on mathematics education. *Journal of Computer Assisted Learning*, 30(5), 465–78.

Heid, M.K. (2018). Digital tools in lower secondary school mathematics education: A review of qualitative research on mathematics learning of lower secondary school students. In L. Ball, P. Drijvers, S. Ladel, H.-S. Siller, M. Tabach & C. Vale (eds), *Uses of technology in primary and secondary mathematics education* (177–201). Cham: Springer.

Hoyles, C. (2010). *Mathematics education and technology: Rethinking the terrain*. Springer.

International Society for Technology in Education (2008). *National Educational Technology Standards for Teachers*. Retrieved from http://www.iste.org/standards/nets-for-teachers.aspx

Jiang, Z., White, A. & Rosenwasser, A. (2011). Randomized control trials on the dynamic geometry approach. *Journal of Mathematics Education at Teachers College*, 2(2).

Koehler, M. & Mishra, P. (2009). What is technological pedagogical content knowledge (TPACK)? *Contemporary Issues in Technology and Teacher Education*, 9(1), 60–70.

Ladel, S. & Kortenkamp, U. (2016). Artifact-centric activity theory: A framework for the analysis of the design and use of virtual manipulatives. In P.S. Moyer-Packenham (ed.), *International perspectives on teaching and learning mathematics with virtual manipulatives* (25–40). Cham: Springer.

Miller, D. & Glover, D. (2010). Presentation or mediation: Is there a need for 'interactive whiteboard technology-proficient' teachers in secondary mathematics? *Technology, Pedagogy and Education*, 19(2), 253–9.

National Council of Teachers of Mathematics (2011). *Strategic use of technology in teaching and learning mathematics*. Retrieved from https://www.nctm.org/Standards-and-Positions/Position-Statements/Strategic-Use-of-Technology-in-Teaching-and-Learning-Mathematics/

National Council of Teachers of Mathematics (2014). *Principles to actions: Ensuring mathematical success for all*. Retrieved from http://www.nctm.org/PtA/

Neal, D., Muir, T., Manuel, K., Livy, S. & Chick, H. (2014). Desperately seeking birthday mates! Or what maths teachers get up to on Saturday nights! *The Australian Mathematics Teacher*, 70(1), 36–40.

Pierce, R. & Bardini, C. (2015). Computer algebra systems: Permitted but are they used? *Australian Senior Mathematics Journal*, 29(1), 32.

Pierce, R. & Stacey, K. (2010). Mapping pedagogical opportunities provided by mathematics analysis software. *International Journal of Computers for Mathematical Learning*, 15(1), 1–20.

Pierce, R., Stacey, K., Wander, R. & Ball, L. (2011). The design of lessons using mathematics analysis software to support multiple representations in secondary school mathematics. *Technology, Pedagogy and Education*, 20(1), 95–112.

Rakes, C.R., Valentine, J.C., McGatha, M.B. & Ronau, R.N. (2010). Methods of instructional improvement in algebra: A systematic review and meta-analysis. *Review of Educational Research*, 80(3), 372–400.

Scootle (2021). *About*. Retrieved from https://www.scootle.edu.au/ec/p/about

Shulman, L.S. (1986). Those who understand: Knowledge growth in teaching. *Educational Researcher*, 4–14.

Sinclair, M.P. (2003). Some implications of the results of a case study for the design of pre-constructed, dynamic geometry sketches and accompanying materials. *Educational Studies in Mathematics*, 52(3), 289–317.

Sinclair, N. & Robutti, O. (2013). Technology and the role of proof: The case of dynamic geometry. In M.A. (Ken) Clements, A.J. Bishop, C. Keitel, J. Kilpatrick & F.K.S. Leung (eds), *Third international handbook of mathematics education* (571–96). New York: Springer.

Slavin, R.E., Lake, C. & Groff, C. (2009). Effective programs in middle and high school mathematics: A best-evidence synthesis. *Review of Educational Research*.

Thompson, A. & Mishra, P. (2008). Breaking news: TPCK becomes TPACK! *Journal of Computing for Teacher Educators*, 24(2), 38.

White, B., Barnes, A. & Lawson, M. (2012). *Student views on the value and use of interactive whiteboards in a secondary school*. ACEC2012 Conference paper, Perth.

White, B. & Geer, R. (2010). Learner practice and satisfaction in a blended learning environment. In *World Conference on E-Learning in Corporate, Government, Healthcare, and Higher Education* (vol. 2010, no. 1, 569–78).

Wolfram|Alpha (2020). *About Wolfram|Alpha*. Retrieved from https://www.wolframalpha.com/about/

Zbiek, R. & Hollebrands, K. (2008). A research-informed view of the process of mathematics technology into classroom practice by in-service and prospective teachers. In *Research on technology and the teaching and learning of mathematics* (vol. 1. Research Syntheses, 287–344). USA: Information Age Publishing.

CHAPTER 5

Inquiry-based learning

Judy Anderson

LEARNING OBJECTIVES

The learning objectives of this chapter are directly linked to the Australian Professional Standards for Teachers (APST) (Australian Institute for Teaching and School Leadership [AITSL], 2011). After studying this chapter, you should be able to:

- identify the benefits and challenges of using student-centred (or learner-centred) teaching approaches in secondary mathematics classrooms (APST 2.1, 4.1)
- describe a range of teaching strategies that foster student-centred approaches (APST 2.1, 4.1)
- describe how problem solving can be implemented in mathematics classrooms as well as the similarities and differences between problem solving and mathematical modelling (APST 3.3)
- identify types of inquiry-based learning with appropriate task selection (APST 3.1)
- plan inquiry-based learning approaches using collaborative or cooperative learning (APST 3.2).

Introduction

Mathematics can be taught in different ways. Good teachers show their passion for mathematics and use different approaches to engage and enthuse their students. When describing a favourite teacher at school, many successful mathematics students talk about their teachers' enthusiasm, willingness to explain concepts in different ways, capability to show them how the mathematics they are learning is used to solve problems, and flexibility and patience when they are struggling to understand challenging concepts.

While some students like to learn mathematics by practising questions on their own, others see this approach as repetitive and meaningless and prefer to talk about mathematics with peers as they solve problems together. Such differences in students' preferences for learning mathematics mean teachers must develop a repertoire of teaching approaches to engage and enthuse their students. Some have referred to this repertoire as a 'mathematics teaching toolkit'. Your teaching toolkit will include ideas for demonstrating new concepts, strategies to start lessons in engaging and challenging ways, examples of rich mathematics tasks, ways to organise students into groups to discuss problems, ideas about connecting mathematics concepts to other areas of the school curriculum, and many more. Some of these approaches have been discussed in other chapters in this volume.

In this chapter, the focus is on teaching strategies which can be used to promote mathematical inquiry. Inquiry-based learning allows students to pose questions about the mathematics they are learning or about ideas they have met during an investigation. They may not be sure how to solve a problem but by talking with peers about what they do know and what issues they are confronting, they begin to wonder about things they do not fully understand. Inquiry-based learning is a broad term which can include problem solving, investigating and modelling. The best way to encourage students to inquire is to allow them the opportunity to work on unfamiliar problems with peers, using either collaborative learning or cooperative learning. Developing a repertoire of ways to form small groups of students, to encourage them to participate so that all group members can contribute, and to take responsibility for developing a shared understanding or solution to the problem are all important components of learning to teach for successful inquiry-based learning. Strategies for implementing cooperative learning are discussed with advice for beginning teachers as they start their journey as secondary mathematics teachers.

The chapter is divided into five sections. The first section investigates student-centred learning in mathematics classrooms by examining the differences between the approaches adopted by two mathematics departments in England. The second section describes teaching strategies that foster student-centred approaches while the third section elaborates two approaches – problem solving and mathematical modelling. The fourth section provides advice about how to establish an inquiry-based learning classroom and the types of tasks to choose. Finally, further advice is provided about collaborative and cooperative learning.

Inquiry-based learning: occurs when students observe phenomena, ask questions, use mathematics to find possible solutions, evaluate their strategies and share their ideas.

Student-centred learning in mathematics classrooms

A distinction is frequently made between 'teacher-centred' and 'student-centred' (or 'learner-centred') approaches in secondary mathematics classrooms. The former has been referred to as 'direct instruction' with the teacher delivering content, demonstrating procedures and directing student learning. Students become recipients of knowledge and rely on the teacher to direct their learning and provide practice questions typically requiring standard procedures. The focus in this type of classroom is on student performance and answering questions correctly (Boaler, 2015).

If this style of teaching was viewed as one end of a continuum of practice, the other end would include students taking more ownership of their learning by negotiating meaning, discussing possibilities and sharing understanding with fellow students and the teacher; trying a range of different types of problems; and assessing their own understanding. The focus in this type of classroom is on students accepting the challenge, being willing to try a task or problem they have not seen before, valuing mistakes as learning opportunities, and not giving up. Students may even pose their own questions and instigate their own inquiries. Contrary to the views of some critics, this alternative does not mean the teacher leaves the learning to students without guidance or advice.

The teacher's role in a student-centred classroom is still critical because to initiate such practices they need to determine what types of learning situations or tasks might stimulate engagement with mathematical ideas, promote discussion and reveal misunderstandings leading to deeper learning, and then they need to be able to summarise the important mathematical ideas for students. Boaler (2002) compared the two approaches described here by collecting data over a three-year period from teachers and students in two secondary mathematics departments in England in the 1990s. Her research describes some of the benefits and challenges of using student-centred approaches in secondary mathematics contexts. The year levels referred to in her research have been renamed here to align with the Australian education system.

CONNECTION **Jo Boaler**

Jo Boaler (1964–) is the Nomellini-Olivier Professor of Mathematics Education at the Stanford Graduate School of Education in California. She rose to fame with her longitudinal study comparing two comprehensive schools in London which reported the impact of different teaching approaches on students' attitudes towards and understanding of mathematics. She published this work in 1997 as the book, *Experiencing School Mathematics: Teaching Styles, Sex and Setting*. Jo co-founded 'you-cubed' at Stanford to provide mathematics resources to parents, teachers and students. She is the author of many books including *The Elephant in the Classroom: Helping Children Learn & Love Maths* (2010), *What's Math Got to Do with It?* (2015) and *Mathematical Mindsets* (2016). Her work promotes mathematics education reform and a shift to more student-focused mathematics teaching.

Were you ever given the opportunity in your secondary mathematics lessons to create your own questions for your peers to solve? If so, how did you feel about being given this opportunity? If not, think about what this opportunity might have made you feel. How might it have promoted your engagement and sense of ownership?

Comparing the two approaches: a research study

Consider two mathematics departments in two similar secondary schools (based on student performance on standardised tests, socioeconomic status and student background characteristics) with competent and dedicated mathematics teachers but with quite different philosophies of teaching. In the study reported here (Boaler, 2002), both schools followed the National Curriculum in England, which included mathematics content strands and a strand referred to as 'using and applying' that incorporated the processes of application, communication, reasoning, logic and proof. One significant difference between the two approaches adopted by these schools was the level of integration of the using and applying strand and its processes into mathematics lessons. Amber Hill school adopted more teacher-centred teaching while Phoenix Park school adopted more student-centred approaches (school names are pseudonyms).

At Amber Hill, students were assigned to classes based on ability and teachers believed students would learn mathematics if they were presented with clear explanations of mathematical concepts, shown mathematical procedures on the board, and given large sets of similar questions to practise – an approach frequently referred to as teacher-centred teaching. Based on the textbook used at the school, some open-ended questions and investigations were occasionally given to the students to meet the requirements of the using and applying strand, although teachers complained about being expected to do this. The school was well managed, with qualified, competent, committed and hard-working mathematics teachers, and students were compliant, maintaining a high level of on-task behaviour during mathematics lessons. The mathematics teaching approach used at Amber Hill was typical of secondary mathematics departments in England (Boaler, 2002; Swan, 2006) and typical of Year 8 mathematics classrooms in Australia (Hollingsworth, Lokan & McCrae, 2003).

At Phoenix Park a very different approach to learning mathematics was implemented, with teachers using a project-based learning approach, with each project typically lasting two to three weeks – a teaching approach more closely aligned with learner-centred teaching. Beginning with a Year 8 cohort of students in mixed-ability mathematics classes, Boaler tracked them for three years, as they worked on open-ended projects in almost every lesson – the using and applying strand was embedded in all projects. An example of a project was Volume 216 – the students were asked to think about what shape would have a volume of 216 and were expected to build on this idea and pose their own questions and interests, thus encouraging independent thinking and taking ownership of their learning. Sometimes teachers would teach mathematics content before the project began but most of

Teacher-centred teaching: the teacher is the director of learning with teaching comprised mainly of demonstration, explanation and checking that students can reproduce skills and procedures after extensive practice.

Learner-centred teaching: the teacher is a facilitator of learning, guiding student group work but allowing students to take responsibility for problem solving, problem posing and inquiry as they work with partners to find solutions.

the time, new mathematics content would be taught to individuals and small groups as required. The projects were sufficiently open to allow for differentiation, and students had choice about what they might pursue in the project and were given formative feedback as they progressed, with grades only being allocated at the end of each academic year. The project work ceased about four months before the end of Year 10 for final mathematics examination preparation, which was teacher organised and managed.

When the students arrived at Phoenix Park, they were not familiar with this approach to learning mathematics and not all of them enjoyed the open-ended, less-structured approach to learning mathematics, particularly some of the boys. Teachers managed this by offering students different amounts of structure depending on their needs – students were never left to manage on their own – always ensuring that all students understood the problem and had some ideas about how to begin. Students needed to learn a new way of working in this environment, including how to explain and justify their thinking, how to extend their ideas and how to take ownership of their learning – teachers taught the students the mathematics content as outlined in the National Curriculum, but they also taught the students how to learn and to be independent thinkers.

Because students often question the purpose of the mathematics they learn at school, and employees have noted that students are not always able to apply their mathematics in real contexts outside school, Boaler (2002, p. 2) was particularly 'interested to discover whether different teaching approaches would influence the nature of the knowledge that students developed and the ways that students approached new and different situations'. She followed one cohort of students in each school using classroom observations, interviews with teachers and students from Years 8 to 10. Data were also collected through a range of assessment tasks, including traditional examinations and some project work, so that Boaler could compare the students' different learning experiences to their achievement data.

For the national mathematics examination system in England in Year 10, the students were entered at one of three levels, with each level allocated a different set of possible grades – higher (A*, A, B, C or fail), intermediate (C, D, E or fail) and foundation (D, E, F, G or fail). At Amber Hill, the levels were determined by the ability grouping students were placed in, and for many students it remained the same as when they entered the school and were first assigned to a class. At Phoenix Park, teachers made decisions about the examination level late in Year 10, providing students with every opportunity to aim for the highest possible grade throughout their learning experiences. Schools also had the option that 20 per cent of the students' final grade could be determined by a project, which they submitted for external moderation. Both schools chose this option, but the Phoenix Park students could choose their best project from many, whereas Amber Hill students submitted the one and only project they had completed in a three-week allocated time during Year 10.

Differences between the two groups of students

While there were no significant differences between the school achievement data of the two cohorts at the beginning of Year 8, by the Year 10 examinations, Phoenix Park students significantly outperformed Amber Hill students, particularly at the foundation level. Based on her data, Boaler speculated that the difference in attitudes between students at the two schools, as well as the absence of anxiety about mathematics at Phoenix Park, may have also been contributing factors.

There were differences between the two groups of students on engagement and enjoyment of lessons and their views of mathematics. The Amber Hill students reported spending more time on task during mathematics lessons than the Phoenix Park students, but the reverse was the case regarding time spent engaged in learning. More of the Phoenix Park than Amber Hill students enjoyed open-ended learning experiences and they also 'believed mathematics to be an active, inquiry-based discipline' (Boaler, 2002, p. 77). At Amber Hill, students described mathematics as rule-bound, requiring memorisation, and during lessons they were observed basing their mathematical thinking on what they thought was expected of them, using cues from the teacher or the textbook rather than thinking about the mathematics within a question. If required to apply mathematics to a real-world context, students at Amber Hill would ask for help rather than think about what they knew – they were reliant on the teacher rather than being independent thinkers.

Differences in creativity were evident when Year 9 students from both schools were required to design a flat or apartment occupying a given space for a student, a couple, a family or themselves. Typically, students from both schools included a kitchen, a living room, a bathroom and at least one bedroom. However, 33 per cent of the designs at Phoenix Park included unusual rooms (games, studies, hi-fi rooms, children's playrooms, etc.), whereas only 3 per cent of Amber Hill designs included something different. Boaler (2002) suggested that the Phoenix Park students included rooms they would want to have in their flat whereas the Amber Hill students included rooms they thought they should have in a flat – those they thought a teacher would approve of.

Using the data from several different forms of assessment, Boaler (2002) concluded that the students had developed different kinds of mathematical knowledge. She stated that while 'the Phoenix Park students did not have greater knowledge of facts, rules, and procedures, [they] were more able to make use of the knowledge they did have in different situations' (Boaler, 2002, p. 104). She described the mathematical knowledge of the Amber Hill students as 'inflexible and inert' – they had difficulty remembering information after a while and were particularly challenged when problems required knowledge from two different mathematics topics. By learning one mathematical procedure at a time, they did not learn the important connections between mathematical ideas. They became rule followers, had little agency and were unable to develop identities as users and creators of mathematics. Boaler (2002) states:

> For if learning mathematics entails more than the construction of cognitive forms, but the development of practices through which identities with the discipline are formed, then repeated and limited practices of procedure repetition will limit the identities of all students who do not go beyond such practices. (p. 133)

The impact of the approaches on male and female students

One important aspect of Boaler's (2002) research was the impact of the different teaching approaches on male and female students. Based on earlier research noting that female students prefer 'connected knowing characterised by intuition, creativity, and experience' and male students 'value separate knowing, characterised by logic, rigor and rationality' (Boaler, 2002, p. 138), Boaler sought to explore whether the different teaching approaches supported these findings. The differences between female and male students at Amber Hill were described as a 'quest for understanding' versus 'playing a kind of school mathematics game'. When given the opportunity to do the open-ended questions in Amber Hill classrooms, more of the female students valued the chance to discuss ideas, work in groups, think creatively and work at their own pace – Boaler noted that the Amber Hill lessons were frequently fast-paced, with students completing many practice questions in each lesson. The male students also liked doing the occasional open-ended work, not because it was a valued learning experience but because it was a change from their normal routine. One striking finding that was established in subsequent studies, revealed the female students from the top-ability class at Amber Hill did not like the fast-paced, procedural approach and many wanted to 'move down' in class even though they could do and understand the mathematics, and they knew it would impact on their final grades.

Another important finding revealed male students outperformed female students at Amber Hill in the final examinations in Year 10, whereas at Phoenix Park there were no significant gender differences in performance. The girls at Amber Hill were particularly disaffected, not because they thought they were not capable, but because they believed they were unable to improve their situation because of the pedagogical approach at the school. From her data, Boaler suggested that the teacher-centred teaching approach at Amber Hill privileged male over female students, thus presenting an inequitable learning environment – this was most acute for the female students in the top-ability class.

Recommendations from the research

As in all research findings, these are generalisations and Boaler (2002) reiterates that some students at Amber Hill could apply their mathematics to real-world problems. Also, not all students at Phoenix Park could apply their knowledge in unfamiliar contexts – there was a range of student capabilities in both schools, but the differences overall were still striking and evident from all data sources. She notes:

> The two approaches are not at opposite ends of a spectrum of mathematical effectiveness, but the differences between the approaches do serve to illuminate the potential of the different methods of teaching for the development of different forms of knowledge and the cultivation of different identities as learners and users of mathematics. (Boaler, 2002, p. 136)

Boaler concluded with advice based on the approach at Phoenix Park. First, teachers of mixed-ability classes must provide differentiated work – this can be achieved by differentiating by task or outcome. She indicated both approaches could be successful, but she recommended differentiating by outcome, whereby all students begin with the same task, but the teacher provides appropriate scaffolding and adapts the task for students as they progress. This is not a trivial exercise and requires careful planning by teachers, with judicious selection of open-ended tasks to meet the needs of students. Not all tasks need to have a real-world focus, but all should involve important mathematical concepts.

Second, students need to develop appropriate practices of doing mathematics just as much as they need to learn mathematics content. If one of the outcomes of schooling is for students to be able to think independently, be creative, choose appropriate methods, connect mathematical ideas and use mathematics to solve real-world problems, then they need to have these experiences within the mathematics classroom, and on a regular basis. Third, teachers need to believe that all students can learn mathematics at a conceptual level and must be challenged with probing questions and not 'spoon fed' – it can be tempting to 'tell' too soon and do the thinking for the students. Fourth, teachers should focus on student engagement through worthwhile activities or tasks that students find interesting. Boaler's study revealed that it is not the amount of work that is completed which leads to the most improved mathematics learning outcomes but the types of tasks and student-centred practices that are undertaken.

Tensions between using the two approaches

It must be acknowledged that very few secondary mathematics classrooms fit neatly into either one or the other of those described above. Most mathematics teachers do implement some teacher-centred teaching approaches and some student-centred approaches when teaching mathematics. However, the balance of each in any one classroom might be quite different based on several factors, including the perceived level of ability or behaviour of the students; type of school; year level; impending large-scale or high-stakes assessments; parent expectations; the knowledge, experience and confidence of the teacher; and the teacher's beliefs about the nature of mathematics as well as about mathematics teaching and learning (Anderson, White & Sullivan, 2005). Boaler's study has provided substantial evidence of the outcomes of different approaches to teaching mathematics and encourages us to reflect on the types of approaches we implement when we teach secondary students.

Further, when teachers are committed to using both approaches in the classroom, tensions or incompatibilities occur which may be difficult to resolve. Swan (2006) describes three that need to be considered – creativity versus coverage, openness versus convergence, and autonomy versus challenge. Teachers frequently feel the pressure of 'covering' the curriculum, particularly when high-stakes examinations are looming, whereas providing students with opportunities to explore, explain and reason, as well as to consider a range of alternative approaches to solving problems, takes time. Open problems may lead to a range of procedures as students choose their preferred methods whereas teachers usually teach a preferred method, one that is expected to be used in high-stakes examinations. Finally, teachers may want their students to be challenged, whereas when given a choice, students may only choose simpler procedures and strategies that they feel more confident using. These tensions can be resolved, but they require careful planning and deliberate action. Swan (2006) advises we explain these tensions to students as we try to implement more open-ended problems that have several pathways to solution. The following section provides advice about implementing some commonly used teaching strategies in student-centred classrooms.

SHORT-ANSWER QUESTIONS

1. Why is it important to use open-ended tasks in the teaching and learning of mathematics?
2. What strategies could teachers use to convince students (and parents) that such tasks are an important part of learning mathematics?

Teaching strategies that foster student-centred approaches

While the literature is replete with examples of different types of student-centred teaching approaches, there are five that help to foster student autonomy in secondary mathematics classrooms – discussion, small group work, cooperative or collaborative learning, problem solving and modelling, and student inquiry. While these strategies are clearly connected, and overlapping, each will be discussed briefly, with problem solving and modelling, inquiry-based learning and cooperative learning considered in detail in subsequent sections of this chapter.

Discussion

Cooperative learning: involves students working in small groups to solve problems with structure and guidance provided by the teacher.

Collaborative learning: involves students working in small groups to solve problems where they organise and negotiate strategies and approaches between themselves.

Problem solving: involves students making choices, interpreting, formulating, modelling and investigating problem situations, and communicating solutions effectively.

Discussion in classrooms may involve whole class discussion or it may take place in small groups (Gillies, 2016). Whole class discussion is a useful teaching strategy to encourage students to listen to a variety of approaches to solving mathematics problems, to reflect on their own understanding and develop useful metacognitive strategies, and to evaluate different processes and procedures. Such discussions need to be carefully orchestrated by the teacher with purposeful selection of student responses so that the solutions offered can be compared, thus highlighting either misconceptions or key mathematical ideas for further elaboration. In addition, whole class discussion enables the teacher to evaluate student thinking, affirm important ideas, highlight useful strategies and identify where the next learning focus should be. However, for whole class discussion to be successful, students need time to think about the problem and prepare a possible solution first, teachers need to guard against a small number of students dominating, and the ideas offered by students need to be focused and on topic – this can be time-consuming if lots of students wish to share their ideas (Killen, 2015). Teachers can prepare for successful whole class discussion if they allow for small group discussion first. By listening to the conversations, selecting the groups to provide input, and determining the ideal order of presentation of ideas to build up from simple strategies to more complex thinking, successful whole class discussion is more likely to occur. Finally, teachers need to summarise the discussion to highlight key points.

Small group discussion can be organised as two or more students work together to share understandings, develop strategies for proceeding on a mathematical task, evaluate their ideas and consider the best ways of communicating their solution. For small group discussion to be effective, it is critical to select mathematics tasks which will generate ideas, that no one student can answer quickly and hence stifle conversations, and which ideally have either several solutions or several methods of solution. For example, the tasks known as 'always, sometimes, or never true' that encourage the students to evaluate the validity of statements and generalisations provide good discussion starters for small groups (Swan, 2006). The following examples are content specific and should be used when appropriate:

- multiplication makes bigger
- $a^2 > a$
- parallelograms have no axes of symmetry (Bills et al., 2004).

These statements can be presented to students to discuss, with teachers encouraging students to determine the conditions when they are true, with counter examples demonstrating when they are false. Ultimately, we want students to be able to generalise and justify their thinking.

PAUSE AND THINK

Consider the statement 'multiplication makes bigger' and develop a response that would make sense to a) a Year 3 student; b) a Year 6 student; and c) a Year 9 student. Think about how the 'spiral curriculum' builds knowledge and understanding by revisiting some mathematical ideas at each year level.

Small group work

Students working in small groups shifts the focus from passive learning to more active participation, teaches students to be less reliant on the teacher, encourages students to verbalise their thinking, enables sharing of different strategies and approaches to solving problems, and encourages cooperation and the learning of important classroom norms (Killen, 2015). Such social norms might include:

1. Explain and justify your solutions and methods.
2. Attempt to make sense of others' explanations.
3. Indicate agreement or disagreement.
4. Ask clarifying questions when the need arises.

For small group work to be effective, teachers need to select tasks that require discussion and negotiation, that will take some time to complete and that will be easier to do with more participants sharing the load. The size of the group should reflect the complexity of the task or the amount of work required in solving the problem – some teachers find allocating group roles beneficial (e.g. recorder, reporter, resource manager). Monitoring group efforts is essential to providing feedback, encouraging further exploration and explanation, extending and challenging when necessary, and identifying misconceptions and resolving disagreements. After listening to group solutions, teachers are able to plan the sharing session for the whole class to benefit from the different approaches and solution strategies, thus building from simple to more complex solutions. Some schools have whiteboards on classroom walls so that groups can display their strategies for sharing.

A more structured form of small group work involves collaborative or cooperative learning (Horn, 2012). For cooperative learning to be effective, teachers need to determine the structure of the groups (mixed ability is preferable) and how group members will interact, the level of challenge of the task, group roles and mutual accountability, and possible assessment approaches so that learning outcomes are made clear. Careful selection of the mathematics task or project is required so that all members can make an active contribution. Further advice about using collaborative and cooperative learning is provided later in this chapter.

Problem solving

There are three approaches to teaching problem solving – teaching *for* problem solving, teaching *about* problem solving, and teaching *through* problem solving (Borbas, 1988, p. 587). The first two approaches are more teacher centred – in the first approach, the teacher focuses on the mathematics before students attempt problems (usually connected to the content) and in the second approach, the teacher teaches the students about the problem-solving process as well as about useful problem-solving strategies (e.g. draw a diagram or table, work backwards, try a simpler example). The third approach is more student centred, since the students can learn new mathematics by confronting problematic situations. This approach is most difficult to implement, but if challenging tasks are used and the challenge can be sustained, conceptual understanding is more likely to develop. The teacher's role is critical, with necessary actions including:

- scaffolding students' thinking
- pressing for students' explanations
- probing of students' strategies and solutions
- helping students accept responsibility for learning in a more open way
- attending to issues of equity.

It is important to understand that students are not left to figure things out on their own or to 'discover' mathematics without assistance.

The Australian Curriculum: Mathematics F–10 (Australian Curriculum, Assessment and Reporting Authority [ACARA], 2020) includes problem solving as one of four proficiencies and recommends that students 'make choices, interpret, formulate, model and investigate problem situations'. Clearly, students need opportunities to solve problems in mathematics classrooms if this proficiency is to be developed (Anderson, 2014). Whether they are asked to apply mathematics they have already learned to problem situations or whether they are asked to struggle with unfamiliar problem contexts to develop new mathematical ideas, it is important that students are given as many opportunities as possible to develop problem-solving skills and competencies. Further information about the Australian Curriculum is presented in Chapter 9.

Mathematical modelling: the process of describing a system in the real world using mathematical concepts and language. A model may help to explain a system, to study the effects of different components, and to make predictions about behaviour.

In addition to problem solving, the curriculum recommends that students have opportunities to model real-world situations. Mathematical modelling involves solving problems set in real-world contexts whereby the real-world problem is simplified to build a real model of the situation – this process is often referred to as 'mathematisation'. The mathematical modelling process is usually represented as a cycle of stages, and like problem solving, it requires a range of competencies for students to be successful. Further information about problem solving and modelling is presented later in this chapter.

Another approach which may be used to refer to implementing problem solving in the classroom is problem-based learning. This teaching method is an approach to curriculum design whereby a comprehensive set of real, complex problems are used that enable students to learn the required knowledge, skills and understandings outlined in curriculum outcomes and objectives – the problems become the curriculum. This approach requires the careful selection and sequencing of suitable problems that allow students to engage in sustained inquiry and thinking using collaboration (Killen, 2015).

Problem-based learning: a teaching method in which real-world problems are used to promote student learning of concepts and principles, rather than direct explanation of facts and concepts. It is like project-based learning, which involves extended investigations where students typically produce a final product.

Inquiry-based teaching

The use of inquiry-based teaching approaches incorporates many of the teaching methods already discussed in this section. By using small-scale student research projects, teachers can blend a combination of discussion (whole class and small group), collaborative learning, and problem solving and modelling into mathematics learning experiences for students. Inquiry-based learning projects may be structured or guided so that students learn mathematics content as well as inquiry skills, but they are typically much more open-ended, with students encouraged to pose their own inquiry questions (Makar, 2012).

Initially attributed to John Dewey (1859–1952), inquiry-based learning was considered to tap into children's natural curiosity, fertile imagination and willingness to play and try to see how things work (Dewey, 1910). By using projects which students find interesting, inquiry-based learning engages students in experimental or scientific inquiry that requires creating questions, giving priority to evidence, formulating explanations from evidence, connecting explanations to what is known, and communicating and justifying conclusions. While initially applied to scientific inquiry, inquiry-based learning has more recently been used in mathematics education as it connects to problem solving and reasoning. While the current research into inquiry-based learning in secondary mathematics education is inconclusive, students report positive experiences when encouraged to pose and investigate complex problems (Huang, Doorman & van Joolingen, 2020).

One form of inquiry-based learning providing links to real-world mathematics applications is project-based learning, which involves cross-disciplinary, multifaceted, open-ended tasks, usually set in a real-world context, with results presented via oral or written presentation (Anderson et al., 2019). Such tasks may take several lessons as students need to define the task, plan the project, collect data, analyse and draw conclusions, and determine how best to present the results. The next three sections of this chapter elaborate on problem solving and modelling, inquiry-based and collaborative learning approaches, and provide examples of tasks for secondary mathematics classrooms.

SHORT-ANSWER QUESTIONS

1. What are the differences between teaching *for* problem solving, teaching *about* problem solving and teaching *through* problem solving?
2. Why should we provide links to real-world problems in mathematics lessons?

IN PRACTICE

Features of inquiry-based learning in the mathematics classroom

Calleja (2016) describes four features of inquiry-based learning in mathematics lessons:

1. Mathematical tasks – need to be accessible and achievable, and have multiple entry points and value process
2. Collaborative learning – supports sharing, values diverse ideas and promotes discussion
3. Purposeful questioning – teacher questions foster reasoning and stimulate exchange of ideas
4. Student agency and responsibility – students can select problems, make decisions, present ideas and be critical.

Problem solving and modelling in mathematics classrooms

Problem solving and modelling are similar, but different tasks are usually used to develop students' problem-solving and modelling skills and dispositions. This section outlines the wide range of skills required for problem-solving and modelling success and presents the characteristics of modelling tasks.

Problem solving

Problem solving in mathematics requires doing a question which is typically unfamiliar and which the problem solver cannot answer immediately from known facts or procedures (Anderson, 2014). In addition, problem solvers are usually motivated to find a solution. If problem solving is to reflect a 'real-world' approach, then problem solvers will usually know why they need to solve the problem, they may not have all the skills required to solve the problem, and the problem may be ill-defined (Killen, 2015). Real problems do not come labelled with the appropriate mathematics topics to be used and they may also contain insufficient information or too much information.

A challenge for teachers is that problem solving requires time and effort for students to understand the problem, devise a plan to solve the problem, carry out the plan and look back to evaluate the solution (Polya, 1957). Problem-solving heuristics, such as draw a diagram, work backwards and think of a simpler example, frequently help students to solve unfamiliar problems. But being familiar with the problem-solving process advocated by Polya (1957) and having had practice in using a range of heuristics is not sufficient to become a competent problem solver. Stacey (2005) suggests there are many skills required to become a competent problem solver (see Figure 5.1). According to Williams (2010), 'students need deep mathematical knowledge and general reasoning ability as well as helpful beliefs and personal attributes for organising and directing their efforts' (p. 9). Coupled with this, students need good communication skills and the ability to work with others (Williams, 2010).

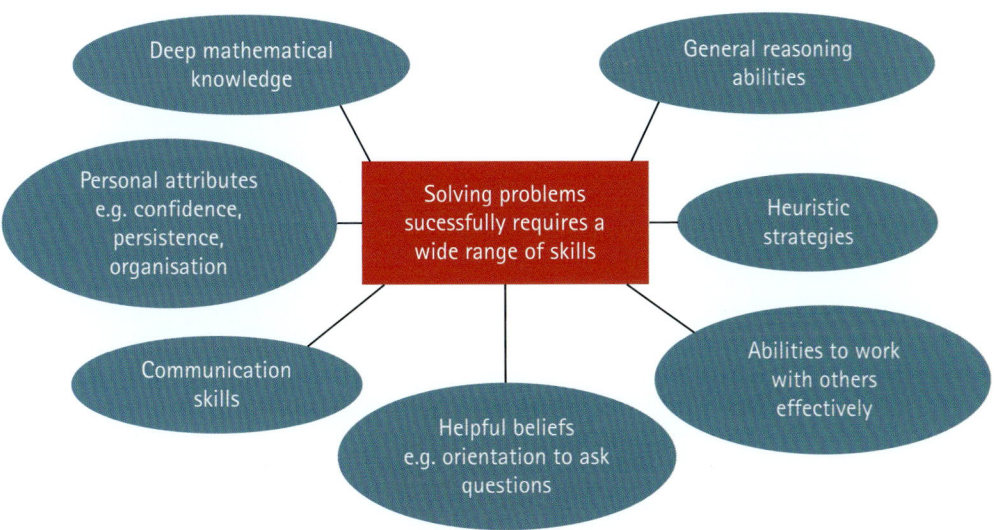

Figure 5.1 Factors contributing to successful problem solving
Source: Stacey, 2005, p. 342.

Another challenge for teachers is choosing the most appropriate problems for the full range of students in the classroom. Sullivan (2011) recommends we choose challenging tasks that are accessible to all students by offering enabling prompts to those who have difficulty starting and extending prompts to those who may find a solution more quickly. In a similar manner, others have described the need to use tasks that have a 'low floor' and 'high ceiling' (e.g. Boaler, 2015).

Some authors classify mathematics tasks according to particular characteristics, including level of complexity. For example, Sullivan (2011) describes tasks that focus on developing procedural fluency, those that use a model or representation, those that use authentic contexts and those which are more open-ended. An example of a challenging, open-ended problem with enabling and extending prompts is:

> Seven people went to the library. The mean number of books rented from the library was 4 and median number of books rented was 2. How many books might each person have rented?

An enabling prompt might be:

> Work on this problem: 'Seven people went to the library. The mean number of books rented from the library was 4. How many books might each person have rented?'

Some extending prompts include:

> How many different answers are possible?
> What if the mode number of books rented was 1?
> If only six people went to the library, what difference would that make to the mean?

Fermi problems are open-ended, challenging tasks that are usually based in real-world contexts but are ill-defined and ideal for small groups working together to share knowledge and understanding. They have been defined as 'the estimation of rough but quantitative answers to unexpected questions about many aspects of the natural world' (Taggart et al., 2007, p. 165), and hence they encourage conjecture, estimation, critical communication and evaluation skills. Some examples include:

- How many piano tuners are there in Sydney?
- How many drops of water are there in a nearby dam?
- How many balloons would fill the school hall?

Mathematical modelling

In mathematical modelling, the nature of the task is the focus. To support students' modelling efforts, approaches have been developed to guide student thinking and promote metacognitive processes. For example, students could use the following four stages in the modelling process, but they are non-linear so students may need to revisit stages as they evaluate their efforts:

- observe and identify the problem
- conjecture how factors are related and interpret them mathematically (mathematising)
- apply mathematical processes and procedures to the model
- obtain and interpret results in the context of the problem.

While there are some similarities between problem solving and modelling, the development of a mathematical model from a situation, which seems to be seemingly non-mathematical in context, is unique to mathematical modelling (Swetz & Hartzler, 1991, p. 2). At the same time, there are sufficient common skills and dispositions to support the development of successful problem solving. English and Watson (2018) suggest common skills include teamwork, breaking complex situations into simpler parts, communication and planning, and monitoring and evaluating. These are all necessary outcomes of the mathematics curriculum, so teachers need to select appropriate modelling tasks to support the development of these skills and dispositions.

Rich model-eliciting tasks have characteristics that involve collaborative problem solving but also require students to mathematise a situation within a real-life context (Hernandez et al., 2017). Finding an appropriate context relevant and meaningful to students is another consideration when choosing tasks (Anderson, 2010). Some teachers have chosen situations within the school or local community to design tasks for students. Examples include:

- designing a new parking area to maximise the number of parking spaces
- examining the recycling in the school and designing ways to improve and increase recycling opportunities
- planning an event for the whole school with timetabling and budgetary requirements (Anderson, 2010).

These situations require the gathering of information, modelling through mathematising, drawing conclusions and making predictions. As noted in Anderson (2010), while the teacher may provide some of the information and place restrictions on what the situation requires, the best opportunities are created when students work in cooperative groups to ask important questions, determine the type of information they require and then work together to formulate a model.

SHORT-ANSWER QUESTIONS

1. Describe some key factors impacting on students' problem-solving capabilities.
2. List some differences between problem solving and mathematical modelling.

Inquiry-based learning in mathematics classrooms

It can be argued that if the task is sufficiently open and there is limited guidance from the teacher about strategies and procedures, problem solving and modelling are examples of inquiry-based learning. Typically, the inquiry begins with a compelling question, which, ideally, is posed by the students (Makar, 2012). By considering how they might go about answering the question and carefully considering their assumptions, students design an investigation and plan the data they may need to collect and how it might be analysed to answer the question. As the investigation progresses, students enter a creative phase where they are encouraged to evaluate what they are discovering, adapt their strategy to incorporate what they are learning and communicate their thinking. Many investigations lead to new questions, which need to be accommodated throughout the inquiry. Students need to develop strategies to discuss and communicate their findings with others and finally reflect on the initial question, their strategies and their findings, and draw conclusions (Anderson, 2014).

Types of inquiry-based learning

Inquiry-based learning may vary depending on the level of support given to students. If using inquiry-based learning with students who are not familiar with this form of learning mathematics, it is recommended you begin with a more structured approach. You will need to carefully plan the inquiry to accommodate students' needs, including support for reading, writing, collaboration, communication, self-direction and the mathematics content. As in the Boaler (2002) study, some students may not like the openness of the approach and will require additional support and a gradual introduction to this way of working mathematically. Three types of inquiry-based learning have been described, including:

- structured inquiry – students are given the problem, as well as the method and materials
- guided inquiry – students are given the problem and materials but not the strategies or methods
- open inquiry – students find their own problems as well as methods and materials (this approach appears to be more common in science classrooms).

Makar (2012) prefers to use ill-structured and open-ended tasks for which the initial conditions and perhaps even the goals are ambiguous. She suggests the teaching of mathematical inquiry requires the need to embrace uncertainty, support student decision making, encourage flexible thinking and tolerate noise and disorganisation. Teachers need to be adaptable and able to balance innovation with efficiency, and it is also helpful to be able to balance mathematical knowledge with contextual knowledge. This approach may be the antithesis of what teachers believe to be a well-organised, orderly mathematics-learning environment.

Many of the types of inquiry questions used in mathematics classrooms require the use of statistics because of the natural link to contextual problems. For example:

- Are athletes getting faster over time?
- Is there a typical Year 8 student?
- Do left-handed students have faster reflexes than right-handed students?
- Does talking on your mobile phone affect concentration on other activities?
- Can a person's height be predicted from body part measurements?
- How much water does fixing dripping taps save?
- How much food is 'thrown away' in the canteen in a year?

An inquiry-based task that requires the application of other mathematics topics and allows for important connections to be made between topics is the cereal box problem. The problem is presented in two parts as follows:

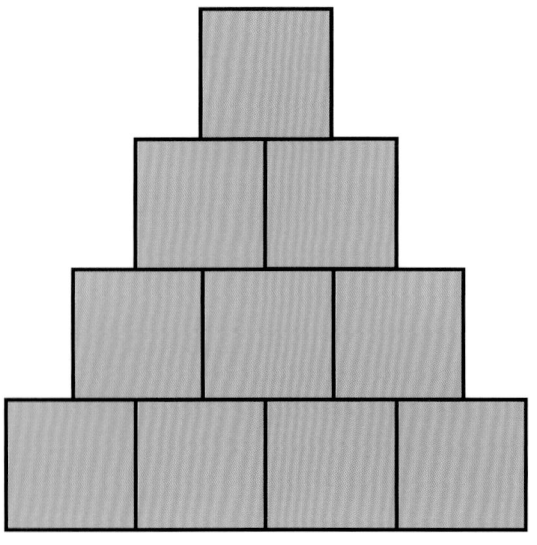

Figure 5.2 Cereal box display

Part 1: A store manager told a sales clerk that 45 cereal boxes had to be stacked in a display window and that all the boxes had to be used. The manager also told the clerk that all the boxes had to be set up in a triangle as shown in [Figure 5.2]. The sales clerk wondered how many boxes would have to be placed on the bottom row to build a triangle that would use all the boxes.

Part 2: What if the clerk had to use 200 boxes in the display? How many boxes would have to be placed on the bottom row to build the triangle? (Manouchehri, 2007, p. 291)

Manouchehri (2007) reported on the use of this problem in a Year 9 class with the expectation that students would use a variety of problem-solving heuristics (use a table, look for a pattern) to solve the problem followed by generalisation of their results. She had planned to use the results to discuss Gauss's method for finding the sums of consecutive positive integers and to introduce triangular numbers, but what followed led to a complete change of plans. The students interpreted the problem in unexpected ways, leading to quite different solutions and the opportunity for the problem to be extended into new investigations. The task led to combining patterns and functions, multiple representations, geometry, graphing and rates of change, and algebraic reasoning and proof. If Manouchehri had stayed with her original plan, the temptation would have been to dismiss students' different interpretations and limit the possibility of new inquiries – she needed to be flexible and responsive and know when to 'tell' or let group conversations continue. This approach can be threatening for teachers, particularly if students pose questions or develop mathematical ideas with which they are unfamiliar. While Manouchehri allowed the inquiries to continue for two weeks, she argued that students learned far more than she anticipated and covered many more mathematical ideas from the curriculum in a deep and meaningful way – they were also highly engaged.

How do I address all curriculum requirements and still have time to do authentic inquiry?

SCENARIO

Having read about the importance of problem solving and reasoning from the curriculum, a student teacher worried about how she could meet all curriculum requirements and yet still provide opportunities for her students to engage in authentic inquiry.

Questions

1. Using the cereal box problem, what topics from the Australian Curriculum for Year 9 can be addressed as students work on this problem for several lessons?
2. What other types of tasks would be necessary for students to complete so that they develop deep understanding of the topics you listed for Question 1?

Choosing tasks for inquiry-based learning

Task choice is critical for successful inquiry-based learning, particularly in mathematics. Suitable tasks must provide challenge and hence have a high cognitive demand (Sullivan, 2020). Table 5.1 summarises the characteristics of high cognitive demand tasks and compares them with the characteristics of low cognitive demand tasks (Henningsen & Stein, 1997).

SHORT-ANSWER QUESTIONS

1. Choose one of the inquiry questions listed in this section. What challenges might students have in planning the inquiry?
2. Now choose a year level suitable for the task and use the curriculum to determine the key mathematical understandings required to conduct the inquiry. What support might be required to encourage student collaboration to do such a task?

Table 5.1 Classifying mathematical tasks by level of cognitive demand

Low cognitive demand tasks and characteristics
Memorisation tasks: <ul><li>Reproducing or memorising facts, rules, formulas or definitions</li><li>Cannot be solved using procedures</li><li>Non-ambiguous, involves exact reproduction of previously seen material</li><li>No connection with the concepts or meaning that underlie procedure being used.</li></ul>
Procedures without connections tasks: <ul><li>Entirely algorithmic</li><li>Use of procedure is called for or evident from prior instruction</li><li>Limited cognitive demand, little ambiguity about what needs to be done and how</li><li>No connection to the concepts or meaning that underlie procedure being used</li><li>Focused on producing correct answers</li><li>Requires no explanations or explanations focus on describing a used procedure.</li></ul>

High cognitive demand tasks and characteristics
Procedures with connections tasks: <ul><li>Focuses attention on the use of procedures for developing deeper levels of understanding of mathematical concepts and ideas</li><li>Suggests pathways to follow that have close connections to underlying conceptual ideas instead of narrow algorithms</li><li>Uses multiple representations</li><li>Requires some degree of cognitive effort, procedures cannot be followed mindlessly</li><li>Students need to engage with conceptual ideas and deep content understandings.</li></ul>
Doing mathematics tasks: <ul><li>Requires complex and non-algorithmic thinking, conjecturing, justifying and reasoning</li><li>Requires students to explore and understand the nature of mathematical concepts, processes or relationships</li><li>Demands self-monitoring or self-regulation of one's own cognitive processes</li><li>Requires students to access and decide on appropriate use of relevant knowledge</li><li>Requires analysis and interpretation of the task</li><li>Requires considerable cognitive effort</li><li>Solution process is unpredictable or ambiguous.</li></ul>

Source: Adapted from Henningsen & Stein, 1997.

Collaborative and cooperative learning

Since inquiry-based learning tasks typically require time, can be comprised of several steps or stages, and have many possible approaches to solution, working with peers helps managing sub-tasks and evaluation of solution methods. Organising students into groups, assigning group roles and helping them to learn how to work collaboratively also takes time but needs to be explicitly addressed before implementing inquiry-based learning tasks. For groups to successfully cooperate, they need to be structured so that:

- all group members work together to complete the task
- there are face-to-face discussions between students
- individual accountability is required
- interpersonal and small-group skills are employed
- group members reflect on their work to monitor their progress (Gillies, 2016).

Ideally, groups should include three to four students with a variety of knowledge, skills and understandings. If the students are not used to working in cooperative groups, begin with pairs of students working together on shorter tasks and spend time identifying key 'rules' or social norms for effective group work, which includes members taking responsibility to share information, reach agreement, provide reasons, challenge ideas, discuss alternatives and include everyone. The teacher's role is critical while students are working in groups – they need to listen carefully to group discussions, probe understandings and offer suggestions when necessary, but without providing answers, which discourages group autonomy (Gillies, 2016).

Some authors distinguish between collaborative and cooperative learning that occurs when groups of students work together. The similarities between these include:

- stressing the importance of active learning
- facilitating learning by the teacher
- teaching and learning shared by both students and teacher
- enhancing higher-order cognitive skills
- emphasising student responsibility for their own learning
- involving situations where students must articulate ideas to the group
- helping students develop social and teambuilding skills
- utilising student diversity.

The differences between collaborative and cooperative learning are summarised in Table 5.2.

Table 5.2 The differences between cooperative learning and collaborative learning

Cooperative learning	Collaborative learning
Students receive training in social skills in small groups.	Students already have the necessary social skills which they use to reach goals.
Activities are structured with each student taking a role.	Students organise and negotiate their efforts themselves.
The teacher observes, listens and intervenes when necessary.	When questions arise, the teacher guides, usually by posing further questions.
Students submit work at end of class for evaluation.	Students retain drafts to complete further work.
Students assess individual and group performance.	Students assess individual and group work.

Source: Adapted from Matthews et al., 1995.

Collaborative and cooperative learning best occur when teachers choose 'group-worthy' tasks. If the task only requires students to exchange information and explanations or request assistance, this will lead to low levels of cooperation, lower levels of discussion and debate, and potentially lower levels of thinking. Group-worthy tasks are more open-ended, multi-dimensional, have high cognitive demand, require the efforts of several students working together and have clear outcomes – see Table 5.3.

Table 5.3 Features of group-worthy tasks

Task features	Characteristics
Open	• Genuine dilemmas and authentic problems with real-life uncertainties and ambiguities. • Many possible solutions. • Accessible problem-solving context.
Multi-dimensional	• Many ways to approach the problem. • Many ways to model the problem (picture, table, graph, etc.). • Requires a variety of skills and mathematical competencies to solve the problem. • Many ways to demonstrate mathematical competence or 'smartness'.
High cognitive demand	• Meets criteria for high cognitive demand tasks. • Highlights a big mathematical idea. • Opportunities for justification and reasoning.
Positive interdependence	• Requires students to work together to complete the tasks. • Group and individual accountability. • Multiple abilities and mathematical competencies needed.
Clear evaluation criteria	• Clearly stated outcomes or product. • Clear deliverables. • Could use a rubric or performance assessment.

Source: Adapted from Lotan, 2002.

In summary, inquiry-based learning encourages students to develop an important set of 'mathematical habits of mind' (Horn, 2012). To develop such habits of mind, students need to be given challenging mathematics tasks that require them to engage in exploring important mathematical ideas, orienting and organising their thinking, representing, justifying, generalising, checking for reasonableness and using mathematical language.

SHORT-ANSWER QUESTIONS

1. What is one key difference between collaborative learning and cooperative learning?
2. What types of roles could be assigned to students working in groups?
3. What strategies should teachers use to help groups stay focused on a task?

Conclusion

The chapter began with an overview of an important and highly regarded research project conducted in the 1990s by Professor Jo Boaler in England. Her research highlighted some key differences in outcomes when using different approaches to the teaching and learning of mathematics. While some approaches are more student centred and others are more teacher centred, most teachers use a combination of approaches in secondary mathematics classrooms. There are several student-centred teaching approaches which support inquiry-based learning, including whole class and small group discussion, problem solving and modelling. Regardless of approach, teachers have a key role to play in organising the learning of their students and in choosing appropriate mathematical tasks.

Task selection is a critical component of teachers' work, as they need to select questions or problems that require considerable thought and effort, and this may vary considerably depending on the group of students they are teaching. Suitable tasks are usually open-ended and may be contextualised to enable students to see links to real-life contexts. It is best that tasks have a high cognitive demand and as such are best attempted in collaborative or cooperative groups. If students are struggling, they may require enabling prompts, and if they finish quickly, it is important that you have prepared extending prompts.

BRINGING IT TOGETHER

1. What are the main differences between teacher-centred teaching and learner-centered teaching?
2. Which mathematics tasks require more teacher-centred teaching and which mathematics tasks require more student-centred teaching?
3. What is the difference between problem solving and mathematical modelling?
4. What are some characteristics of group-worthy tasks?
5. If a student's parent asked you why you were allowing students to work in groups to answer mathematics problems rather than have them work on exercises from their textbook, how would you justify your approach?
6. Select a lesson you recently taught during practicum and create a new lesson which involves students working on an unfamiliar problem in small groups. Ensure the problem satisfies the group-worthy tasks characteristics and develop enabling and extending prompts for the problem.

REFERENCES

Anderson, J. (2010). Collaborative problem solving as modelling in the primary years of schooling. In B. Kaur & J. Dindyal (eds), *Mathematical applications and modelling* (78–94). Singapore: World Scientific.

Anderson, J. (2014). Forging new opportunities for problem solving in Australian mathematics classrooms through the first national mathematics curriculum. In Y. Li & G. Lappan (eds), *Mathematics curriculum in school education* (209–29). Dordrecht: Springer.

Further resources

Anderson, J., White, P. & Sullivan, P. (2005). Using a schematic model to represent influences on, and relationships between, teachers' problem-solving beliefs and practices. *Mathematics Education Research Journal*, 17(2), 9–38.

Anderson, J., Wilson, K., Tully, D. & Way, J. (2019). 'Can we build the wind powered car again?' Students' and teachers' responses to a new integrated STEM curriculum. *Journal of Research in STEM Education*, 5(1), 20–39.

Australian Curriculum, Assessment and Reporting Authority (ACARA) (2020). *Foundation–Year 10 Curriculum*. Sydney: ACARA. Retrieved from https://www.acara.edu.au/curriculum/foundation-year-10

Australian Institute for Teaching and School Leadership (AITSL) (2011). *Australian Professional Standards for Teachers*. Melbourne: AITSL.

Bills, C., Bills, L., Watson, A. & Mason, J. (2004). *Thinkers*. Derby, UK: Association of Teachers of Mathematics.

Boaler, J. (2002). *Experiencing school mathematics: Traditional and reform approaches to teaching and their impact on student learning*. Mahwah, NJ: Lawrence Erlbaum Associates.

Boaler, J. (2015). *What's math got to do with it? How teachers and parents can transform mathematics learning and inspire success* (Revised edn). New York: Penguin Books.

Borbas, A. (1988) *Proceedings of the Annual Conference of the International Group for the Psychology of Mathematics Education* (12th, Veszprem, Hungary, July 20–25, 1988), 2, 1–359.

Calleja, J. (2016). Teaching mathematics through inquiry: A continuing professional development programme design. *Educational Designer*, 3(9), 1–29.

Dewey, J. (1910). *How we think*. Lexington, MA: D.C. Heath. (reprinted in 1991 by Prometheus Books, Buffalo).

English, L. & Watson, J.M. (2018). Modelling with authentic data in sixth grade. *ZDM – International Journal on Mathematics Education*, 50(1–2), 103–15.

Gillies, R.M. (2016). *Enhancing classroom-based talk: Blending practice, research and theory*. Abingdon: Routledge.

Henningsen, M. & Stein, M.K. (1997). Mathematical tasks and student cognition: Classroom-based factors that support and inhibit high-level mathematical thinking and reasoning. *Journal for Research in Mathematics Education*, 28, 524–49.

Hernandez, M.L., Levy, R., Felton-Koestler, M.D. & Zbiek, R.M. (2017). Mathematical modelling in the high school curriculum. *The Mathematics Teacher*, 110(5), 336–42.

Hollingsworth, H., Lokan, J. & McCrae, B., (2003). *Teaching mathematics in Australia: Results from the TIMSS 1999 video study*. Camberwell: Australian Council of Educational Research.

Horn, I.S. (2012). *Strength in numbers: Collaborative learning in secondary mathematics*. Reston, VA: National Council of Teachers of Mathematics.

Huang, L., Doorman, M. & van Joolingen, W. (2020). Inquiry-based learning in lower secondary mathematics education reported by students from China and the Netherlands. *International Journal of Science and Mathematics Education*. Online first https://doi.org/10.1007/s10763-020-10122-5

Killen, R. (2015). *Effective teaching strategies: Lessons from research and practice* (7th edn). South Melbourne: Cengage Learning.

Lotan, R.A. (2002). Group-worthy tasks: Carefully constructed group learning activities can foster students' academic and social growth and help close the achievement gap. *Educational Leadership*, 60(6), 72–5.

Makar, K. (2012). The pedagogy of mathematical inquiry. In R.M. Gillies (ed.), *Pedagogy: New developments in the learning sciences* (371–97). New York: Nova Science Publishers.

Manouchehri, A. (2007). Inquiry-discourse: Mathematics instruction. *Mathematics Teacher*, 101(4), 290–300.

Matthews, R.S., Cooper, J.L., Davidson, N. & Hawkes, P. (1995). Building bridges between cooperative and collaborative learning. *Change*, 29(4), 35–40.

Polya, G. (1957). *How to solve it*. Princeton, NJ: Princeton University Press.

Stacey, K. (2005). The place of problem solving in contemporary mathematics curriculum documents. *Journal of Mathematical Behaviour*, 24, 341–50.

Sullivan, P. (2011). *Teaching mathematics: Using research-informed strategies*. Camberwell: Australian Council for Education Research. Retrieved from http://research.acer.edu.au/aer/13/

Sullivan, P. (2020). *Leading improvement in mathematics teaching and learning*. Camberwell: Australian Council for Education Research Press.

Swan, M. (2006). *Collaborative learning in mathematics: A challenge to our beliefs and practices*. London: National Research and Development Centre for Adult Literacy and Numeracy.

Swetz, F. & Hartzler, J.S. (1991). *Mathematical modeling in the secondary school curriculum*. Reston, VA: National Council of Teachers of Mathematics.

Taggart, G.L., Adams, P.E., Eltze, E., Heinrichs, J., Hohman, J. & Hickman, K. (2007). Fermi questions. *Mathematics Teaching in the Middle School*, 13(3), 164–7.

Williams, G. (2010). MERGA response to Australian Curriculum (Mathematics). *Mathematical Education Research Group of Australasia*. 1–52. Retrieved from https://studylib.net/doc/15418636/merga-s-response-to-the-acara-k-10-curriculum-draft

CHAPTER 6

Gender, culture and diversity in the mathematics classroom

Gregory Hine

LEARNING OBJECTIVES

The learning objectives of this chapter are directly linked to the Australian Professional Standards for Teachers (APST) (Australian Institute for Teaching and School Leadership [AITSL], 2011). After studying this chapter, you should be able to:

- offer explanations to account for gender differences in motivation and achievement in mathematics (APST 1.1, 1.2)
- identify instructional approaches to be used in making a mathematics classroom gender equitable (APST 1.1, 1.2, 4.1)
- outline instructional guidelines for working with students from different cultural backgrounds (APST 1.3, 1.4)
- justify the need for diverse learners to receive instruction according to their particular learning needs (APST 1.3, 1.4)
- delineate instructional approaches to be used with diverse learners (APST 1.3, 1.4, 4.1).

Introduction

In this chapter, the topics of gender, culture and learner diversity are explored in the context of the Australian secondary mathematics classroom. Each of these topics is underpinned by the goals of the Alice Springs (Mparntwe) Education Declaration (Council of Australian Governments Education Council, 2019), which, in turn, informs the development of the Australian Curriculum framework. First, the topic of gender is unpacked with regard to pervasive belief systems, achievement and motivation, teacher behaviour and the promotion of gender equity. Second, the issue of culture is investigated through sub-topics of cultural diversity and cultural awareness, and best instructional practices for English language learners (ELLs) and Indigenous Australian learners. Third, learner diversity is defined and two groups of diverse learners receive particular attention: gifted and talented students, and students with learning difficulties. In light of this attention, theoretical and practical advice concerning the identification and education of such diverse learners is outlined.

> **Culture:** the manifested and integrated pattern of human knowledge, beliefs and behaviour (also understood as the 'way of life' for any people within a setting).

Gender, culture and diversity in the Australian Curriculum

The Alice Springs (Mparntwe) Education Declaration (Council of Australian Governments Education Council, 2019) provides the policy framework for the Australian Curriculum. This framework includes two distinct but interconnected goals:

> Goal 1: The Australian education system promotes excellence and equity.
> Goal 2: All young Australians become confident and creative individuals, successful lifelong learners, and active and informed members of the community.

In addressing these goals, the Australian Curriculum has been developed to be appropriate and accessible for all students (Australian Curriculum, Assessment and Reporting Authority [ACARA], 2020). The propositions that have shaped the development of the Australian Curriculum are:

- that each student can learn and that the needs of every student are important
- that each student is entitled to knowledge, understanding and skills that provide a foundation for successful and lifelong learning and participation in the Australian community
- that high expectations should be set for each student, as teachers account for the current level of learning of individual students and the different rates at which students develop
- that the needs and interests of students will vary, and that schools and teachers will plan from the curriculum in ways that respond to those needs and interests.

With specific reference to student diversity, ACARA demonstrates a clear commitment to the development of a high-quality curriculum for all Australian students that promotes excellence and equity in education. The Teacher Education Ministerial Advisory Group (TEMAG) (2014) recommends that teachers be suitably equipped with the pedagogical knowledge that will allow them to effectively address the learning and development needs of all students in their class. Moreover, TEMAG (2014, p. 16) cited that 'a growing body of research acknowledges that teachers need a broad range of skills and strategies to maximise the learning of diverse student populations'. As a corollary to this research, TEMAG has suggested that an:

> ability to work effectively with special needs students, and in particular students with disability and learning difficulties, needs to be considered a core requirement of all teachers rather than a specialisation. (2014, p. 17)

In Australian schools, the term *student diversity* can include those students who are culturally and linguistically diverse, or those who have specific learning needs, difficulties or disabilities (ACARA, 2020). The work in this chapter is underpinned by the notion that all students require opportunities to advance their mathematical knowledge through teaching approaches that are attentive and responsive to their learning needs. Such approaches strive to provide educational equity, which has been described by scholars as much more than providing students with an equal opportunity to learn mathematics. Rather, educational equity 'attempts to attain equal outcomes for all students by being sensitive to individual differences' (Van de Walle, Karp & Bay-Williams, 2014, p. 100).

Gender in the secondary mathematics classroom

Over the last three decades, research has contributed towards a significant shift in male and female students' secondary mathematics achievement and motivation (Forgasz & Leder, 2001; Huetinck & Munshin, 2008). Before the 1990s a widely held societal belief was that boys were better at mathematics than girls, and up until this time in Australian schools the proportion of boys was higher than girls in advanced levels of mathematics in the senior years (Siemon et al., 2011). In the early 1980s researchers and teachers began to discern decreasing gender differences in mathematics and science, where previously female students consistently had performed more poorly than male students (Huetinck & Munshin, 2008). Such discernment prompted researchers and practitioners to address the question: 'What is wrong with the girls?' In the late 1980s researchers began to realise that the ways in which mathematics classes were taught was affecting the performance of girls, and that curriculum content and delivery could be modified to accommodate girls (Siemon et al., 2011). Moreover, Vale and Bartholomew (2008) have argued that male and female students may be shaped by the contexts they are in. Specifically, these authors contend that for teachers 'paying attention to the relationships within the classroom, the different identities, and hence the different needs of students in the mathematics classroom, are central to equity' (Vale & Bartholomew, 2008, p. 273). As a result of heightened attention on issues affecting gender performance, various approaches have become standard practice in creating greater equality of performance. Siemon et al. (2011) noted that these approaches include an emphasis on:

> collaborative learning, discussion in small and whole-class groups, using applications of mathematics to social contexts, making contexts clear (for example, descriptions of sporting contexts made overt, rather than assuming learners have played particular sports), and including contexts that would appeal to girls and boys. (pp. 156–7)

More recently, the notion that mathematics is a 'male domain' has dissipated significantly and the mathematical potential of girls is no longer regarded as inferior to boys (Forgasz & Leder, 2001; Forgasz, Leder & Tan, 2014). Perhaps as a corollary to this dissipation, attention has been focused on the apparent underperformance of adolescent boys (Booker et al., 2014) where scholars posit that the most successful boys continue to achieve scholastic success, but increasingly other boys are not (Burton, 2001; Siemon et al., 2011). Following an inquiry into the education of boys, the Australian House of Representatives Standing Committee on Education and Training concluded in its final report:

> While it is dangerous to generalise, boys and girls do tend to prefer different learning styles. Boys tend to respond better to structured activity, clearly defined objectives and instructions, short-term challenging tasks and visual, logical and analytical approaches to learning. They tend not to respond as well as girls to verbal, linguistic approaches. Good teachers respond to the different learning styles of their students and utilise students' preferred learning styles while also aiming to develop the full range of capacities in each student. (House of Representatives Standing Committee on Education and Training, 2002, p. xviii)

Despite acknowledgement of gender-preferred learning styles, more male than female students continue to study the most demanding mathematics courses offered, and male students still dominate in careers related to science, technology, engineering and mathematics (STEM) (Siemon et al., 2011; Thomson, 2014). As such, there remains a continuing need to encourage girls to participate in mathematics at all levels, and for educators to engage girls with the science and technology courses that lead to more high-status and influential careers (Burton, 2001; Thomson, 2014).

Figure 6.1 Closing the gender gap in Australian mathematics education

PAUSE AND THINK

Research has suggested that 'higher proportions of boys in single-sex and co-educational schools than girls in single-sex and co-educational schools are enrolled in specialist mathematics courses' (Forgasz & Leder, 2017, p. 259). Looking between the gendered setting of the school, there was also a higher proportion of girls and boys from single-sex than co-educational schools enrolled in advanced courses. Higher proportions of girls and boys in single-sex schools than in co-educational schools were enrolled in intermediate courses, and the proportions of boy and girls in both gendered school settings for elementary mathematics is virtually identical (Forgasz & Leder, 2017).

Think back to the most demanding Year 11 and Year 12 mathematics classes when you were in secondary school. Can you recall how many girls and boys were in these classes? What approaches can teachers and school leaders take to ensure that boys and girls have every opportunity to study these demanding courses?

The gender gap

For the past 20 years, results from large-scale international testing, such as the Trends in International Mathematics and Science Study (TIMSS), have revealed consistently no gender differences in mathematics achievement in Australia, with the exception of TIMSS 2007 (Thomson, 2014). In a similar vein, Thomson noted that there were no gender differences apparent during the measurement of mathematical literacy in the 2003 Programme for International Student Assessment (PISA). However, this author has emphasised that PISA 2012 found that:

> while average scores in mathematics had declined in Australia, males in Australia were significantly outperforming females, and females had significantly higher average levels of anxiety about and significantly lower levels of confidence in mathematics. (Thomson, 2014, p. 59)

Thomson et al. (2019) indicated that between 2015 and 2018, the gap between male and female performance had not changed, with males outperforming females by 6 points (equivalent to approximately one-fifth of a school year). Several years earlier Vale and Bartholomew (2008) analysed the 2006 PISA survey, concluding that boys were more often among the highest achievers, and:

> there remains a difference in the ways that male and female students respond to their own mathematical experiences . . . boys reported higher levels of enjoyment, interest and self-efficacy in mathematics than girls, and boys more highly valued the use of technology in mathematics. (p. 286)

Vale and Bartholomew (2008) also suggested that these findings led to boys' enrolments in higher level mathematics outnumbering those of girls, largely due to a recognised positive relationship between affective factors and enrolment.

Typically, the gender gap has been attributed to outmoded social stereotypes. However, neurological discoveries have been used as potential explanations to assist in mathematics learning as it pertains to gender. Sousa (2008) drew attention to these neurological findings, stating:

> male brains are about 6 to 8 percent larger than female brains. But males are on the average about 6 to 8 percent taller than females, which could also explain the similar differences in brain sizes. And brain imaging studies show that males seem to have an advantage in visual-spatial ability (the ability to rotate objects in their heads) while females are more adept at language processing. (p. 65)

Additionally, Sousa (2008) pointed out that the *corpus callosum* in women is proportionally larger than in men, resulting in more efficient communication between brain hemispheres. However, the male brain appears to communicate more efficiently within a hemisphere. Despite these neurological differences, no genetic advantage for mathematical processing or learning has yet been determined (Sousa, 2008). In a meta-analysis of over 100 academic studies and papers, Spelke (2005) found that most suggested that men and women have an equal aptitude for mathematics and science. As such, it is important for teachers to know about gender differences – especially neurological development – as a factor related to students' perceptions, participation and achievement in mathematics (Booker et al., 2014; Sousa, 2008).

Corpus callosum: the bundle of nerves that connects the cerebral hemispheres of the brain.

Looking at gender issues abroad: the United States

Consistent with findings drawn from an Australian context, a general international trend has been that gender differences in mathematics achievement are declining (Hyde et al., 2008). In the United States, the gap between boys' and girls' high school mathematics course enrolment has narrowed, as has the difference on standardised tests (Ellison & Swanson, 2010). Huetinck and Munshin (2008) commented on the summary of more than 600 programs funded by the National Science Foundation and the American Association for the Advancement of Science from 1966 to 1982. This summary revealed real gains in helping female students to excel in science and mathematics, where common elements of the more effective programs included: academic emphasis, multiple strategies and systems approaches. Huetinck and Munshin noted that in particular, 'achievement in mathematics was nearly equal for males and females until the fifth grade, when females began falling behind' (2008, p. 340). Additionally, the gender differences became more pronounced in high school (Huetinck, 1990). A large-scale study analysing standardised test scores from more than 7.2 million students in the United States in grades 2–11 revealed that there were no differences in mathematics scores between girls and boys (Hyde et al., 2008).

Despite the acknowledged decline of gender differences in mathematics achievement, there are still significant differences at the advanced course levels (American Association of University Women, 1992; Sousa, 2008), university entrance examinations (Sousa, 2008; Wai et al., 2010), large-scale international testing (Guiso et al., 2008) and mathematics competitions (Ellison & Swanson, 2010). Following high school, more male than female students enter fields of study that emphasise STEM areas (Ceci & Williams, 2010). To this end, Tortolani (2007) cited the president of the Society of Women Engineers as asking rhetorically 'Why, while girls comprise 55 per cent of undergraduate students, do they account for only 20 per cent of engineering majors, and boys remain four times more likely to enrol in undergraduate engineering programs?' Looking at high-achieving students, Ellison and Swanson stated that:

> there is a 2.1 to 1 male–female ratio among students scoring 800 on the math SAT, and a ratio of at least 1.6 to 1 among students scoring in the 99th percentile on the PISA test in 36 of the 40 countries. (2010, p. 109)

An analysis of the American Mathematics Competition data provides three key findings concerning the magnitude of the gender gap at very high performance levels (Ellison & Swanson, 2010). First, the gender gap appears to widen substantially at percentiles beyond the 99th, and at the very high end of analysed data, the male–female ratio exceeds 10 to 1. Second, and although some gender variation was found across all participating schools, there was enough variation:

> from school to school to suggest that the number of girls reaching high performance levels would increase substantially if all school environments could somehow be made to resemble those where girls are currently doing relatively well. (Ellison & Swanson, 2010, p. 110)

Third, an examination of extreme high-achieving students chosen to represent the United States in international competitions revealed that the highest-scoring boys and the highest-scoring girls appeared to be drawn from very different pools (Ellison & Swanson, 2010). While the boys came from a variety of backgrounds, the top-scoring girls were almost exclusively drawn from a remarkably small set of super-elite schools (Ellison & Swanson, 2010). According to Ellison and Swanson (2010), this finding suggests that almost all American girls with extreme mathematical ability are not developing their mathematical talents to the degree necessary to reach the extreme top percentiles of these contests. To assist with this development – particularly in the formative years – teachers are reminded to challenge pervasive gender stereotypes, create gender-friendly learning environments, and stimulate all students' interest in pursuing university studies and careers in mathematics-related fields (Van de Walle, Karp & Bay-Williams, 2014).

Gender differences explained: achievement and motivation

Current literature suggests several possible explanations to account for gender differences in motivation and achievement in mathematics. These explanations are socially and culturally constructed, and include pervasive belief systems, teacher behaviour and attitudes, and student attitudes. In examining the social and cultural causes of gender differences, teachers can create gender-equitable mathematics instruction for boys and girls (Van de Walle, Karp & Bay-Williams, 2014).

Attitude: a semi-permanent way of thinking, feeling or acting towards someone or something.

Pervasive belief systems

A pervasive, stereotypical view held by parents and society generally is that mathematics is a male activity which is incongruous with femininity (Else-Quest, Hyde & Linn, 2010; Hall, 2012). According to Hall (2012, p. 70), these 'deeply entrenched societal views and stereotypes will be difficult to change and may take many years to slowly evolve; however, change can begin now from within the classroom'. Within the classroom, Nosek, Banaji and

Greenwald (2002) contended that stereotypical views (i.e. boys are better in mathematics) shape girls' self-perceptions and motivations. As a consequence of these views being upheld, Van de Walle, Karp and Bay-Williams (2014) conjectured that girls' emerging interest in mathematics would decrease. Research from Stevens et al. (2007) indicated that female students report subject interest as a very influential factor when deciding to pursue higher level mathematics courses. In a Victorian context, Year 12 girls enrolled in either Further Mathematics or Mathematical Methods courses were significantly less likely than boys to perceive mathematics as relevant and useful for the future (Helme & Teese, 2011). Furthermore, these commentators found:

> Female Further Mathematics students were significantly less likely than their male counterparts to agree that their teacher understands how they learn, and significantly more likely to report that the pace of learning is too fast. Female Mathematical Methods students were significantly less likely than their male counterparts to agree that they enjoy the subject. (Helme & Teese, 2011, n.p.)

At a later milestone in life, Sousa (2008, pp. 65–6) underscored how differences in career choices are made not due to 'differing abilities in mathematics but to cultural factors, such as subtle but pervasive gender expectations that emerge in high school'. Interestingly, female students often express that they are less proficient than their male classmates at mathematics, even when they perform at similar levels (Correll, 2001). Perhaps an examination of stereotype threat can illuminate the basis for such a belief system. For instance, one research project was conducted to determine whether merely telling female students that a mathematics test often shows gender differences was enough to hurt their performance (Spencer, Steele & Quinn, 1999). After giving a mathematics test to male and female students, the researcher told half of the female students that the test would reveal gender differences and the remaining half that the test would find none. There were two key findings to this research: first, those female students who expected gender differences on the test performed significantly worse than the male students. Second, the female students who were told the test would reveal no gender disparity performed equally to the male students. Moreover, the experiment was conducted with high-performing female mathematics students (Spencer, Steele & Quinn, 1999).

Stereotype threat: this phenomenon occurs when people believe they will be evaluated on societal stereotypes about their particular group.

Teacher behaviour and attitudes

Although teachers may not intentionally set out to stereotype students by gender, the gender-based biases of wider society can affect teacher–student interactions (Van de Walle, Karp & Bay-Williams, 2014). In describing observations of teachers' gender-specific interactions in the classroom, Campbell (1995) noted that boys receive both more attention and different kinds of attention than girls. Furthermore, boys tend to be more involved in discipline-related attention (Campbell, 1995). Another study concluded that female students in mathematics classes go unobserved and are known as 'quiet achievers' (Clarke et al., 2001). Research from the United Kingdom revealed that most mathematics teachers held different beliefs about students, based on students' gender (Soro, 2002). Although some teachers did not hold gendered beliefs, girls were perceived to use inferior cognitive skills and succeed because of their diligence, while boys were seen to be talented in mathematics but lacking in effort

(Soro, 2002). In California, Marshall (1984) hypothesised that girls outperformed boys in elementary school only if there were algorithms to follow in problem solving. After enrolling in higher-level courses, girls were less able to use novel approaches – perhaps due to internalising rules so well their creativity suffered. These gender-based perceptions are incongruent with findings suggesting that female students achieve as good or better grades in mathematics than male students (Gallagher & Kaufmann, 2005; Riegle-Crumb, 2006). However, female teachers with mathematics anxiety can negatively influence female students' mathematics achievement – even over the course of one year (Beilock et al., 2010).

Student attitudes

Although the gender gap concerning mathematics achievement is narrowing, it is still considerably wide in terms of students' attitudes towards mathematics (Hall, 2012; Hannula, 2009). Studies conducted over time and across levels of education have generally found that boys hold a more positive attitude towards mathematics (Fennema & Sherman, 1977; Hall, 2012; Saranen, 1992). In particular, scholars from Finland have discerned that gender differences are pronounced with regard to how difficult mathematics is perceived (Kangasniemi, 1989) and how these perceptions affect students' self-confidence, with boys displaying remarkably higher self-confidence than girls (Hannula & Malmivuori, 1997; Hannula et al., 2005). Hall (2012) examined data presented on students' mathematics achievement, attitudes and participation from the United States, Australia, New Zealand and Canada. Over time and across all educational levels (from primary school to university), Hall (2012) challenged the notion that gender issues in mathematics have been 'solved' in those countries. To support this challenge, gender gaps in achievement, positive attitude and greater self-confidence favoured boys at primary and secondary school levels. Furthermore, Hall noted that scholarly work has shown that student attitudes, achievement and participation are highly inter-related factors: 'Students with positive attitudes toward mathematics have greater achievement, and both of these factors are related to participation in mathematics' (2012, p. 70).

SCENARIO   ## Finding a way to make a geometry lesson more gender equitable

During an internship, a preservice teacher (PST) received feedback from his mentor teacher about some Year 9 lessons he had just taught on geometric reasoning (ACMMG220). From the PST's point of view, he had taught the concepts well (using a traditional method) and all students seemed to understand and then complete prescribed textbook exercises independently afterwards. Few questions were asked by students in the lessons, which reinforced the PST's idea that key concepts had been taught and learned well. But in her feedback, the mentor teacher suggested that more could be done to more actively engage the girls in the lessons.

Questions

1. Read the sub-section *Instructional practices to involve all students*. What do you think some of the specific suggestions were that the mentor teacher made to the PST?

2. Looking at the suggestions, are there any instructional practices you feel your mathematics learning could have benefited from? How?

Promoting gender equity in secondary mathematics

Researchers and practitioners offer suggestions for secondary teachers to promote gender equity within mathematics classrooms. Specifically, these suggestions require teachers to develop a heightened awareness of treating male and female students equitably (Van de Walle, Karp & Bay-Williams, 2014) so that instructional practices to involve all students can be implemented (Hall, 2012; Sousa, 2008).

Awareness of gender equity

Van de Walle, Karp and Bay-Williams (2014, p. 114) suggest that when interacting with students, teachers should be sensitive to the following:

- number and type of questions asked
- ability of students to act out or model mathematical situations or concepts with movements and gestures
- amount of attention given to disturbances
- kinds and topics of projects and activities assigned
- praise given in response to students' participation
- makeup and use of groups
- context of problems
- discussions of STEM careers to increase students' interest in these fields.

Additionally, Koontz (1997) recommended that teachers promote gender equity by modelling an attitude of acceptance for both sexes' participation in lessons (e.g. acknowledging all responses, giving supportive comments to incorrect answers) and by using gender-fair language (e.g. men and women scientists/mathematicians/doctors). Sensitivity to these issues can heighten teachers' awareness of their own gender-specific actions. Although Van de Walle, Karp and Bay-Williams (2014) acknowledge that detecting this awareness may be difficult initially, analysing video-recorded instructional lessons may be an instructive exercise. During an analysis, these authors recommend looking at the number of questions asked of male and female students, noting which students ask questions and what kinds of questions are being asked, and examining the types of feedback given (Van de Walle, Karp & Bay-Williams, 2014).

Instructional practices to involve all students

To minimise gender differences in learning, a variety of teaching approaches and strategies should be used. For instance, Sousa (2008) proposed that educators take into account the learning styles of students, address multiple intelligences, consider teaching styles available and examine how students think about mathematics in an attempt to make mathematics education equitable for all learners. Van de Walle, Karp and Bay-Williams (2014) recommend teachers find ways to involve all students in their classes, and not just those who appear to be eager. These commentators acknowledge that boys and girls alike may tend to avoid class involvement, lack motivation or seem reluctant in requesting teacher assistance. In a similar vein, Mau and Leitze (2001) offered that when teachers are in a 'show-and-tell mode', there are significantly more opportunities to reinforce boys' more overt behaviours and girls' more passive behaviours. Instead of reinforcing these behaviours, teachers must expect all students to speak, listen and share their thinking with others (Mau & Leitze, 2001). Hall (2012) suggested that making mathematics accessible to both genders can be achieved by avoiding traditional teaching approaches such as memorisation and rule following. Alternatively, planning lessons that focus more on conceptual understanding and connecting mathematics to students' lives may help both male and female students relate better and understand the subject (Becker, 1995; Belenky et al., 1986; Hall, 2012, p. 70; Morrow & Morrow, 1995). According to Hall (2012, p. 70), 'these approaches have been shown to have positive outcomes with respect to all students' achievement, attitudes, and participation in mathematics, but particularly for girls and women'. Additionally, Hall noted that various researchers (Boaler, 1997; Burton, 1995) advocate for creating supportive, inquiry-focused mathematics classroom environments in which female voices are heard and many approaches to solving problems are valued. Various instructional strategies teachers can employ to assist girls and boys in learning mathematics can be found in Table 6.1. It should be noted that these strategies are empirically supported but may not be universally applied.

Table 6.1 Strategies for assisting boys and girls to learn mathematics

Girls	Boys
Include group work.	Structure activities.
Include opportunities for discussion in both small groups and whole-class settings.	Assist learners to break tasks into achievable steps, particularly extended tasks such as investigations.
Include contexts of interest to girls.	Structure assignments with long deadlines (e.g. over several weeks) to include regular class work and milestones that are achievable (this also assists a teacher to be aware of ownership of the work).
Provide assessment opportunities that allow for a variety of responses (e.g. open-ended investigations). Avoid multiple choice questions as the only assessment approach.	If homework is set, ensure it is checked regularly.
Ensure contexts are understood (e.g. if football is used as a context, ensure all students are familiar with the context).	Employ clearly defined objectives and instructions.
	Ensure all students in the class are clear about what they are required to do.
	Put tasks on the whiteboard so they are retrievable. If a student has not heard the instructions, or joined the class after the task was set, the student will be able to catch up if the task has been recorded clearly.
	Use the 'tour to be sure' strategy: when a task has been set, wait for a few minutes for students to begin and then move around the students to check for understanding of the task.
Use contextual tasks. Include tasks that show the value of mathematics to the solution of social problems. These can include looking at environmental and social justice issues.	Set short-term challenging tasks.
	Include tasks with a focus on mathematical thinking for all learners in the class. Open-ended tasks allow challenge for all learners.
Use verbal and linguistic approaches.	Use visual, logical and analytical approaches.
Communication is an important part of mathematics and needs to be taught explicitly and emphasised for all learners.	A great deal of mathematics can be accessed using visual representations.
Girls may particularly enjoy opportunities to communicate mathematics in linguistic forms.	Allow students opportunities to explain their thinking and to record it in a variety of ways.
Engage learners with problems set in real contexts.	

Source: Siemon et al., 2011, p. 157.

PAUSE AND THINK

For those who attended a co-educational secondary school

Reflect on your secondary school experience and write down some recollections as they pertain to how teachers treated both genders in the mathematics classroom. After listing some thoughts, analyse the extent to which boys and girls were treated in terms of the following: praise and encouragement, selection to offer answers (verbally + demonstrations), amount and quality of attention given, instructional strategies and feedback. In your experience, and in light of what you have already read in this chapter, were boys and girls treated equitably?

For those who attended a single-gender secondary school

Reflect on your secondary school experience and write down some recollections as they pertain to how teachers treated students in the mathematics classroom. After listing some thoughts, analyse the extent to which boys or girls were treated in terms of: praise and encouragement, selection to offer answers (verbally + demonstrations), amount and quality of attention given, instructional strategies and feedback. In your experience, and in light of what you have already read in this chapter, were the instructional approaches used by teachers the best for your gender?

SHORT-ANSWER QUESTIONS

1. Define the two major goals provided for the policy framework of the Australian Curriculum by the Alice Springs (Mparntwe) Education Declaration.
2. List and describe the four propositions that have been developed to make the Australian Curriculum appropriate and accessible for all students.
3. What are some possible explanations to account for gender differences in motivation and achievement in mathematics?
4. Outline several instructional strategies for assisting boys and girls to learn mathematics.

Culture in the secondary mathematics classroom

Australia is considered one of the most culturally diverse nations in the world. Churchill et al. (2013) assert that migration is a powerful force in Australia's history, where 'waves of migration from Europe, Asia and Africa have brought many benefits to everyday life in both urban and regional communities in Australia' (p. 147). The 2016 Census of Population and Housing of the Australian Bureau of Statistics (ABS) found that approximately 22 per cent of the population speak a language other than English at home. This statistic represents an approximate increase of 1.2 per cent from the 2011 Census. The five most commonly spoken languages other than English were Mandarin, Arabic, Cantonese, Vietnamese and Italian, with speakers comprising approximately 7.5 per cent of the total population (ABS, 2016). Furthermore, approximately 21 per cent of households speak two or more languages at home (ABS, 2016). The educational implications for this linguistic diversity are considerable, and Marsh (2004, p. 238) has stated earlier that 'teachers working in Australian schools require a broad understanding of our cultural diversity – diversity due to race, ethnicity, socio-economic status, geographic region, religion and gender'. And while the Australian student population is linguistically and culturally diverse, 'it is significant that the Australian teaching profession is overwhelmingly Anglo-Australian and of middle-class background' (Churchill et al., 2013, p. 147). In light of these data, it must be acknowledged that there are significant regional differences within Australia regarding language and cultural diversity. Within this section, attention is given to teaching mathematics with cultural awareness, and teaching mathematics to ELLs.

PAUSE AND THINK

Are you aware of any methods of teaching and/or learning mathematics which are particular to a specific country or culture? How are these methods different to how you learned mathematics?

Teaching mathematics with cultural awareness

While mathematics is commonly heralded as a 'universal language', the accuracy of this statement is questionable. For instance, conceptual knowledge (e.g. what multiplication is) is universal, while procedural knowledge (e.g. how to multiply) and symbols are culturally determined, and are non-universal (Van de Walle, Karp & Bay-Williams, 2014). Following research into culturally responsive instructional models within New Zealand mathematics classrooms, Averill et al. (2009) deemed the following conditions as necessary for teachers of culturally diverse students:

> [a] deep mathematical understanding; effective and open relationships, cultural knowledge, opportunities for flexibility of approach and for implementing change, many accessible and non-threatening mathematics learning contexts, involvement of a responsive learning community, and most important, working within a cross-cultural teaching partnership. (p. 180)

Van de Walle, Karp and Bay-Williams (2014) argue that culturally relevant mathematics instruction is for all students, including students from different ethnic groups and socioeconomic status. These authors suggest such instruction requires teachers to focus on 'important mathematics', make content relevant, incorporate students' identities and ensure there is shared power between teacher and student. Teachers should also be careful to select learning materials (e.g. textbooks) that do not exhibit cultural bias (Marsh, 2004).

IN PRACTICE

Working with students from different cultural backgrounds

Marsh, Clarke and Pittaway (2014, p. 261) recommend a number of strategies for teachers working with students from different cultural backgrounds. The strategies are:

1. Respect the ethnic and racial backgrounds of students – encourage and support.
2. Use role plays to provide student empathy for different cultures.
3. Ensure students from different cultures have opportunities to work with others in small group activities.
4. Use curriculum resources that highlight multicultural perspectives.
5. Provide opportunities for students to examine, in depth, particular values, beliefs and points of view relating to cultures.
6. Use media examples to highlight undesirable bias and discrimination.
7. Encourage students to be open and willing to evaluate their values.

Teaching mathematics to English language learners (ELLs)

English language learners (ELLs) enter the mathematics classroom at different ages, and at different stages of learning the English language. ACARA (2020) acknowledges that English as an additional language or dialect (EAL/D) students must achieve the aims of the Australian Curriculum: Mathematics 'while simultaneously learning a new language and learning content and skills through that new language'. To compound the matter for ELLs, there are unique features of the 'mathematics language' that make learning difficult; for instance, certain specific terminology may not be translated easily into another language (Van de Walle, Karp & Bay-Williams, 2014). Moreover, Janzen (2008) highlights that worded problems are particularly difficult for ELLs to complete as the sentences are constructed differently from sentences in conversational English. Khisty (1997) points out that ELLs develop conversational skills in English much faster than their proficiency for benefiting from content lessons taught in English. The corollary of this assertion is that teachers should not assume that because a student can converse competently in English, the same student can 'follow complex, multi-faceted instruction in English requiring her to listen, read, write, speak, and employ other communication structures (e.g. interpret body language and illustrations)' (Cangelosi, 2003, p. 79). In fact, it may take up to seven years for an ELL to learn an academic language, such as mathematics (Cummins, 1994).

When teaching ELLs, one approach is content-based instruction (CBI), which uses specific mathematical content on which to base language instruction. Said another way, in CBI 'the language is taught within the context of a specific academic subject' (Teaching English to Speakers of Other Languages, 2008, p. 1). According to DelliCarpini and Alonso (2014), CBI can enhance the acquisition of both language and content. However, these authors concede that at a secondary level 'to date most of the CBI practice that occurs, whether content-driven or language-driven, does so in the English as a second language classroom exclusively, and mainstream content teachers are often unprepared or underprepared to work with ELLs in their classrooms' (DelliCarpini & Alonso, 2014, p. 158). Other approaches to assist in language acquisition for ELLs recommended in the literature have been included within the In Practice section below.

IN PRACTICE

Making the Australian Curriculum accessible to EAL/D students

A national EAL/D framework has been developed to support teachers in making the Australian Curriculum: Foundation to Year 10 in mathematics accessible to EAL/D students (ACARA, 2020). Presented in this framework are the following suggestions:

1. Allow students more time to complete tasks.
2. Allow students to work in their native language as required.
3. Provide input and directions at current level of comprehension.
4. Provide additional support to develop students' language needs.
5. Teach specific vocabulary (i.e. the language of mathematics) explicitly.
6. Consider carefully cooperative learning groups to support language acquisition.

Cultural considerations for Indigenous learners

Over the past three decades there have been a number of large-scale literacy and numeracy programs developed to enhance the learning opportunities for Indigenous Australian students (Cooke & Howard, 2009). Some of the more recent programs have included Mathematics in Indigenous Contexts (1999–2005), What Works (2005–2008), Turn the Page: Indigenous Mathematics and Numeracy (2009–2012), and Make it Count (2009–2012). For instance, the Make it Count program was undertaken by the Australian Association of Mathematics Teachers (AAMT) to address factors contributing to the differences in achievement between Indigenous and non-Indigenous students. The program 'developed evidence-based, responsive pedagogies and resources to improve learning outcomes of Aboriginal and Torres Strait Islander learners across Australia' (AAMT, 2013, p. 1). Additionally, seven key findings were synthesised as a result of university researchers and eminent Indigenous and mathematics educators working with over 1500 Indigenous students from 35 schools across five states. These summary findings reflect the APST (AITSL, 2011) and underscore the importance of improving current practices used in teaching mathematics to Indigenous students. Herein, the AAMT (2013, p. 2) presented the findings as:

Professional knowledge

1. Know Indigenous learners and know how they learn.
2. Know the mathematics content and know the different ways to teach it effectively to Indigenous learners.

Professional practice

3. Plan for and implement Responsive Mathematics Pedagogy for Indigenous learners that is culturally, academically and socially inclusive.
4. Create and maintain learning environments in which Indigenous learners feel safe and supported.
5. Develop and use tools that assess both affective and cognitive learning outcomes specific to Indigenous learners, provide feedback, and report on student learning.

Professional engagement

6. Engage with colleagues – in professional learning communities in ongoing, action oriented, professional learning – who are prepared to push the boundaries and move outside their comfort zone. Strive for collegial innovation in both Indigenous education and mathematics and numeracy education.
7. Engage with Indigenous parents, families and community in two-way dialogue.

Concomitant with these findings, scholars outline that to successfully teach Aboriginal and Torres Strait Islander students a culturally responsive approach is required (Parkin & Hayes, 2006; Perry & Howard, 2008; Warren & DeVries, 2010). Such a response moves away from a transmission model of learning and towards one that requires a 'creative and thoughtful use of each teacher's repertoire of professional skills, and a careful consideration of context' (AAMT, 2013, p. 1). To illustrate, Parkin and Hayes (2006) contend that the linguistic demands of the mathematics curriculum (e.g. mathematics textbooks, worded problems) render mathematics difficult for Indigenous students to access. Teachers can facilitate students' accessibility through several key linguistic strategies; broadly speaking, these strategies assist students in 'sorting out the mathematically important from the unimportant, and in recognising what constituted the mathematical task hidden somewhere in this mass of seemingly irrelevant verbiage' (Parkin & Hayes, 2006, p. 27). Specific strategies include scaffolded literacy, lower book orientation, higher book orientation and text patterning. Treacy and Frid (2008) conducted research that illustrated how Western mathematics (which is taught in Australian schools) is distinctly different from how traditional Indigenous cultures make sense of, organise and act in their environments. For the research, Treacy and Frid (2008) created three mathematical 'counting' tasks for 18 Aboriginal students (Years 1–11) to complete and examined the extent to which students used the strategy of counting to arrive at their answers. The researchers concluded that while most students demonstrated specific counting knowledge and skills in two of the tasks, there was a preference not to use counting for the third task. Acknowledging that Aboriginal languages do not have many counting words – and as a consequence, many Aboriginal people tend not to count in their everyday situations (Treacy & Frid, 2008, p. 536) – the findings from this research suggest that students did not view the third task as one that required counting. Instead, students preferred to use an estimation strategy rather than obtain an exact answer, which supports the idea that precision is much more central to Western society than in most Aboriginal contexts (Malcolm et al., 1999).

PAUSE AND THINK

Suppose that a student – who is from a culture you know little about, and who speaks a language that you do not understand – is the newest member of your secondary mathematics classroom. After having read this section on culture, what would you do in order to teach this student effectively? Furthermore, what factors will influence the steps you will take in teaching this student?

SHORT-ANSWER QUESTIONS

1. List and describe some instructional guidelines for teaching students from different cultural backgrounds.
2. What are some practical approaches to support the language acquisition of English language learners (ELLs)?
3. Outline some evidence-based, responsive pedagogies and resources to improve learning outcomes of Indigenous learners across Australia.

Special needs learners in the secondary mathematics classroom

Since the mid-1970s the literature base for special needs education has grown considerably. Concomitant to legislation steadily securing the educational rights of students with special needs, research has provided much insight into best instructional practices for these students. Aligning with the belief that all learners – who, with good teaching and the appropriate support – are able to learn mathematics, this section examines two broad groups of special needs learners. These groups are gifted and talented students and students with learning difficulties.

PAUSE AND THINK

Who do you consider to be learners with special needs? What are the special educational needs of these learners, and how should teachers and school leaders respond to these needs?

Gifted and talented students

Gifted and talented students generally demonstrate above-average ability, high levels of task commitment, creativeness (Huetinck & Munshin, 2008), and an exceptional level of performance in one or more areas of expression (National Association for Gifted Children, 2007). In addition, students who are mathematically gifted and talented may possess 'an intuitive knowledge of mathematical concepts, whereas others have a passion for the subject even though they may have to work hard to learn it' (Van de Walle, Karp & Bay-Williams, 2014, pp. 115–16). Diezmann and Watters (2002, p. 220) have listed characteristics of gifted learners of mathematics to include:

- exceptional reasoning ability
- exceptional memory and concentration span
- preference for abstract and self-directed work
- ability to solve problems in unexpected ways
- ability to identify patterns and relationships
- preference for mathematical activities and puzzles, including enjoyment in posing problems.

Researchers have also identified spatial ability as a characteristic of gifted and talented mathematics students (Perry, Anthony & Diezman, 2004). Students' giftedness can become apparent to parents and teachers through a demonstrated grasp or articulation of mathematical concepts at an age earlier than expected (Van de Walle, Karp & Bay-Williams, 2014). Rotigel and Fello (2005) have noted that these students are often found to easily make connections between topics of study, and frequently are unable to explain how they quickly got an answer. Teachers who have a keen ability to detect giftedness in mathematics note they are students who possess strong number sense or visual/spatial sense (Gavin & Sheffield, 2010). However, gifted mathematics students should not be confused solely with those who are fast with number facts, but those who possess an ability to reason and make sense of mathematical tasks.

Despite the gifts and talents of such learners being recognised by teachers and parents, the students themselves may not embrace comfortably the label of mathematically gifted (Huetinck & Munshin, 2008; Van de Walle, Karp & Bay-Williams, 2014). Popular culture and the media consistently portray talented learners of mathematics as looking strange or acting weird (Sheffield, 1997), socially inept outcasts (Van de Walle, Karp & Bay-Williams, 2014) or as eccentrics with prodigious calculating abilities (Siemon et al., 2011). Such portrayals are unhelpful for gifted adolescent students who 'absorb these powerful negative messages about showing their intelligence in public settings' (Van de Walle, Karp & Bay-Williams, 2014, p. 115). Nevertheless, it is important for teachers to encourage and support mathematically gifted students by showing the class appropriate, successful, 'real-world' mathematics role models who appear in popular culture, the media and the real world (Van de Walle, Karp & Bay-Williams, 2014).

Educational strategies for gifted and talented students

According to Huetinck and Munshin (2008), it is generally assumed within education that good teaching can respond to the varying needs of diverse learners – including the gifted and talented. However, educators may tend to feel less inclined to provide for these learners than for students of low ability. A considerably worse outcome is where mathematically talented learners have experienced classrooms where giftedness is not recognised or cultivated appropriately. For instance, such learners may have been given 'more of the same' work for practice and consolidation, given 'free time' or independent time on a technological device, or routinely assigned to assist struggling learners – all of which are considered unsupportive strategies for talented learners (Huetinck & Munshin, 2008; Siemon et al., 2011). Irrespective, teachers should not wait for students to demonstrate their mathematical content; these students require instructional programs designed to assist them reach their full potential (Huetinck & Munshin, 2008). Such programs can include challenging, rich extension tasks (Siemon et al., 2011) and an inquiry-based instructional approach (Van Tassel-Baska & Brown, 2007) which can both promote creative and divergent thinking. To this end, Sheffield (1999) contended that gifted students need to experience the:

> joys and frustrations of thinking deeply about a wide range of original, open-ended, or complex problems that encourage them to respond creatively in ways that are original, fluent, flexible, and elegant. (p. 46)

ACARA recommends that teachers enrich gifted students' learning through the provision of opportunities to work with learning area content in further depth or breadth. This enrichment can take place through teachers:

> emphasising specific aspects of the general capabilities learning continua (for example, the higher order cognitive skills of the Critical and Creative thinking capability); and/or focusing on cross-curriculum priorities. Teachers can also accelerate student learning by drawing on content from later levels in the Australian Curriculum: Mathematics and/or from local state and territory teaching and learning materials. (ACARA, 2020)

Gallagher and Gallagher (1994) suggest four strategies for adapting mathematics content for gifted students: acceleration, enrichment, sophistication and novelty. Within each strategy, these authors contend that planned activities require students to apply learned information rather than simply acquire it. Additionally, an emphasis on the implementation and extension of ideas must eclipse the acquisition of facts and concepts (Gallagher & Gallagher, 1994).

CONNECTION　Carl Friedrich Gauss

Considered one of the greatest mathematicians of all time, Carl Friedrich Gauss (1777–1855) made significant contributions to mathematics, physics, astronomy and statistics. One famous story (NRICH, 2012) recalls how one of Carl's primary school teachers asked the class to add together all the integers from 1 to 100 inclusively. After only a few moments, Carl shocked his teacher by correctly calculating the sum of 5050 in his head! But Carl was able to outline his method of calculation rather easily, which forms a general rule for the sum of the first n integers, such that:

$$1 + 2 + 3 + \cdots + n = \frac{1}{2}n(n+1)$$

Question

How did Carl Gauss add all the integers from 1 to 100 inclusively so quickly? Try using the general rule for the sum of the first n integers to add all of the integers from 1 to 10 inclusively, and then all of the integers from 1 to 50 inclusively.

Four strategies: acceleration, enrichment, sophistication and novelty

Acceleration involves providing students with access to curriculum content different from their year level while demanding more learning independence, recognising that 'students may already understand the mathematics content that will be presented'; consequently, 'teachers can either reduce the amount of time these students spend on aspects of the topic or move altogether to more advanced and complex content' (Van de Walle, Karp & Bay-Williams, 2014). According to Chambers and Timlin (2013, p. 159), the provision of acceleration 'allows able pupils to progress through the school system in fewer years than is normal'. *Enrichment* activities take further the prescribed, curriculum-based topics of study to extend original mathematical tasks. Alternatively, enrichment activities 'can involve studying the same topic as the rest of the class while differing on the means and outcomes of the work' (Van de Walle, Karp & Bay-Williams, 2014). Van de Walle, Karp and Bay-Williams (2014, p. 116) suggest that examples of enrichment can include 'group investigations, solving real problems in the community, writing letters to outside audiences, or identifying applications of the mathematics learned'. In contrast to acceleration programs, which are designed for students to learn mathematics in settings with older peers (or independently in age-appropriate classes), enrichment can benefit many in the class if it leads to more interesting mathematical activities being offered to all (Foster, 2013).

When educators raise the level of complexity of a topic or have students pursue a task at greater depth, they are using the strategy of *sophistication*. Said another way, students have the opportunity to explore a larger set of ideas within which a mathematics topic exists. For instance, a teacher may ask mathematically gifted students to develop a method for calculating reducible interest on a car loan when the class is studying arithmetic and geometric sequences. *Novelty* requires educators to introduce gifted students to completely different material from the prescribed curriculum, and within their developmental grasp. Alternatively, the novelty approach provides students with various options to demonstrate their mathematical skills and knowledge through demonstrations, inventions, experiments and oral presentations (Gallagher & Gallagher, 1994).

Teaching a gifted student

SCENARIO

During her internship, a PST learns from her mentor teacher that a new student will arrive in their Year 7 class. From her previous school in another state, this new student is already working proficiently at a Year 10 level (i.e. three years' ahead of her age-appropriate peers) and her parents are strictly opposed to out-of-class acceleration practices. The PST will be teaching a series of lessons on ACMNA153 (Solve problems involving addition and subtraction of fractions, including those with unrelated denominators).

Questions

1. Read the sub-section 'Educational strategies for gifted and talented students'. Based on your reading, what approach do you feel will best suit this student's educational needs? What inferences do you think the mentor teacher and the PST made about the parents of this student strictly opposing out-of-class acceleration practices?

2. Outline some of the key factors you feel the PST will need to take into account when planning learning activities for her class as well as for this gifted and talented student. For teaching the first lesson on ACMNA153, what are some possible activities the PST could plan to engage all learners in the class?

Students with learning difficulties

The identification and remediation of students whose low ability in mathematics is based on learning difficulties is a complex process (Huetinck & Munshin, 2008). Furthermore, these authors offer that learning difficulties can comprise intellectual disability, serious emotional disturbances, autism, traumatic brain injury, attention deficit disorder, attention deficit hyperactivity disorders or other specific learning disabilities (Huetinck & Munshin, 2008, p. 343). In turn, students may have very specific difficulties with perceptual or cognitive processing which may affect 'memory; general strategy use; attention; the ability to speak or express ideas in writing; the ability to perceive auditory, visual, or written information; or the ability to integrate abstract ideas' (Van de Walle, Karp & Bay-Williams, 2014, p. 104). Sousa (2008) has identified environmental (e.g. mathematics anxiety) and neurological (e.g. dyscalculia) factors that can impede mathematics learning – as well as mathematics disorders (e.g. procedural, memory), difficulties (arithmetic, number concepts) and deficits (counting skills, visual-spatial). In 1975, a landmark event for special needs education occurred in the United States, where under federal law, the Education for All Handicapped Children Act mandated that special education students be placed in the least restrictive environment for success in learning. This meant that special education students were to remain in educational environments with their non-identified peers unless there was a compelling reason to withdraw them. Since then, Australia has enacted through legislation the *Disability Discrimination Act 1992* and, more importantly for education, the Disability Standards for Education (Department of Education, Science and Training [DEST], 2006). These standards outline how students with a disability have the right to 'education and training opportunities on the same basis as students without disabilities' (DEST, 2006, p. 42). Over the past three decades the conceptualisation of disability has been described as one of 'deficit', where emphasis was placed on identifying students' learning difficulties in order to 'fix them up' (Siemon et al., 2011). A more recently developed view positions Australia's educational system to:

Least restrictive environment: the requirement that, where appropriate, students with disabilities be educated in settings with students without disabilities.

> promote excellence and equity . . . improve outcomes for educationally disadvantaged young Australians . . . Targeted support can help learners such as . . . children with disabilities reach their potential. This means tailoring to the needs of individuals across a system that prioritises equity of opportunity and that supports achievement. (Alice Springs (Mparntwe) Education Declaration, 2019, p. 17)

Irrespective of how disability is conceptualised, the strategies used to teach students with disabilities should be based upon the tenets of sound teaching practices that help all students. Some strategies to be used in a secondary mathematics classroom will now be considered.

CONNECTION | Maria Montessori (1870–1952)

After completing medical school in 1896, Dr Maria Montessori voluntarily joined a research program in the psychiatric clinic at the University of Rome. Her work here initiated a deep interest within her in the needs of children with learning difficulties. Maria advocated strongly for greater educational support for mentally and developmentally disabled children, and in 1907 she opened the first *Casa dei Bambini* (Children's House). Maria came to realise that children were able to educate themselves when placed in an environment where activities were designed to support their natural development. Now known globally, the

Montessori Method of Education provides a nurturing, supportive and individualised environment for children of all abilities and learning styles.

Question

Conduct an internet search on Maria Montessori. From your wider reading, how has the Montessori Method of Education contributed to instructional approaches used today, and, in particular, within special needs education?

Teaching students with learning disabilities

At all stages in their learning, students should be able to engage with all topics within the Australian Curriculum: Mathematics. One instructional challenge for the classroom teacher, therefore, is to ensure that all learners – whatever their capability or prior learning – have access to mathematical topics in a meaningful and developmentally appropriate manner. For this access, teachers must make necessary curriculum adjustments by drawing from:

> content at different levels along the Foundation to Year 10 sequence. Teachers can draw from content at different levels along the Foundation to Year 10 sequence. Teachers can also use the extended general capabilities learning continua in Literacy, Numeracy and Personal and Social capability to adjust the focus of learning according to individual student need. (ACARA, 2020)

Additionally, teachers can enact considerations for instructional planning, delivery and assessment, including identification of learning disabilities, using an Individualised Educational Plan (IEP) and pedagogical strategies.

Individualised Educational Plan (IEP): a written education plan that is created with a child's specific learning needs in mind.

Identification of learning disabilities

Initially, working with students who have learning disabilities may be a difficult task. As beginning teachers gain experience in interacting with students of varying ages, maturity levels and mathematical abilities, they become increasingly aware and sensitive to students who may have learning disabilities (Huetinck & Munshin, 2008). While a low assessment result may arouse a teacher's curiosity – and, at the same time, provide valuable information about student weaknesses or possible learning difficulties – there are numerous non-cognitive factors to be considered. Factors including low motivation, prior mathematics instruction, cultural bias of tests and immaturity can be investigated by the teacher before referring a student to a special needs coordinator (SNC) (Huetinck & Munshin, 2008). Miller and Mercer (1997) posited that students with learning disabilities exhibit problems that contribute to poor mathematics achievement and that are linked to information processing. Following their review of research, these authors suggested various ways in which weaknesses in selected components of information processing may affect mathematics performance (Miller & Mercer, 1997). The general problems and accompanying specific weaknesses listed in Table 6.2 could be used by a teacher who suspects a student has learning difficulties in mathematics.

Table 6.2 How components of information processing may affect mathematics performance

Attention deficits	1. Student has difficulty maintaining attention to steps in algorithm of problem solving.
	2. Student has difficulty sustaining attention to critical instruction (e.g. teacher modelling).
Visual–spatial difficulties	1. Student loses place on the worksheet.
	2. Student has difficulty differentiating between numbers, coins, operation symbols, clock hands.
	3. Student has difficulty writing across the paper in a straight line.
	4. Student has difficulty relating to directional aspects of mathematics and number alignment.
	5. Student has difficulty using a number line.
Auditory processing difficulties	1. Student has difficulty doing oral skills.
	2. Student is unable to count on from within a sequence.
Memory problems	1. Student is unable to retain mathematical facts or new information.
	2. Student forgets steps in an algorithm.
	3. Student performs poorly on review lessons or mixed probes.
	4. Student has difficulty telling time.
	5. Student has difficulty solving multi-step word problems.
Motor difficulties/Disabilities	1. Student writes numbers illegibly, slowly and inaccurately.
	2. Student has difficulty writing numbers in small spaces.

Source: Miller & Mercer, 1997, p. 50.

Using an individualised education program (IEP)

Once a student's learning disability has been diagnosed by a SNC accurately, an IEP will be prepared by educators (usually classroom teachers in conjunction with a SNC), parents and, where appropriate, the student. This document outlines specifically the services the student will receive, including 'positive appropriate behaviour interventions when necessary, assigning certain supplementary learning aids and services, and possibly modifying programs for the youngster' (Huetinck & Munshin, 2008, p. 343). While implementing strategies prescribed within the IEP, teachers are encouraged to actively contribute to the development of the document, seek further advice from relevant professionals (e.g. educational psychologist, SNC) and work closely with the student's parents (Siemon et al., 2011). Doing so will allow the student's profile of educational strengths to be developed over time.

Pedagogical strategies

Teaching students with learning disabilities is a challenging task, as the classroom teacher requires additional and special skills (Tannock & Martinussen, 2001). The strategies listed below (Marsh, 2004, p. 236) focus on structured, effective approaches for teaching students with learning disabilities. In offering these general instructional strategies, Marsh acknowledges that learning goals for these students 'tend to be based on basic reading, writing and arithmetic, and on social, vocational and domestic skills' (Marsh, 2004, p. 236). Nonetheless, the strategies provide a solid foundation for any secondary teacher looking to make mathematics accessible to all learners:

- Carefully develop readiness for each learning task.
- Present material in small steps.
- Develop ideas with concrete, manipulative and visually oriented materials.
- Be prepared for large amounts of practice on the same idea or skill.
- Relate learnings to familiar experiences and surroundings.
- Focus on a small number of target behaviours so that students can experience success.
- Motivate work carefully.
- Ensure that the material used is appropriate for the physical age of the student and is not demeaning.
- Every time students complete a task successfully, they should be rewarded (Marsh, 2004, p. 236).

Key principles of effective instructional design

IN PRACTICE

Specific to the discipline of mathematics, Carnine (1997) recommends five key principles of effective instructional design for teachers of students with learning disabilities. These principles include:

1. Teach big ideas

These concepts can make learning subordinate concepts easier and more meaningful, and represent central ideas within a discipline (Carnine, 1997). For instance, in the Measurement and Geometry strand students learn various formulas for calculating volume. Using a big idea reduces the number of formulas students must learn through using a single formula (i.e. Volume = Area of Base × Height).

2. Teach conspicuous strategies

Acknowledge that a strategy is 'a series of steps that students follow to achieve some goal' (Carnine, 1997, p. 134), as it is an instructional challenge to develop strategies that are 'just right' interventions for students who have difficulty forming them independently. Such strategies are intermediate in generality and facilitate students' use of knowledge to address the *what*, *how* and *when*.

3. Use time efficiently

This principle is centred on teaching students 'all they need to know (in both quantitative and qualitative terms) without "losing" them by trying to do too much, too quickly' (Carnine, 1997, p. 137). To achieve this, teachers should abandon low-priority learning objectives in favour of focusing on big ideas, ease into the instruction of complex problem-solving strategies, use a strand organisation for lessons and use manipulatives in a time-efficient manner.

4. Clearly and explicitly communicate strategies

Teachers should explain new concepts and strategies in a clear, concise, accurate and comprehensible manner. To achieve this, they must accommodate differences in students' prior knowledge,

▶ ▶

▶ ▶ provide carefully scaffolded examples that will lead to self-directed learning and use corrective feedback consistently.

5. Facilitate retention through practice and review

Keeping in mind that an important goal of any 'mathematics instructional program is to remember and apply increasingly complex concepts and strategies' (Carnine, 1997, p. 138), students need opportunities to carefully practise and review mathematical skills. To facilitate retention, teachers can help students develop automaticity through the provision of concisely delivered demonstrations and the reception of carefully scaffolded practice.

Additionally, Van de Walle, Karp and Bay-Williams (2014) suggest that *explicit instruction*, the *concrete-representational-abstract (CRA) teaching sequence* and *peer-assisted learning* may also be effective in teaching students with learning disabilities. First, explicit instruction involves highly structured, step-by-step, teacher-led explanations of concepts and strategies that focus on 'critical connection building and meaning making that help learners relate new knowledge with concepts they know' (Van de Walle, Karp & Bay-Williams, 2014, p. 104). Using this instructional approach, the teacher uses a tightly scripted sequence that commences with modelling, and then prompts students through the model towards independent practice. Second, the CRA teaching sequence enables students to move away from concrete representations (manipulative materials), through representations (drawings or pictures) and towards abstraction (mathematical symbols) when solving problems. Third, peer-assisted learning allows students to benefit from classmates' modelling and support on an 'as needed' basis. Importantly, tutors and tutees can interchange roles, providing special needs students with a valuable opportunity to explain to another student (Van de Walle, Karp & Bay-Williams, 2014). Additionally, Huetinck and Munshin (2008) note that to succeed, special needs students require considerable encouragement and frequent reminders from the teacher about their progress.

PAUSE AND THINK

Examine your personal philosophy of teaching and learning for statements of how you plan to make mathematics accessible to all students in your class. As a guide, read the Department of Education and Training's NCCD Guidelines (DET, 2019). From this document, what key points have you included in your personal philosophy?

SHORT-ANSWER QUESTIONS

1. Justify the need for diverse learners to receive instruction according to their particular learning needs.

2. Outline some instructional approaches for teaching gifted and talented students in the secondary mathematics classroom.

3. How is an individualised educational plan (IEP) developed and implemented?

4. Carnine (1997) recommends five key principles of effective instructional design for teaching students with learning disabilities. What are these principles?

Conclusion

Several possible explanations can account for perceived gender differences in motivation and achievement in mathematics. These explanations are socially and culturally constructed and include pervasive belief systems, teacher behaviour and attitudes, and student attitudes.

To make a classroom gender equitable, a teacher can take into account the learning styles of students, address multiple intelligences, consider teaching styles available and examine how students think about mathematics.

When working with students from different cultural backgrounds, it is important to provide instruction for culturally relevant mathematics. Such instruction requires teachers to focus on 'important mathematics', make content relevant, incorporate students' identities and ensure there is shared power between teacher and student.

To make learning accessible for diverse learners, teachers should modify their approaches towards instructional planning, delivery and assessment. For students with learning difficulties, teachers can plan activities focused on big ideas, teach conspicuous strategies, use time efficiently, communicate strategies clearly and explicitly, and facilitate retention through practice and review. Teachers of gifted and talented learners can promote creative and divergent thinking through the use of strategies including acceleration, enrichment, sophistication and novelty.

BRINGING IT TOGETHER

1. Using the two goals from the Alice Springs (Mparntwe) Education Declaration, summarise your instructional approach for a secondary mathematics classroom.

2. Describe what you think is meant by the term *stereotype threat*. How do you plan to minimise the possibility of stereotype threat emerging within your classroom?

3. Paraphrase what you feel are considered best principles of instructional practice for working with (i) students with intellectual disabilities, and (ii) gifted and talented students.

4. Recall and explain some instructional approaches suggested by the AAMT when teaching mathematics to Aboriginal and Torres Strait Islander students.

5. A student arrives in your Year 10 class with minimal English-speaking capabilities. Using the content of this chapter as a starting point, what are some of the things you will do to ensure that this student can engage with the content you are currently teaching?

6. Select a lesson plan you have developed recently for your teaching practicum. Modify the content and pedagogical approach of the main body of the lesson for a student who (i) has auditory processing difficulties, and (ii) has an exceptional ability to identify patterns and relationships.

Further resources

REFERENCES

American Association of University Women (1992). *How schools shortchange girls: A study of major findings on girls and education*. AAUW Educational Foundation, The Wellesley College Center for Research on Women.

Australian Association of Mathematics Teachers (AAMT) (2013). *Make it Count: Numeracy, mathematics and Indigenous learners project summary*.

Australian Bureau of Statistics (ABS) (2016). *2016 census quick stats: People – cultural and language diversity*. Retrieved from https://quickstats.censusdata.abs.gov.au/census_services/getproduct/census/2016/quickstat/036

Australian Curriculum, Assessment and Reporting Authority (ACARA) (2020). *Student diversity*. Retrieved from https://www.acara.edu.au/curriculum/student-diversity

Australian Institute for Teaching and School Leadership (AITSL) (2011). *Australian Professional Standards for Teachers*. Melbourne: AITSL.

Averill, R., Anderson, D., Easton, H., Maro, P.T., Smith, D. & Hynds, A. (2009). Culturally responsive teaching of mathematics: Three models from linked studies. *Journal for Research in Mathematics Education*, 40(2), 157–86.

Becker, J.R. (1995). Women's ways of knowing in mathematics. In P. Rogers & G. Kaiser (eds), *Equity in mathematics education: Influences of feminism and culture* (163–74). London: Falmer.

Beilock, S.L., Gunderson, E.A., Ramirez, G. & Levine, S.C. (2010). Female teachers' math anxiety affects girls' math achievement. *Proceedings of the National Academy of Sciences*, 107(5), 1860–963.

Belenky, M.F., Clinchy, B.M., Goldberger, N.R. & Tarule, J.M. (1986). *Women's ways of knowing: The development of self, voice, and mind*. New York, NY: Basic Books.

Boaler, J. (1997). Reclaiming school mathematics: The girls fight back. *Gender and Education*, 9(3), 285–305.

Booker, G., Bond, D., Sparrow, L. & Swan, P. (2014). *Teaching primary mathematics* (5th edn). Frenchs Forest: Pearson.

Burton, L. (1995). Moving towards a feminist epistemology of mathematics. In P. Rogers & G. Kaiser (eds), *Equity in mathematics education: Influences of feminism and culture* (209–26). London: Falmer.

Burton, L. (2001). Fables: The tortoise? The hare? The mathematically underachieving male. In B. Atweh, H. Forgasz & B. Nebres (eds), *Sociocultural research on mathematics education: An international perspective* (379–92). Mahwah, NJ: Lawrence Erlbaum.

Campbell, P.B. (1995). Redefining the 'girl problem in mathematics'. In W.G. Secede & L.B. Adajian (eds), *New directions for equity in mathematics education* (225–41). New York: Cambridge University Press.

Cangelosi, J.S. (2003). *Teaching mathematics in secondary and middle school: An interactive approach* (3rd edn). Upper Saddle River, NJ: Pearson Education.

Carnine, D. (1997). Instructional design in mathematics for students with learning disabilities. *Journal of Learning Disabilities*, 30(2), 130–41.

Ceci, S.J. & Williams, W.M. (2010). Sex differences in math-intensive fields. *Current Directions in Psychological Science*, 19(5), 275–9.

Chambers, P. & Timlin, R. (2013). *Teaching mathematics in the secondary school* (2nd edn). Los Angeles, CA: SAGE Publications.

Churchill, R., Ferguson, P, Godinho, S., Johnson, N., Keddie, A., Letts, W., Mackay, J., McGill, M., Moss, J., Nagel, M., Nicholson, P. & Vick, M. (2013). *Teaching: Making a difference* (2nd edn). Milton: John Wiley & Sons.

Clarke, D., Cheeseman, J., Clarke, B., Gervasoni, A., Gronn, D., Horne, M. & Sullivan, P. (2001, July). *Understanding, assessing and developing young children's mathematical thinking: Research as a powerful tool for professional growth.* Keynote paper presented at the Annual Conference for the Mathematical Education Research Group of Australasia, Sydney.

Cooke, S. & Howard, P. (2009). '*Can WE address the issues surrounding Aboriginal education?' 'Yes we can!!!! Together!*' Paper presented to the Dare to Lead conference, Canberra. Retrieved from http://makeitcount.aamt.edu.au/Resources/Indigenous-education

Correll, S.J. (2001). Gender and the career choice process: The role of biased self-assessment. *American Journal of Sociology*, 106(6), 1691–730.

Council of Australian Governments Education Council (2019). *Alice Springs (Mparntwe) Education Declaration.* Retrieved from https://docs.education.gov.au/system/files/doc/other/final_-_alice_springs_declaration_-_17_february_2020_security_removed.pdf

Cummins, J. (1994). Knowledge, power, and identity in teaching English as a second language. In F. Genesee & J.C. Richards (eds), *Educating second language children: The whole child, the whole curriculum, the whole community* (39–57). Cambridge: Cambridge University Press.

DelliCarpini, M. & Alonso, O.B. (2014). Teacher education that works: Preparing secondary-level math and science teachers for success with English language learners through content-based instruction. *Global Education Review*, 1(4), 155–78.

Department of Education, Science and Training (DEST) (2006). *Disability standards for education.* Retrieved from https://www.dese.gov.au/disability-standards-education-2005

Department of Education and Training (DET) (2019). *Nationally consistent collection of data on school students with disability [NCCD]. 2019 guidelines.* Retrieved from https://www.nccd.edu.au/sites/default/files/nccd_guidelines.pdf

Diezmann, C. & Watters, J. (2002). Summing up the education of mathematically gifted students. In B. Barton, K.C. Irwin, M. Pfannkuch & M.O.J. Thomas (eds), *Mathematics education in the South Pacific. Proceedings of the 25th Annual Conference of Mathematics Education Research Group of Australasia (MERGA)* (219–26). Auckland: MERGA.

Ellison, G. & Swanson, A. (2010). The gender gap in secondary school mathematics at high achievement levels: Evidence from the American mathematics competitions. *Journal of Economic Perspectives*, 24(2), 109–28.

Else-Quest, N.M., Hyde, J.S. & Linn, M. (2010). Cross-national patterns of gender differences in mathematics: A meta-analysis. *Psychological Bulletin*, 136(1), 103–27.

Fennema, E. & Sherman, J. (1977). Sex-related differences in mathematics achievement, spatial visualization, and affective factors. *American Educational Research Journal*, 14, 51–71.

Forgasz, H. & Leder, G. (2001). A+ for girls, B for boys: Changing perspectives on gender equity and mathematics. In B. Atweh, H. Forgasz & B. Nebres (eds), *Sociocultural research on mathematics education: An international perspective* (347–66). Mahwah, NJ: Lawrence Erlbaum.

Forgasz, H. & Leder, G. (2017). Gender and VCE mathematics subject enrolments 2001–2015 in co-educational and single-sex schools. In A. Downton, S. Livy & J. Hall (eds), *40 years on: We are still learning! Proceedings of the 40th Annual Conference of the Mathematics Education Research Group of Australasia (MERGA)* (253–60). Melbourne: MERGA.

Forgasz, H., Leder, G. & Tan, H. (2014). Public views on the gendering of mathematics and related careers: International comparisons. *Educational Studies in Mathematics*, 87(3), 369–88.

Foster, C. (2013). *The essential guide to secondary mathematics: Successful and enjoyable teaching and learning.* London: Routledge.

Gallagher, A.M. & Kaufmann, J.C. (2005). *Gender differences in mathematics: An integrative psychological approach.* Cambridge: Cambridge University Press.

Gallagher, J. & Gallagher, S. (1994). *Teaching the gifted child* (4th edn). Boston, MA: Allyn & Bacon.

Gavin, M.K. & Sheffield, L.J. (2010). Using curriculum to develop mathematical promise in the middle grades. In M. Saul, S. Assouline & L.J. Sheffield (eds), *The peak in the middle.* Reston, VA: National Council of Teachers of Mathematics.

Guiso, L., Monte, F., Sapienza, P. & Zingales, L. (2008). Culture, gender, and math. *Science*, 320(5880), 1164–5.

Hall, J. (2012). Gender issues in mathematics: An Ontario perspective. *Journal of Teaching and Learning*, 8(1), 59–72.

Hannula, M.S. (2009). *The effect of achievement, gender and classroom context on upper secondary students' mathematical beliefs*. Paper presented at CREME Conference, Lyon, France. Retrieved from http://ife.ens-lyon.fr/publications/edition-electronique/cerme6/wg1-01-hannula.pdf

Hannula, M.S., Maijala, H., Pehkonen, E. & Nurmi, A. (2005). Gender comparisons of pupils' self-confidence in mathematics learning. *Nordic Studies in Mathematics Education*, 10(3–4), 29–42.

Hannula, M. & Malmivuori, M.L. (1997). Gender differences and their relation to mathematics classroom context. In E. Pehkonen (ed.), *Proceedings of the 21st Conference of the International Group for the Psychology of Mathematics Education (PME)* (vol. 3, pp. 33–40). Lahti: PME.

Helme, S. & Teese, R. (2011). *How inclusive is year 12 mathematics?* Paper presented at the 2011 MERGA Conference. Retrieved from http://www.merga.net.au/node/38?year=2011

House of Representatives Standing Committee on Education and Training (2002). *Boys: Getting it right: Report on the inquiry into the education of boys*. Retrieved from http://www.ph.gov/house/committee/edt/eofb/report/front.pdf

Huetinck, L. (1990). *Gender differences on science exams with respect to item type, format, student interest and experience*. Doctoral dissertation. Los Angeles, CA: UCLA.

Huetinck, L. & Munshin, S.N. (2008). *Teaching mathematics in the 21st century: Methods and activities for grades 6–12* (3rd edn). Upper Saddle River, NJ: Pearson Education.

Hyde, J., Ellis, A., Linn, M.C. & Williams, C. (2008). Gender similarities characterize math performance. *Science*, 321(5888), 494–5.

Janzen, J. (2008). Teaching English language learners in the content areas. *Review of Educational Research*, 78(4), 1010–38.

Kangasniemi, E. (1989). *Opetussuunnitelma ja matematiikan koulusaavutukset*. [Curriculum and mathematics achievement.] Kasvatustieteiden tutkimuslaitoksen julkaisusarja A 28. Finland: University of Jyväskylä.

Khisty, L.L. (1997). Making mathematics accessible to Latino students: Rethinking instructional practice. In J. Trentacosta & M.J. Kenney (eds), *Multicultural and gender equity in the mathematics classroom: The gift of diversity*, 1997 Yearbook. Reston, VA: National Council of Teachers of Mathematics.

Koontz, T. (1997). Know thyself, the evolution of an intervention gender-equity program. In J. Trentacosta & M.J. Kenney (eds), *Multicultural and gender equity in the mathematics classroom: The gift of diversity*, 1997 Yearbook. Reston, VA: National Council of Teachers of Mathematics.

Malcolm, I., Haig, Y., Konigsberg, P., Rochecouste, J., Collard, G., Hill, A. & Cahill, R. (1999). *Towards more user-friendly education for speakers of Aboriginal English*. Perth: Edith Cowan University.

Marsh, C.J. (2004). *Key concepts for understanding curriculum*. Psychology Press.

Marsh, C., Clarke, M. & Pittaway, S. (2014). *Marsh's becoming a teacher* (6th edn). Frenchs Forest: Pearson.

Marshall, S. (1984). Sex differences in children's mathematics achievement: Solving, computations and story-problems. *Journal of Educational Psychology*, 76(2), 194–204.

Mau, T.S. & Leitze, A.R. (2001). Powerless gender or gender-less power: The promise of constructivism for females in the mathematics classroom. In J.E. Jacobs, J.R. Becker & G.F. Kilmer (eds), *Changing the faces of mathematics: Perspectives on gender* (37–41). Reston, VA: National Council of Teachers of Mathematics.

Miller, S.P. & Mercer, C.D. (1997). Educational aspects of mathematics disabilities. *Journal of Learning Disabilities*, 30(1), 47–56.

Morrow, C. & Morrow, J. (1995). Connecting women with mathematics. In P. Rogers & G. Kaiser (eds), *Equity in mathematics education: Influences of feminism and culture* (13–26). London: Falmer Press.

National Association for Gifted Children (2007). *What is giftedness?* Retrieved from http://www.nag.org

Nosek, B.A., Banaji, M.R. & Greenwald, A.G. (2002). Harvesting implicit group attitudes and beliefs from a demonstration web site. *Group Dynamics: Theory, Research, and Practice*, 6(1), 101–15. https://doi.org/10.1037/1089-2699.6.1.101

NRICH (2012). *Clever Carl.* Retrieved from https://nrich.maths.org/2478

Parkin, B. & Hayes, J. (2006). Scaffolding the language of maths. *Literacy Learning: The Middle Years*, 14(1), 23–35.

Perry, B., Anthony, G. & Diezmann, C. (2004). *Research in mathematics education in Australasia 2000–2003.* Flaxton, QLD.

Perry, J. & Howard, P. (2008). Mathematics in Indigenous contexts. *Australian Primary Mathematics Classroom.* 13(4), 4–9.

Riegle-Crumb, C. (2006). The path through math: Course-taking trajectories and student performance at the intersection of gender and race/ethnicity. *American Journal of Education*, 113(1), 101–22.

Rotigel, J. & Fello, S. (2005). Mathematically gifted students: How can we meet their needs? *Gifted Child Today*, 27(4), 46–65.

Saranen, E. (1992). *Lukion yleisen oppimäärän opiskelijoiden matematiikan taidot ja käsitykset matematiikasta.* [Upper secondary school general course mathematics students' skills in and conceptions about mathematics]. Kasvatustieteiden tutkimuslaitoksen julkaisusarja A, Tutkimuksia; 38. Jyväskylä: Kasvatusteiteen tutkimuslaitos.

Sheffield, L.J. (1997). From Doogie Howser to dweebs – or how we went in search of Bobby Fischer and found that we are dumb and dumber. *Mathematics Teaching in the Middle School*, 2(6), 376–9.

Sheffield, L.J. (1999). *Developing mathematically promising students.* Reston, VA: National Council of Teachers of Mathematics.

Siemon, D., Beswick, K., Brady, K., Clark, J., Faragher, R. & Warren, E. (2011). *Teaching mathematics: Foundations to middle years.* South Melbourne: Oxford University Press.

Soro, R. (2002). Teachers' beliefs about gender differences in mathematics: 'Girls or boys?' scale. In A.D. Cockburn & E. Nardi (eds), *Proceedings of the 26th Conference of the International Group for the Psychology of Mathematics Education (PME)* (vol. 4, 225–32). Norwich: PME.

Sousa, D.A. (2008). *How the brain learns mathematics.* Thousand Oaks, CA: Sage Publications.

Spelke, E.S. (2005). Sex differences in intrinsic aptitude for mathematics and science? A critical review. *American Psychologist*, 60, 950–8.

Spencer, S.J., Steele, C.M. & Quinn, D.M. (1999). Stereotype threat and women's math performance. *Journal of Experimental Social Psychology*, 35, 4–28.

Stevens, T., Wang, K., Olivarez, A., Jr. & Hamman, D. (2007). Use of self-perceptions and their sources to predict the mathematics enrolment intentions of girls and boys. *Sex Roles*, 56(3), 51–63.

Tannock, R. & Martinussen, R. (2001). Reconceptualizing ADHD. *Educational Leadership*, 59, 20. Retrieved from http://www.ascd.org/publications/educational-leadership/nov01/vol59/num03/Reconceptualizing–ADHD.aspx

Teacher Education Ministerial Advisory Group (TEMAG) (2014). *Action now: Classroom ready teachers.* Retrieved from http://www.studentsfirst.gov.au/teacher-education-ministerial-advisory-group

Teaching English to Speakers of Other Languages (2008). *Position statement on teacher preparation for content based instruction (CBI).* Alexandria, VA: Author.

Thomson, S. (2014). *Gender and mathematics: Quality and equity.* Retrieved from http://research.acer.edu.au/cgi/viewcontent.cgiarticle=1226&context=research_conference

Thomson, S., De Bortoli, L., Underwood, C. & Schmid, M. (2019). *PISA 2018: Reporting Australia's results. Volume I student performance.* Melbourne: Australian Council for Educational Research.

Tortolani, M. (2007). *Presentation at the 2007 National Association of Multicultural Engineering Program Advocates National Conference.* Baltimore, MD.

Treacy, K. & Frid, S. (2008). *Recognising different starting points in Aboriginal students' learning of number. Proceedings of the 31st Annual Conference of Mathematics Education Research Group of Australasia (MERGA)* (vol. 2, 531–9). MERGA.

Vale, C. & Bartholomew, H. (2008). Gender and mathematics: Theoretical frameworks and findings. In H. Forgasz, A. Barkatssas, A. Bishop, B. Clarke, S. Keast, W.T. Seah & P. Sullivan (eds), *Research in mathematics education in Australasia 2004–2007* (271–90). Rotterdam: Sense Publishers.

Van de Walle, J.A., Karp, K.S. & Bay-Williams, J.M. (2014). *Elementary and middle school mathematics: Teaching developmentally* (8th international edn). Essex: Pearson Education.

Van Tassel-Baska, J. & Brown, E.F. (2007). Toward best practice: An analysis of the efficacy of curriculum models in gifted education. *Gifted Child Quarterly*, 51(4), 342–58.

Wai, J., Cacchio, M., Putallaz, M.C. & Mackel, M.C. (2010). Sex differences in the right tail of cognitive abilities: A 20-year examination. *Intelligence*, 38(4), 412–23.

Warren, E. & DeVries, E. (2009). Young Australian Indigenous students' engagement with numeracy: Actions that assist to bridge the gap. *Australian Journal of Education*, 53(2), 159–75.

Aboriginal and Torres Strait Islander learners and mathematics

Judy Anderson

LEARNING OBJECTIVES

The learning objectives of this chapter are directly linked to the Australian Professional Standards for Teachers (APST) (Australian Institute for Teaching and School Leadership [AITSL], 2011). After studying this chapter, you should be able to:

- define cultural competence and outline the features of culturally responsive teaching (APST 1.4, 3.7)
- describe teaching approaches which support Aboriginal and Torres Strait Islander students as learners of mathematics (APST 2.4, 4.1)
- present the outcomes of at least one successful mathematics program designed to support Aboriginal and Torres Strait Islander students' learning (APST 1.4, 2.4, 3.6)
- identify strategies mathematics teachers could use to address the Australian Curriculum cross-curriculum priority of Aboriginal and Torres Strait Islander Histories and Cultures (APST 2.3, 2.4).

Introduction

I would like to respectfully begin by acknowledging the ancestors of the Country in which I work and live, the Gadigal people of the Eora nation. I pay my respects to Elders past, present and emerging and acknowledge the many Aboriginal colleagues I have worked with, and learned from.

As mathematics teachers, our attitudes, values and beliefs determine how we teach (Gay, 2009). As culturally responsive teachers, we need to consider our attitudes and values if we are to address some commonly held beliefs about mathematics. For example, mathematics holds a position of status, power and privilege and is frequently referred to as a 'gatekeeper' to further study. As such, mathematics can be considered challenging and learnable by only the 'brightest' (Matthews, 2015). In addition, mathematics can be perceived as depersonalised and hence lacking in relevance (Morris & Matthews, 2011) and the technical language of mathematics can become a barrier for learners (Gay, 2009). These beliefs can impact students' engagement and should be addressed if we are to support the mathematics learning of *all* students and meet the goals for *all* Australian students.

As teachers in the Australian context, we need to learn about and understand the rich and diverse Aboriginal and Torres Strait Islander histories and cultures. As noted by Mousley and Matthews (2019):

> ... at least 60,000 years ago – Australian Aboriginal peoples had developed a sophisticated culture with, for example, its own versions of education, medicine, astronomy, agriculture, and other fields. Australian Aboriginal peoples are the world's oldest continuing cultures ... [They used] complex concepts of measurement, space, time and direction [that] were all based on extensive mathematical thinking that was different from the 'western' thinking of colonists. (pp. 113–14)

Country: often used by Aboriginal peoples and Torres Strait Islander peoples to describe family origins and associations with particular parts of Australia. Relationships to Country are complex and interrelated. Within traditional Aboriginal and Torres Strait Islander societies, each language group has a defined area of land or country that each group is connected to, both geographically and spiritually (Queensland Curriculum and Assessment Authority [QCAA], 2008, pp. 1–2).

Identity: who you are and what defines you.

We need to understand Aboriginal and Torres Strait Islander peoples' sense of belonging to family, community and Country as a part of their culture and identity. For non-Indigenous people, understanding this culture can be difficult but we must make that effort if we are to support all learners in all our classrooms. As noted by the Honourable Linda Burney, a member of the Australian national parliament, 'being Aboriginal has nothing to do with the colour of your skin or the shape of your nose. It is a spiritual feeling, an identity you know in your heart. It is a unique feeling that may be difficult for non-Aboriginal people to understand' (cited in Bamblett, 2005, p. 20).

While non-Aboriginal teachers may not understand the nature of this 'spiritual feeling', we must make the effort to understand how Aboriginal and Torres Strait Islander students learn and how they use mathematics to interpret their world. We must develop pedagogies which are inclusive, which are respectful and which are culturally responsive (Mukhopadhyay, Powell & Frankenstein, 2009). We must recognise the importance of community, and endeavour to design lessons which are relevant and engage all learners.

We should recognise, acknowledge and teach our students about the contribution Aboriginal and Torres Strait Islander peoples have made to our country, for example, the inventor David Unaipon.

David Unaipon

CONNECTION

David Unaipon (born David Ngunaitponi, 1872–1967) of the Ngarrindjeri people was the son of a preacher – he became a preacher, an inventor, an author and a political activist. He invented the modern shears that revolutionised the Australian sheep shearing industry. Likened to Leonardo da Vinci, Unaipon described the principle of the helicopter based on the aerodynamics of the boomerang, he predicted the development of lasers, and took out patents for a centrifugal motor, a multi-radial wheel and a mechanical propulsion device. He appeared before royal commissions into Aboriginal welfare and was awarded a Coronation medal in 1953. He lacked recognition for his written work throughout his lifetime with some of his writings plagiarised. He is currently commemorated on the Australian $50 note.

Acknowledging the importance of preparing to teach Aboriginal and Torres Strait Islander students, the APST (AITSL, 2011) make specific reference to these students in two of the standards for graduate teachers:

> 1.4. Demonstrate broad knowledge and understanding of the impact of culture, cultural identity and linguistic background on the education of students from Aboriginal and Torres Strait Islander backgrounds.
>
> 2.4. Demonstrate broad knowledge of, understanding of, and respect for Aboriginal and Torres Strait Islander histories, cultures and languages.

In addition to acknowledging First Nations peoples in the APST, the Australian Curriculum includes Aboriginal and Torres Strait Islander Histories and Cultures as one of three cross-curriculum priorities (see Chapter 9 for further detail).

Teachers need to understand and know their students, their backgrounds and their learning needs in every class they teach. By doing so, they will be able to develop respectful relationships, and avoid stereotyping students and limiting students' learning opportunities based on inappropriate judgements.

The Alice Springs (Mparntwe) Education Declaration (Council of Australian Governments Education Council, 2019) presents two distinct but interconnected goals for young Australians:

> Goal 1: The Australian education system promotes excellence and equity
>
> Goal 2: All young Australians become:
> - confident and creative individuals
> - successful lifelong learners
> - active and informed members of the community. (p. 4)

This latest Declaration by the Australian state and territory governments sets a clear agenda that every student deserves a quality education.

According to the Australian Bureau of Statistics (2018), Aboriginal and Torres Strait Islander students comprised 5.6 per cent of all student enrolments in Australian schools in 2017. These students are present in many of our classrooms from large metropolitan locations to rural, regional and remote schools. As First Nations peoples, their history and culture provides a rich history of mathematical understandings. This chapter builds on Chapter 6, beginning with the notions of cultural competence and culturally responsive teachers and pedagogies. Knowing and understanding our students' backgrounds and learning needs is promoted as well as approaches to support Aboriginal and Torres Strait Islander students' mathematics learning. Examples of successful programs designed to support Indigenous students' mathematics learning are presented, highlighting key features of success including the role of storytelling. Finally, the Australian Curriculum cross-curriculum priority of Aboriginal and Torres Strait Islander Histories and Cultures is discussed in relation to mathematics teaching and learning.

Following the advice of Perso and Hayward (2015), I wish to clarify my use of terminology in this chapter. As they suggest, 'the use of terminology is complex' (p. xv). The term 'Indigenous' is frequently used to include all people who are Aboriginal and/or Torres Strait Islander whereas the terms 'Aboriginal' and 'Torres Strait Islander' refer to particular groups of First Nations people. While acknowledging that some Aboriginal and Torres Strait Islander people do not like the term 'Indigenous', I use it in this chapter to aid readability.

Culturally responsive teaching and cultural competence

Culturally responsive teaching: enacted cultural competence. As a personal capability, cultural responsiveness involves 'respect for cultural differences, empathy in being about to see the world from the other's perspective, and perception of yourself as a humble learner rather than an all-knowing teacher' (Perso & Hayward, 2015, p. xxvi).

Culturally responsive teaching involves designing learning based on awareness of difference, respect for students, as well as listening to, and learning, from students (Mukhopadhyay, Powell & Frankenstein, 2009). Cultural competence 'involves the capacity to understand, interact and communicate effectively and sensitively with people from different cultural backgrounds' (Perso, 2012). Just as your students come to your class with different knowledge and understanding of mathematics, they also come to your class with different cultural backgrounds. It is important not to make assumptions about your students based on their culture, language, social and/or economic backgrounds as that can lead to incorrect stereotyping and inappropriate judgement-making. You need to be aware that some 'children are raised differently, taught differently, learn differently and see the world differently' (Perso & Hayward, 2015, p. xvii), but regardless of these differences, they are all entitled to a quality mathematics education.

The three critical dimensions of responsive mathematics pedagogy are cultural inclusion, academic inclusion and social inclusion (Morris et al., 2015). An awareness of cultural difference allows teachers to adopt culturally responsive pedagogical practices including:

- wanting to know more about your students, including their backgrounds
- respecting your students
- nurturing caring relationships and wanting the best for your students
- designing lessons that maximise their learning
- designing quality assessment tasks that give your students opportunities to demonstrate the full extent of their learning (Perso & Hayward, 2015, p. xv).

Culturally responsive teachers also understand that diversity exists within cultural groups and the importance of having some knowledge of local communities within which they work. Understanding how to teach Aboriginal and Torres Strait Islander students in a large metropolitan school may not be the same as teaching students in a remote region in the Northern Territory. Teaching in a different context will require learning about the students in that new location, understanding the local community and appreciating local cultural beliefs and practices.

When learning about your students, it is critical to acknowledge individual strengths, including knowledge, skills and values – this is referred to as a 'strengths-based' approach rather than focusing on what students do not know, which is a 'deficit' approach. To build relationships and respect, you need to begin with what your students know and understand, and the skills they bring to the classroom. It is also important not to have lower expectations of some students because they belong to a particular cultural group – this has been referred to as 'deficit thinking' (Burton, 2017). Sarra (2011) and Pearson (2014) warn against labelling all Indigenous students 'underachievers' and having low expectations – views reinforced by funding from government agencies under the umbrella of 'closing the gap' initiatives.

Students can tell when teachers have low expectations and give them 'busy work' in lessons, typically with an overreliance on worksheets. Rather than motivate and engage, this approach of low-level, repetitive tasks can demotivate students. It is possible to scaffold learning if students require further support without 'dumbing down' the curriculum. See Chapters 1 and 5 for suggested approaches to scaffolding learning. A better approach to motivate and engage students is to acknowledge culture and its impact on learning, to provide appropriate contexts for learning mathematics, and to empower students to use mathematics to critically examine their world and the inequities which might confront them. An approach which provides opportunities for problem solving and critical and creative thinking by connecting mathematics and culture is referred to as ethnomathematics.

> **Ethnomathematics:** a term used to express the relationship between mathematics and culture. Ethnomathematics characterises mathematics as a human activity as opposed to the typical school mathematics curriculum which does not connect with people's lives.

How do I use Aboriginal kinship in my lessons?

SCENARIO

With academic colleagues, community members and educators, Dr Lynette Riley, from the University of Sydney, has designed an online kinship program that involved developing a teaching framework for cultural competence. The online module is a free-access resource designed to promote cross-cultural understanding by explaining the intricacies of the Aboriginal Kinship system.

The online 'Kinship Module' can be found on the University of Sydney website.

The module contains eight themes:

1. Welcome and acknowledgements
2. Nations, clans and family groups
3. First level of Kinship – Moeity
4. Second level of Kinship – Totems
5. Third level of Kinship – Skin Names
6. Language and traditional affiliations
7. Lines of communication
8. Disconnected themes.

> **Kinship:** 'a feature of Aboriginal social organisation and family relationships across Central Australia' (Central Land Council, n.d.). According to the Central Land Council, kinship 'is a complex system that determines how people relate to each other and their roles, responsibilities and obligations in relation to one another, ceremonial business and land' (Central Land Council, n.d.).

Questions

1. By listening to the Community Narratives, what did you learn about Aboriginal people's experiences of education?
2. How could you use knowledge of Aboriginal Kinship in mathematics lessons on patterns and relationships?

An ethnomathematical perspective on culturally responsive mathematics education

D'Ambrosio (2001, p. 309) stated 'we can help students realise their full mathematical potential by acknowledging the importance of culture to the identity of the child and how culture affects how children think and learn'. This suggests we can support students' mathematics learning if we learn about the connections between mathematics and Aboriginal and Torres Strait Islander cultures and use this to inform our teaching. See, for example, the guide published by NarraGunnaWali (n.d.) which includes information about Aboriginal and Torres Strait Islander mathematics, noting it is intricately linked to many other subject areas and that by delineating certain cultural practices using the European term of 'mathematics' we lose some of the intricacy and complexity of their culture. Further, mathematics is 'very much a social construct, involving abstracted natural patterns and symbolic expressions of relationships. It is a cultural practice, and the symbols and language of mathematics itself are cultural products' (p. 2). Further information about the mathematics used by Aboriginal and Torres Strait Islander peoples can be found in Meaney and Evans (2013).

In addition, further concerns are offered by Nakata (2003) about repositioning the notion of what constitutes mathematics for Aboriginal and Torres Strait Islander students by focusing on studying mathematics that is embedded in their culture (i.e. using an ethnographic approach). He suggests there are instances 'where the infiltration of anthropological schemes into Indigenous educational practices have been counterproductive, giving the anthropological discourse primacy over the educational context' (p. 9). Such concerns remind us of the complexity of embedding cultural perspectives into what is predominantly a Western education system with Western approaches to teaching, learning and assessment.

However, the link between culture and mathematics has been achieved in other contexts with other cultural groups. Gutstein and Peterson (2013, p. 3) suggest that rethinking mathematics means 'using culturally relevant practices and experiences of students and their communities' and that 'teachers should view students' home cultures and languages as strengths upon which to build' knowledge and understanding. The teaching ideas provided in their edited volume demonstrate how mathematics can be used to help students more clearly understand their lives and to see mathematics 'as a tool to help make the world more equal and just' (p. 1).

To address a potential tension between the embedding of cultural or community knowledge into the school mathematics curriculum, which is largely described as Western knowledge, Gutstein (2006) advocates a balance between three forms of mathematical practice – classical mathematics knowledge, community knowledge and critical knowledge – suggesting all three are important for mathematics learners. Warren and Miller (2016) also recommend a balanced perspective for school mathematics learning to recognise 'different ways of knowing' (p. 4). In marginalised contexts, teacher assistants with high levels of cultural knowledge have been employed to work alongside mathematics teachers to ensure both ways are embedded in lessons (Warren & Miller, 2016).

Some of the challenges or issues for the mathematical investigations in Gutstein and Peterson (2013) address all three of Gutstein's (2006) mathematical practices – some examples include:

- disparities in wealth distribution
- examining unemployment rates and how they are determined
- environmental hazards
- investigating overcrowding in schools and local communities.

These investigations were designed as appropriate for the students' cultural backgrounds and relevant to their communities in marginalised contexts in the United States.

Bringing a social justice perspective to your classroom means you are committed to equitable outcomes for all students. As Bell (2007, p. 3) puts it:

> Social justice is both a process and a goal. The goal of social justice is full and equal participation of all groups in a society that is mutually shaped to meet their needs. Social justice includes a vision of society in which the distribution of resources is equitable and all members are physically and psychologically safe and secure . . .

Social justice: the equal access to wealth, opportunities and privileges within a society. It is both a process and a goal.

PAUSE AND THINK

Does equity mean we should treat all mathematics learners the same? Why or why not?

As noted by Ewing (2017, p. 788), 'schools and teachers play a critical role in providing a climate of learning and effective practices that encourage engagement and active student participation'. She draws on social theory to explore what constitutes participation, engagement and community in mathematics classrooms and highlights the importance of addressing issues of access as well as inclusion and exclusion. A culturally responsive approach to teaching supports a socially just orientation which has the potential to be inclusive and to develop trust and caring relationships with students (Buckskin, 2015). Learning about your Aboriginal and Torres Strait Islander students' cultures and backgrounds so that you can connect and communicate more effectively, will support the development of cultural competence. Connecting culture and mathematics is a key component of effective mathematics teaching in multicultural classrooms.

SHORT-ANSWER QUESTIONS

1. List some culturally responsive teaching strategies.
2. Why is cultural competence so important when working with Aboriginal and Torres Strait Islander students?
3. How could you connect the curriculum to the ways Australian Indigenous peoples used mathematics?

Aboriginal and Torres Strait Islander students as mathematics learners

The previous section described the need to have high expectations of learners and to provide students with opportunities to problem solve and develop skills in critical and creative thinking rather than busy work. It is essential to set the 'tone' of the classroom and encourage students to do their best with support from you through scaffolding when necessary (Perso & Hayward, 2015). This section of the chapter, Aboriginal and Torres Strait Islander students as mathematics learners, begins with a description of the eight ways of learning before presenting further research into appropriate pedagogical strategies for learning mathematics.

Eight Ways of Learning for Indigenous students

There are strong links between culture and how people learn. In his PhD, Yunkaporta (2009) conducted research into the ways Aboriginal people in Western NSW used patterns in stories, songs and kinship to develop a spirit of learning. He created a framework for 'eight ways of learning' and represented each with a symbol, as in Figure 7.1.

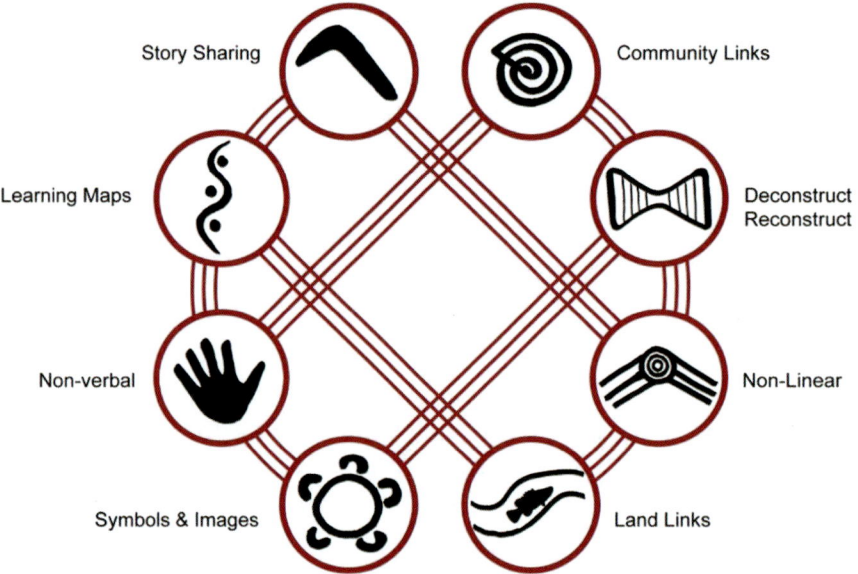

Figure 7.1 Eight Ways of Aboriginal Learning
Source: Yunkaporta, 2009, p. 46.

In presenting this pedagogical framework, Yunkaporta (2009, p. 50) states:

> The main contentions here have been that culture impacts on optimal pedagogy for all learners, that explicit Aboriginal pedagogy is needed to improve outcomes for Indigenous learners, that there is common ground between Aboriginal pedagogies and the optimal pedagogies for all learners, that the eight ways of learning developed

in my research are supported in the international literature, that such a practical framework is needed for teachers to be able to organise and access this knowledge in cultural safety, and finally that a reconciling interface approach is needed to harmonise the relationship between the two pedagogical systems.

The common ground between 'mainstream' pedagogies and Aboriginal pedagogies can be summarised as follows:

- learning through narrative
- planning and visualising explicit processes
- working non-verbally with self-reflective, hands-on methods
- learning through images, symbols and metaphors
- learning through place-responsive, environmental practice
- using indirect, innovative and interdisciplinary approaches
- modelling and scaffolding by working from wholes to parts
- connecting learning to local values, needs and knowledge (Yunkaporta, 2009).

Each of these pedagogical approaches can be used to support the learning of mathematics and can be used to help design effective learning experiences. An example is provided on the '8 Ways Maths' website, which states 'working with Fibonacci sequences and spirals is a good example of how to teach basic maths using Land Links, Symbols and Images, etc.' (8 Ways, n.d.). However, if the Aboriginal content is tokenistic, separated from core content, or treated as a 'fun' activity, this will marginalise Aboriginal learners even further (Yunkaporta, 2009).

PAUSE AND THINK

Consider a lesson you might have to plan to help students understand different ways to represent data. How could you embed some of the '8 ways of learning' into the lesson?

Strategies to embed mathematics in cultural practices

Maths as Storytelling

This section of the chapter is informed by the work of applied mathematician Professor Chris Matthews. In much of his writing Matthews described his school experiences as alienating and developing a feeling of 'not belonging' (Matthews, 2009; Morris & Matthews, 2011; Matthews, 2015). He was successful in school mathematics and computing because 'in the mathematics classroom you do not have to deal with notions of race, and behind a computer you do not have to deal with people' (2015, p. 103). Unlike many other learners, Matthews (2009) described learning algebra as the 'turning point' in his mathematics education when he started to understand and find mathematics easy. While he did not have an enjoyable experience at school, he credited his family and friends for their positive influence and for developing in him a strong Aboriginal identity. I encourage you to read his story and reflect on his schooling experiences as a learner of mathematics.

CONNECTION **Chris Matthews**

Professor Chris Matthews is from Quandamooka people of Minjerribah (Stradbroke Island) in Queensland. He is a mathematician, was awarded a PhD in applied mathematics from Griffith University in 2003 and is currently Associate Dean (Indigenous Leadership and Engagement) in the Science Faculty at the University of Technology, Sydney (UTS). As part of this role, Chris is leading a team of academics at UTS to transform the science curriculum to meet the Indigenous Graduate Attribute and develop a community of Indigenous STEM professionals. Throughout his career, Chris has developed a deep interest in mathematics education for Aboriginal and Torres Strait Islander learners and exploring the connections between mathematics and Indigenous knowledge. He is a curriculum advisor to the Australian Curriculum, Assessment and Reporting Authority (ACARA), played a key role in the Make It Count project, and is Chair of the Aboriginal and Torres Strait Islander Mathematics Alliance (ACEMS, 2020).

Drawing on his experiences of learning mathematics, Matthews is passionate about 'allowing students to connect with mathematics through their own social and cultural background' (p. 103). Through his reflections on Praeger's (2008) description of mathematics as having three interconnected essential elements – power, truth and beauty – Matthews (2015) describes a dichotomy of mathematics as objective to enable rigour, precision and logic when solving problems, but also as subjective since it is 'beautiful' and a 'social construct'. He uses a 'Cloud Model' to represent his theory of mathematics knowledge (epistemology) showing 'a cycle in which the development of mathematics starts with the reality of the observer' (p. 107) (see Figure 7.2).

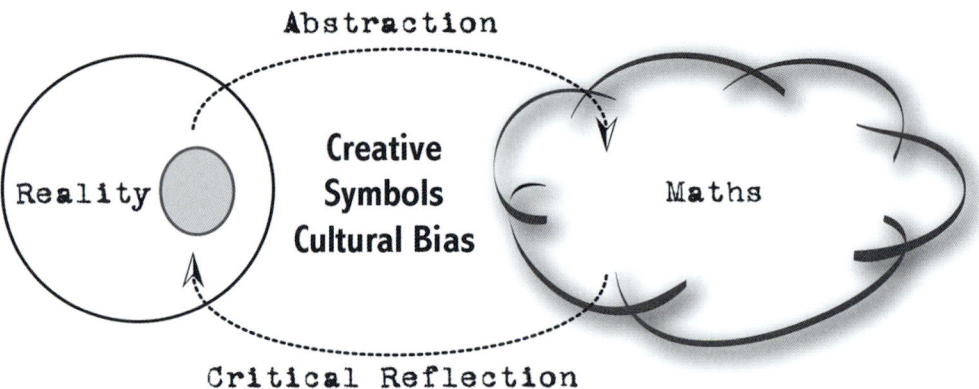

Figure 7.2 The Cloud Model – Matthews' theory of mathematics knowledge
Source: Matthews, 2009, p. 47.

Matthews (2015) argues that school students experience the abstract nature of mathematics without experiencing opportunities for critical reflection. He notes the model has three main interconnecting ideas that inform the processes of abstraction and critical reflection – creativity, symbols and cultural bias. Further, he argues students should be able to express themselves through mathematics (creativity), that the symbolic language of mathematics has a strong connection to oral language which will vary depending on the context of the student (symbols), and understanding mathematics provides opportunities 'to explore the connection between mathematics and people' (culture) (p. 109). Matthews (2009, p. 47) recommends students can develop mathematical understanding by the process of:

- choosing a part of reality (represented by the grey circle)
- creating an abstract representation of that real-life situation, using a range of mathematical symbols which are combined to form a symbolic language we call mathematics
- using mathematics in its abstract form to explore the attributes and behaviours of the real-life situation and mathematically communicating these ideas to others
- critically reflecting on their mathematical representation to ensure that it fits with the observed reality.

Consequently, the abstraction and critical reflection processes form an important cycle in which mathematics and its knowledge are created, developed and refined.

Maths as Storytelling is a pedagogical approach that originated from the model in Figure 7.2 and has been used to teach students pre-algebra using connections to arithmetic as well as more complex algebra – it focuses on stories and explores how symbols and their meanings can communicate these stories. The six steps of the Maths as Storytelling process (Matthews, 2015) are:

1. understanding symbols – students explore the use of symbols in constructing abstract representations
2. a simple maths story – students act out a story which might involve forming groups and connecting groups
3. students' representation (unstructured) – on blank paper, students draw the story using their own symbolism
4. students' representation (structured) – with teacher guidance, concrete materials and then symbols are used to represent the story
5. sharing symbol systems – by choosing a student's or the teacher's symbols, new stories are explored and represented
6. modifying the story – by removing some materials from step 4, the students realise the story no longer makes sense and so there is a need for finding a more general way of representing maths stories (abstraction).

Maths as Dance: an extension of Maths as Storytelling

During the Make It Count program, which is described in the next section of the chapter, the Maths as Dance evolved from a mathematics camp within an Indigenous community aimed at bringing members of the local Indigenous community and the school community together. In contrast to Maths as Storytelling, Maths as Dance starts with the abstract idea and progresses through five steps (Matthews, 2015):

1. Starting in the abstract – in groups of four students, each group was given a different equation such as $4 \times 3 = 12$, and was asked to represent this equation by creating a story and performing a dance.
2. Symbolism – the symbols in the equation are then explained as subjects and actions which are separating $(-, \div)$ or joining $(+, \times)$.

3. Students create a new story – students use the constructs in step 2 to create a new story using symbols.
4. Students create a dance for their story – at the camp, students were supported by a local Song Man from the region who demonstrated how 'various movements and sound can create meaning and provoke feelings and emotions pivotal to the story' (p. 116).
5. Students perform their dance and discuss how their story is represented.

At the mathematics camp, Matthews (2015) described how the students' story contexts were very different to those typically arising in mathematics lessons, including hunting kangaroos, gathering food, and raindrops falling from a cloud (p. 116). Student surveys (Morris & Matthews, 2011) revealed new beliefs about learning mathematics and connecting with culture. Matthews noted the importance of connecting with the community, bringing the local community and the school community together, and the important role of the Indigenous Education Worker to make the connections and support such initiatives.

Further factors supporting Indigenous students as mathematics learners

Like the Eight ways of Learning, Matthews (2015) suggests Indigenous students learn best through contextualised, hands-on tasks; learn best through visual and spatial information rather than verbal; and learn by observation and non-verbal communication. Several studies have revealed improvements in Indigenous students' learning outcomes if programs:

- reinforce pride in culture
- encourage attendance
- highlight potential for success in mathematics
- challenge
- provide overviews of topics
- have Indigenous leadership.

It has been noticed that when Aboriginal students progressed through their schooling years, there was a steady increase in preference for cooperative learning and a steady decline in preference for both competitive and individualistic learning situations. Catering for the needs of all students, including Indigenous learners, requires strong leadership and a culturally responsive school approach that values the learning needs of all students, sets high expectations and supports teachers to become culturally competent.

SHORT-ANSWER QUESTIONS

1. Explain how the '8 Ways of Learning' informs the embedding of Aboriginal and Torres Strait Islanders' learning approaches into lessons.
2. How does Matthews embed cultural practices into learning mathematics?
3. What further factors are critical to supporting the learning of Aboriginal and Torres Strait Islander students?

Practices of schools in remote locations with successful numeracy outcomes

Research investigating 'what builds success in remote numeracy' identified a range of practices that were grouped into three levels – enacted, enabled and envisioned (Jorgensen, 2018). More than 30 remote schools participated in this study from several states and territories across Australia – these schools typically had enrolments comprising at least 80 per cent Indigenous students. These schools were judged successful by NAPLAN results over several years when compared to 'like' schools. Further analyses of the identified practices led to the development of sets of norms which provided an overarching model describing the principles that underpin each level of practice as follows.

Enacted mathematical norms:
- All students can learn mathematics to high levels
- Embedding mathematics is critical for understanding – embedding in the brain as well as embedding in contexts
- Mathematics is as much about language as it is mathematical concepts
- Transparency in learning and teaching mathematics enables students to access the 'secret knowledge' of school mathematics
- Mathematics lessons should engage learners at their levels of understanding, and then extend learning into new levels (p. 63)

Enabled mathematical norms:
- Teacher quality is essential for quality learning
- Recruitment, development, retention of staff is critical
- Teacher support is integral to developing quality mathematical environments
- A key person for mathematical support across the school enables quality teaching and environments – numeracy coach
- Indigenous people are a key resource in teaching and the classroom (p. 64)

Envisaged mathematical norms:
- Leadership is critical for developing a positive mathematics culture, supporting teachers and supporting community
- Establishing a whole-school approach to teaching mathematics ensures consistency and transparency – for students, teachers and community (p. 65)

Mathematics programs designed to support the learning of Aboriginal and Torres Strait Islander students

Several programs have been developed to support Aboriginal and Torres Strait Islander students as learners of mathematics. This section describes three successful programs that have been used in secondary schools and presents some of the key factors which led to their success. The selection of these three programs was based on the authors' familiarity with them and access to resources through their websites. While there are differences between the programs, the similarities suggest there is a set of constructs or key factors which increase the potential for success.

Mathematics in Indigenous Contexts, 1999–2005

The Mathematics in Indigenous Contexts program involved the three key elements of collaborative engagement of Aboriginal parents in curriculum development, the contextualisation of mathematics learning and the professional development of teachers (Howard et al., 2005, p. 2). The NSW Board of Studies (now NSW Education Studies Authority), in collaboration with the NSW Department of Education and Training, implemented the program to support the development and trialling of contextual mathematics units in classrooms from Kindergarten to Year 8. Staff from schools in Western Sydney and in Western NSW regional locations with significant enrolments of Aboriginal students worked in learning teams of teachers, Aboriginal educators and local Aboriginal community members to develop the units of work.

Contextualistion: involves relating to a context. Mathematical ideas and concepts can be more meaningful to students if they are presented in a familiar context. However, contexts can be culturally specific so care must be taken to ensure the students are familiar with a particular context.

The program was 'based upon the principle that the mutually beneficial engagement of people and cultures is essential in building a community's capacity for educating Aboriginal students' (Perry & Howard, 2008, p. 4). Building community capacity involves addressing key challenges, including the need to develop:

- mutual respect between the Aboriginal community and the school community;
- mutual engagement with the community in developing learning approaches based upon alternative and creative discourses;
- evidence-based discourses to inform one's understanding of learners and learning;
- home-school-community alignment for enhancing student learning; and
- personal and collective efficacy for community engagement. (Howard & Perry, 2007, p. 405)

With support from academic mentors, the learning teams in each school developed tasks that investigated local Aboriginal histories and cultures in the community with input from Elders and other community members. For example, in one small rural NSW town, the upper primary and lower secondary students completed a range of mathematics tasks involving mapping changes in land use near the school. Students visited local sites, and heard stories about growing up in the area, how families lived and supported themselves, and the impact Aboriginal peoples had on the land. This is a good example of place-based education where 'respect for local knowledge, and development of social competencies, a strong sense of identity, and a sense of responsibility to the local community are seen as crucial to effective learning' (Burgess et al., 2019, p. 310).

During the contextualised unit of work, teachers noted heightened engagement and enthusiasm by the Aboriginal students and a new appreciation and understanding of the importance of the area by non-Aboriginal students. The context enriched the learning and provided a purpose for the mathematics tasks undertaken by the students. Key features for successful implementation of the program were identified as context, engagement and learning, sustainability, and activities and processes (Perry & Howard, 2008).

Make It Count, 2009–2012

The Make It Count program was instigated by the Australian Association of Mathematics Teachers and funded by the Australian Government as part of the Closing the Gap initiative. The aim of the program was to develop responsive pedagogies informed by research that improved Aboriginal and Torres Strait Islander students' mathematics outcomes. Eight clusters of schools across regional and urban Australia participated in the program (Morris & Matthews, 2011) and involved each cluster working with an academic mentor to develop responsive mathematics pedagogy (RMP) that valued what Indigenous learners brought to the classroom and built learning that was culturally relevant. The learning was focused on support through scaffolding, with explicit instruction when necessary, it was carefully sequenced, and additional tutoring was available. Students were challenged using investigations and problems that were contextualised, allowing for transfer of knowledge and building resilience.

Given the diversity of contexts involved in the program from large metropolitan areas to more remote locations, the focus for each cluster was developed from local needs but needed to be inclusive of cultural, social and academic needs of Aboriginal learners – the three key elements of RMP. One case study described by Thornton, Statton and Mountzouris (2012) noted that in their cluster mathematics tasks involved the 'processes of mathematisation, the use of mathematical models and representations of real world contexts, and contextualisation, [and that] the embedding of mathematical ideas into a meaningful context, [were] key aspects of students' mathematical learning' (p. 730). Through solving rich mathematics problems with an authentic purpose, the students worked in a Deadly Design studio to draw solutions to problems (a preferred learning approach). Teachers developed a more comprehensive approach to planning learning experiences that focused on contexts and connections with culture rather than beginning with the linear arrangement of mathematics concepts from the curriculum. Teachers noted the Aboriginal students' increased engagement and deeper mathematics learning.

> **Mathematisation:** to interpret a situation in a mathematical way. A process which is necessary to solve real-world problems using mathematics.

The Make It Count program enabled whole school communities to work together to improve mathematics pedagogies by building the capacity of school communities to maximise Aboriginal students' learning. Mathematics and numeracy learning were enhanced through teachers:

- developing high expectations of themselves and of their students
- learning more about what their students know, and using this to inform teaching and learning
- using knowledge of students' cultural and home backgrounds to engage students in learning mathematics at school
- giving students opportunities to make decisions about learning of mathematics
- actively fostering positive attitudes towards mathematics, and building Indigenous students' confidence in themselves as mathematicians (Australian Association of Mathematics Teachers, n.d.).

YuMi Deadly Maths

The YuMi Deadly Maths (YDM) program was developed by academics from the Queensland University of Technology (Sarra et al., 2016). The aim was to work with teachers in schools with high proportions of Aboriginal and Torres Strait Islander students which were located in lower socioeconomic status communities, to enhance mathematics learning outcomes and improve employment and life chances for students. The YuMi Deadly Centre (YDC) worked with schools and communities to enhance the capacity of teachers to teach mathematics using life-related, inquiry-based pedagogies that promoted:

- deep learning of powerful mathematics
- positive student engagement
- problem-solving skills, creativity and critical reflection
- symbolic language and structure
- overall disciplinary content knowledge.

Building on Matthews' model of mathematics (see Figure 7.2), the YDM pedagogical framework was developed to help teachers plan instructional sequences. The RAMR cycle has four components – Reality, Abstraction, Maths, Reflection (see Figure 7.3).

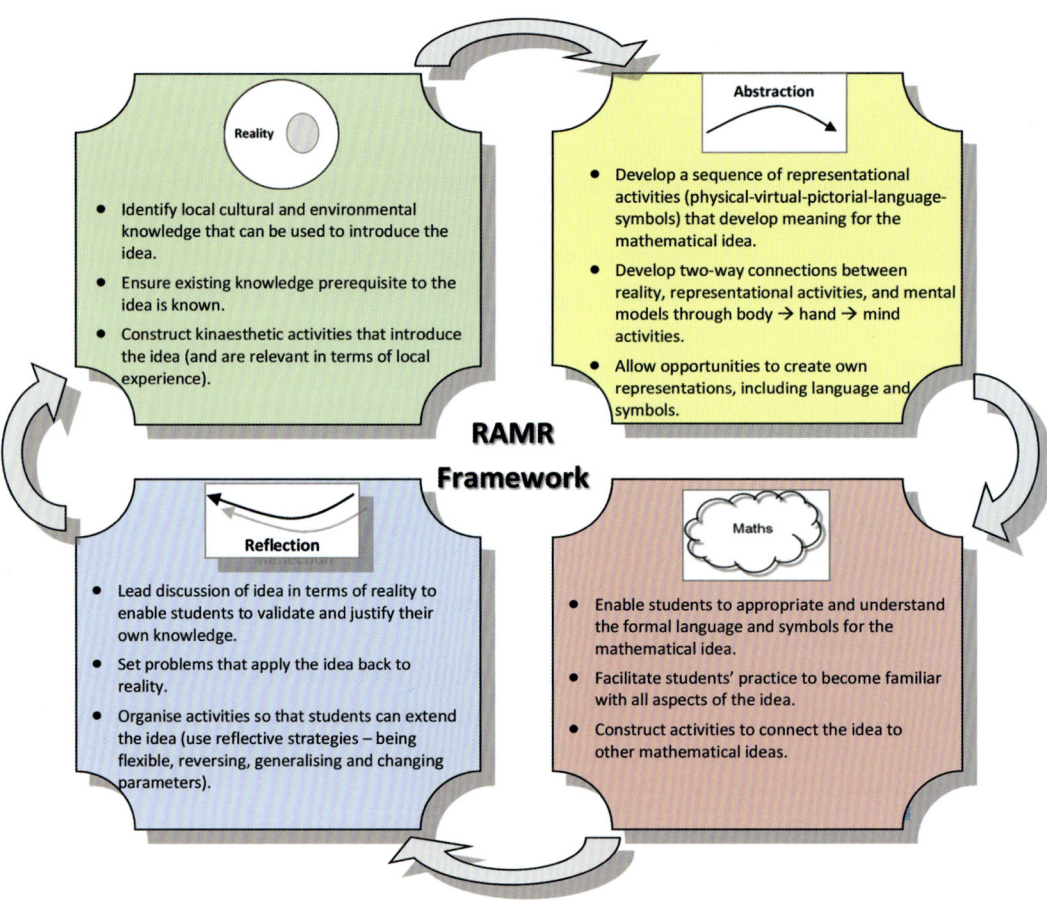

Figure 7.3 The RAMR cycle
Source: YuMi Deadly Centre, 2016, p. 32.

Before using the RAMR cycle, teachers needed to develop an extended plan for teaching as well as identify the central mathematics ideas to be taught – referred to as the 'big ideas' of mathematics (see Chalmers et al., 2017). Rather than the program being prescriptive, teachers were encouraged to use the framework to design lessons that met the needs of their students, schools and communities. As noted by Cooper and Carter (2016, p. 177):

> The RAMR framework begins and ends with the reality of the students' lives. It starts with something that interests the students, and then acts this out with kinaesthetic or whole body activities to build visual images or pictures in the mind of the mathematics idea(s). It then moves to consolidation, which involves making connections as well as practice, and finally reflects back to the students' reality.

A recent initiative of the YDC, the PRIME Futures program, involved 60 schools in 10 geographical locations across Australia to provide professional learning to teachers using the YDM approach. The RAMR framework was used by teachers to scaffold the development of lesson and unit plans. An important component of the program was engaging with the local community, who actively participated in the teacher workshops 'to model to schools how local knowledge should be made legitimate in the classroom and school' as well as 'challenge poor attendance and behaviour, integrate and legitimise local community knowledge, build in practice to support the culture of the students, and change teacher attitudes towards relationships with the students' (Sarra et al., 2016, p. 5).

Acknowledging the importance of school leadership, schools participating in YDC programs were encouraged to adopt change processes such as those developed by the Stronger Smarter Institute (Sarra, 2011). The Stronger Smarter Institute (2017) delivers a leadership program to influence personal, school and community spheres and support schools delivering a high-quality education for every Indigenous student. The Institute uses a framework developed by Sarra that includes four key underlying elements necessary for schools to be able to implement the Stronger Smarter approach:

- responsibility for change (professional accountability)
- taking a strengths-based approach
- embracing a positive Indigenous student identity (Strong and Smart)
- building 'high-expectations relationships'. (2017, p. 4)

School leaders are encouraged to use five interconnected metastrategies appropriate for their local context:

1. Acknowledging, embracing and developing a positive sense of identity in schools.
2. Acknowledging and embracing Indigenous leadership in schools and school communities.
3. 'High expectations' leadership to ensure 'high expectations' classrooms, with 'high expectations' teacher/student relationships.
4. Innovative and dynamic school models in complex social and cultural contexts.
5. Innovative and dynamic school staffing models in complex social and cultural contexts. (2017, p. 4)

High-expectations relationships: behaviours associated with 'understanding personal assumptions, creating spaces for dialogue, and engaging in challenging conversations' (Stronger Smarter Institute, 2017, p. 10).

CONNECTION   **What are high-expectations relationships?**

Sarra and colleagues (2018) state that 'in Australian Indigenous education, recommendations for "what works" for Indigenous student success start with high expectations, along with strong teacher-student relationships, quality teaching and positive cultural acknowledgement' (p. 32). With a focus on strengths-based education, they describe high-expectations relationships as behaviours associated with understanding personal assumptions, creating spaces for dialogue and engaging in challenging conversations. Further, they argue that 'for a high-expectations relationship to occur, it is critical that both socially just relating (fair) and critically reflective relating (firm) are present'. This means that when firm and fair work together, teachers will 'challenge behaviours but also seek to understand the circumstances that caused the incident and work with the child and the parents to discuss expectations and co-create constructive solutions' (p. 37).

Questions

1. Consider the situation presented on p. 38 in Sarra et al. (2018) where students are heard using racist language when talking about another student. How might you respond using a high-expectations relationship approach?

2. Select a situation you have observed during a recent school visit or professional placement and describe how it could have been handled differently using a high-expectations relationship approach.

Proposed constructs: providing successful mathematics education

The three programs in the previous section promote the development of mathematics for Aboriginal and Torres Strait Islander students, and share common approaches that are key constructs for success (Howard et al., 2011):

- enhanced mathematics learning – curriculum and pedagogies that help students see a purpose and meaning in the mathematics they are learning
- social justice – all students are treated with dignity and respect
- empowerment – students gain the necessary mathematical knowledge to achieve authentic educational outcomes
- engagement – students are able to interact purposefully with the mathematics discourse in classrooms as well as being acknowledged as capable learners
- reconciliation – teachers taking the time to listen, learn and care as well as collaborating to bring about enhanced educational outcomes
- connectedness – an individual sense of belonging, of being accepted and knowing you are valued
- relevance – bringing the Aboriginal students' environments into the mathematics classroom.

Reconciliation: the act of bringing people together. For Aboriginal and Torres Strait Islander peoples, reconciliation requires more. Reconciliation is a process of building understanding, relationships and respect, to acknowledge past injustices, and a commitment towards working towards a more equal and respectful future.

This set of constructs provides a framework for the design of mathematics programs which aim to support and enhance the learning of Aboriginal and Torres Strait Islander students. A further challenge for mathematics teachers is to consider how the cross-curriculum priorities could be embedded in mathematics experiences for all students and, in particular, how the Aboriginal and Torres Strait Islander Histories and Cultures could be used to support all mathematics learners. This is discussed in the last section of this chapter.

SHORT-ANSWER QUESTIONS

1. Select one of the three successful mathematics programs for Indigenous students and outline the key success criteria.
2. Which aspects of the YuMi Deadly Maths Program could be used to help you design suitable mathematics lessons for your students?

The Australian Curriculum cross-curriculum priority: Aboriginal and Torres Strait Islander Histories and Cultures

As noted in Chapter 9, the three-dimensional Australian Curriculum model includes learning areas (or subjects), seven general capabilities (e.g. literacy, numeracy, critical and creative thinking) and three cross-curriculum priorities (CCPs). One CCP is Aboriginal and Torres Strait Islander Histories and Cultures (ACARA, 2020), which uses a conceptual framework to provide a context for learning – see Figure 7.4.

For each CCP, a set of organising ideas reflects the essential knowledge, understandings and skills for the priority. The organising ideas are embedded in the content descriptions and elaborations of each learning area as appropriate. For Aboriginal and Torres Strait Islander Histories and Cultures, the organising ideas are presented in Table 7.1 under the three key concepts of country/place, culture and people.

An examination of the Australian Curriculum: Mathematics reveals few links to the Aboriginal and Torres Strait Islander Histories and Cultures CCP. Some have suggested the links are tokenistic (Lowe & Yunkaporta, 2013) so it is important we consider how the advice

Figure 7.4 Conceptual framework for the Aboriginal and Torres Strait Islander Histories and Cultures CCP
Source: ACARA, 2020.

contained in this chapter might provide some useful and purposeful ways secondary mathematics teachers can develop teaching and learning approaches which support Aboriginal and Torres Strait Islander learners and do justice to this CCP.

Some suggestions are listed here in the order they appear in the chapter, although this is not an exhaustive list:

1. Investigate the cultural backgrounds of students in each of your classes through surveys, stories and family histories – link to time, location, and collecting and representing data – a useful resource is *If the World Were a Village of 100 People* (Smith & Armstrong, 2013).

Table 7.1 Organising ideas for each of the key concepts for the cross-curriculum priority – Aboriginal and Torres Strait Islander Histories and Cultures

Code	Organising ideas
Country/Place	
OI.1	Australia has two distinct Indigenous groups: Aboriginal Peoples and Torres Strait Islander Peoples, and within those groups there is significant diversity.
OI.2	Aboriginal and Torres Strait Islander communities maintain a special connection to and responsibility for Country/Place.
OI.3	Aboriginal and Torres Strait Islander Peoples have holistic belief systems and are spiritually and intellectually connected to the land, sea, sky and waterways.
Culture	
OI.4	Aboriginal and Torres Strait Islander societies have many Language Groups.
OI.5	Aboriginal and Torres Strait Islander Peoples' ways of life are uniquely expressed through ways of being, knowing, thinking and doing.
OI.6	Aboriginal and Torres Strait Islander Peoples live in Australia as first peoples of Country or Place and demonstrate resilience in responding to historic and contemporary impacts of colonisation.
People	
OI.7	The broader Aboriginal and Torres Strait Islander societies encompass a diversity of nations across Australia.
OI.8	Aboriginal and Torres Strait Islander Peoples' family and kinship structures are strong and sophisticated.
OI.9	The significant contributions of Aboriginal Peoples and Torres Strait Islander Peoples in the present and past are acknowledged locally, nationally and globally.

Source: ACARA, 2020.

2. Design lessons that highlight the mathematics associated with students' interests and activities including sports, card games and leisure activities popular in the local community – encourage students to share their knowledge of these activities.
3. Design investigations like those from Gutstein and Peterson (2013) to investigate social justice issues related to the local context – local data could be used as a prompt for discussions allowing students to pose questions for further investigation.
4. Encourage students to create stories to represent number and algebra relationships.
5. Present mathematical ideas more holistically showing connections and relationships.
6. Connect with local Aboriginal and/or Torres Strait Islander Elders to learn about their histories and connections to country – organise visits to local sites or have Elders talk to students.

Some have suggested the CCPs might be ignored. In their critique of the approach of 'embedding' CCPs in the Australian Curriculum, and particularly the inclusion of Aboriginal and Torres Strait Islander Histories and Cultures, Salter and Maxwell (2016) suggest 'in the classroom, the CCPs are susceptible to omission by teachers too busy or disinclined to teach them. At a policy level, the CCPs could be interpreted and depicted as permeating every corner of the Australian Curriculum – as pushing out core content and leaving Australian students, and the nation, at the mercy of education fads' (p. 309).

Further, Lowe and Yunkaporta's (2013) analyses of the content descriptions and elaborations contained in the first four subjects developed for the Australian Curriculum (which included mathematics) indicated 'a lack of intention on ACARA's part ... to integrate high-quality learning around the histories and cultures of Aboriginal and Torres Strait Islander peoples' (p. 1).

With only two content elaborations about this CCP in the Years 7 to 10 Australian Curriculum: Mathematics that appear in analyses of statistics and data, the 'embedding' of Aboriginal and Torres Strait Islander Histories and Cultures in mathematics falls short of expectations. While the ACARA website offers some advice for mathematics teachers, it may be difficult to apply without deep engagement with context and connection to local communities. However, Henderson (2018, p. 147) reminds us that 'this priority area was also envisaged as a context for developing the general capability of intercultural understanding'. Without substantial effort by teachers to engage with cultural knowledge and understandings as well as local communities, the CCP is likely to be largely ignored.

It is clearly up to teachers to commit to supporting the mathematics learning of all students, and to provide each student with a learning experience that values their cultural backgrounds and enables opportunities for all students to learn deeply about First Nations peoples. As mathematics teachers, we can use components of the mathematics curriculum to explore 'many of the significant social justice issues that have impacted on the daily lives of Aboriginal and Torres Strait Islander peoples' (Lowe & Yunkaporta, 2013, p. 11) through culturally responsive pedagogies.

SHORT-ANSWER QUESTIONS

1. What are the three cross-curriculum priorities in the Australian Curriculum?
2. What are the three organising ideas on the cross-curriculum priority Aboriginal and Torres Strait Islander Histories and Cultures?
3. What are some of the suggestions in the Australian Curriculum for including the Aboriginal and Torres Strait Islander Histories and Cultures CCP in the mathematics curriculum?

Conclusion

It is very difficult to implement the school mathematics curriculum without learning about our students' backgrounds and cultures as well as their knowledge and understandings. It is very difficult for students to learn the mathematics curriculum without teachers investigating what they already know and understand and being prepared to build on that knowledge. As well as mathematics knowledge, our students have cultural understandings which impact upon their learning. Aboriginal and Torres Strait Islander students bring to school a wealth of knowledge and understanding about their cultures and histories. We have a responsibility to learn from and with our students and to connect with community.

This chapter has been written to support preservice secondary mathematics teachers by providing a summary of advice from the literature. Clearly not all the literature on this topic is referenced but the selection of articles has sought to include many prominent scholars working in this area of mathematics education research. The chapter encourages readers to continue to explore the many facets of mathematics education for Aboriginal and Torres Strait Islander students.

BRINGING IT TOGETHER

1. What teaching and learning strategies are recommended for catering for difference in mathematics classrooms?
2. What approaches are recommended to support the mathematics learning of Aboriginal and Torres Strait Islander students?
3. How can the 8 Ways of Learning be used to design mathematics learning activities for students?
4. What factors appear to be key to designing and implementing a successful mathematics program for Aboriginal and Torres Strait Islander students?
5. Sarra et al. (2018) describe the importance of high-expectations relationships as a foundation to enacting high expectations in Australian schools. In their article, they present examples of the difference between high-expectations rhetoric and a high-expectations relationship.
   a. Consider the situation of a student refusing to participate in or complete a classroom, homework or assessment task that is 'too hard'.
   b. Describe how you would respond to this situation using a high-expectations relationship approach. How does that compare to a high-expectations rhetoric approach?
6. If you were sent to a school with a higher population of Aboriginal and Torres Strait Islander students than the average secondary school, what strategies would you use to begin to learn more about the students' backgrounds and cultures? Who else in the school community would you talk with about the local community and how might you begin to embed connections to culture in your lessons?

REFERENCES

Further resources

8 Ways (n.d.). *8 Way Maths*. Retrieved from https://www.8ways.online/8way-maths

ACEMS (ARC Centre of Excellence for Mathematical & Statistical Frontiers) (2020). *Teaching culture = Deep learning. Teaching mathematics from an Aboriginal perspective* [YouTube video], August 3. Retrieved from https://youtu.be/JnkrKDTufcQ

Australian Association of Mathematics Teachers (n.d). *Make It Count 2009–2012*. Retrieved from https://mic.aamt.edu.au/Resources/Make-It-Count-2009-2012

Australian Bureau of Statistics (2018). *Schools Australia, 2017 – Summary of findings*. Retrieved from https://www.abs.gov.au/ausstats/abs@.nsf/Lookup/4221.0main+features22017

Australian Curriculum, Assessment and Reporting Authority (ACARA) (2020). *Cross-curriculum priorities – Aboriginal and Torres Strait Islander Histories and Cultures*. Retrieved from https://www.australiancurriculum.edu.au/f-10-curriculum/cross-curriculum-priorities/aboriginal-and-torres-strait-islander-histories-and-cultures/

Australian Institute for Teaching and School Leadership (AITSL) (2011). *Australian Professional Standards for Teachers*. Melbourne: AITSL.

Bamblett, M. (2005). Living and learning together: A celebration and appreciation of diversity. *Developing Practice*, 13, 17–30.

Bell, L.A. (2007). Theoretical foundations for social justice education. In M. Adams & L.A. Bell (eds), *Teaching for diversity and social justice* (3rd edn, 3–26). New York: Routledge.

Buckskin, P. (2015). Engaging Indigenous students: The important relationship between Aboriginal and Torres Strait Islander students and their teachers. In K. Price (ed.), *Aboriginal and Torres Strait Islander education: An introduction for the teacher profession* (2nd edn, 174–91). Port Melbourne: Cambridge University Press.

Burgess, C., Tennent, C., Vass, G., Guenther, J., Lowe, K. & Moodie, N. (2019). A systematic review of pedagogies that support, engage and improve the educational outcomes of Aboriginal students. *The Australian Educational Researcher*, 46(2), 297–318.

Burton, L. (2017). Ditching deficit thinking: Changing to a culture of high expectations. *Issues in Educational Research*, 27(2), 198–214.

Central Land Council (n.d.). *Kinship and skin names*. Retrieved from https://www.clc.org.au/articles/info/aboriginal-kinship

Chalmers, C., Carter, M., Cooper, T. & Nason, R. (2017). Implementing the 'Big Ideas' to advance the teaching and learning of science, technology, engineering and mathematics (STEM). *International Journal of Science and Mathematics Education*, 15(Supplement 1), S25–243.

Cooper, T. & Carter, M. (2016). Large-scale professional development towards emancipatory mathematics: The genesis of the YuMi Deadly Maths. In B. White, M. Chinnappan & S. Trenholm (eds), *Opening up mathematics education research. Proceedings of the 39th Annual Conference of the Mathematics Education Research Group of Australasia (MERGA)* (174–81). Adelaide: MERGA.

Council of Australian Governments Education Council (2019). *Alice Springs (Mparntwe) Education Declaration*. Retrieved from https://docs.education.gov.au/documents/alice-springs-mparntwe-education-declaration

D'Ambrosio, U. (2001). In my opinion: What is ethnomathematics, and how can it help children in schools? *Teaching Children Mathematics*, 7(6), 308–10.

Ewing, B. (2017). Theorizing participation, engagement and community for primary and secondary mathematics classrooms. *Creative Education*, 8, 788–812.

Gay, G. (2009). Preparing culturally responsive mathematics teachers. In B. Greer, S. Mukhopadhyay, A.B. Powell & S. Nelson-Barber (eds), *Culturally responsive mathematics education* (189–205). New York: Routledge.

Gutstein, E. (2006). *Reading and writing the world with mathematics: A pedagogy for social justice*. New York: Routledge.

Gutstein, E. & Peterson, B. (eds). (2013). *Rethinking mathematics: Teaching social justice by the numbers* (2nd edn). Milwaukee, WI: Rethinking Schools, Ltd.

Henderson, D. (2018). Cross-curriculum priorities: Stirring the passions and a work in progress? In A. Reid & D. Price (eds), *The Australian Curriculum: Promises, problems and possibilities* (143–62). Canberra: Australian Curriculum Studies Association Inc.

Howard, P., Cooke, S., Lowe, K. & Perry, B. (2011). Enhancing quality and equity in mathematics education for Australian Indigenous students. In B. Atweh, M. Graven, W. Secada & P. Valero (eds), *Mapping equity and quality in mathematics education* (365–78). Dordrecht: Springer.

Howard, P., Feirer, M., Lowe, K., Ziems, S. & Anderson, J. (2005). *Mathematics in Aboriginal contexts: Building school-community capacity*. Sydney: Board of Studies, NSW.

Howard, P. & Perry, B. (2007). A school-community model for enhancing Aboriginal students' mathematics learning. In J. Watson & K. Beswick (eds), *Proceedings of the 30th Annual Conference for the Mathematics Education Group of Australasia (MERGA)* (402–9). Adelaide: MERGA.

Jorgensen, R. (2018). *Celebrating success: Numeracy in remote Indigenous contexts. Final report for the remote numeracy project.* Canberra: University of Canberra. Retrieved from https://www.canberra.edu.au/research/faculty-research-centres/stem-education-research-centre/research-projects/remote-numeracy/tabs/publications/documents/Remote-numeracy-final-repor-31-3-18.pdf

Lowe, K. & Yunkaporta, T. (2013). The inclusion of Aboriginal and Torres Strait Islander content in the Australian National Curriculum: A cultural, cognitive and socio-political evaluation. *Curriculum Perspectives*, 33(1), 1–14.

Matthews, C. (2009). Stories and symbols: Maths as storytelling. *Professional Voice*, 6(3), 45–50. Retrieved from https://www.aeuvic.asn.au/sites/default/files/PV_6-3.pdf

Matthews, C. (2015). Maths as storytelling: Maths is beautiful. In K. Price (ed.), *Aboriginal and Torres Strait Islander education: An introduction for the teacher profession* (2nd edn, 102–20). Port Melbourne: Cambridge University Press.

Meaney, T. & Evans, D. (2013). What is the responsibility of mathematics education to the Indigenous students that it serves? *Educational Studies in Mathematics*, 82(3), 481–96.

Morris, C. & Matthews, C. (2011). Numeracy, mathematics and Indigenous learners: Not the same old thing. In Australian Council for Educational Research (ACER), *Indigenous education: Pathways to success*. Conference Proceedings. Retrieved from https://research.acer.edu.au/cgi/viewcontent.cgi?article=1130&context=research_conference

Morris, C., Toberty, K., Thornton, S. & Statton, J. (2015). *Numeracy, mathematics, and Aboriginal learners: Developing responsive mathematics pedagogy*. Paper presented at the annual conference of ATSIMA, Wollongong, NSW. Retrieved from https://atsimaths.files.wordpress.com/2013/12/mes7_thpaper2013.pdf

Mousley, J.A. & Matthews, C. (2019). Australia: Mathematics and its teaching in Australia. In J. Mack & B. Vogeli (eds), *Mathematics and its teaching in the Asia–Pacific region* (113–56). Singapore: World Scientific.

Mukhopadhyay, S., Powell, A.B. & Frankenstein, M. (2009). An ethnomathematical perspective on culturally responsive mathematics education. In B. Greer, S. Mukhopadhyay, A.B. Powell & S. Nelson-Barber (eds), *Culturally responsive mathematics education* (65–84). New York: Routledge.

Nakata, M. (2003). Some thoughts on literacy issues in Indigenous contexts. *Australian Journal of Indigenous Education*, 1(2), 181–2.

NarraGunnaWali (n.d.). *Resource guide: Mathematics*. Retrieved from https://www.narragunnawali.org.au/storage/media/professional-learning/mathematics-resource-guide-d0d5280776.pdf

Pearson, N. (2014). A rightful place: Race, recognition and a more complete commonwealth. *Quarterly Essay*, 55, 1–72. Retrieved from https://www.quarterlyessay.com.au/essay/2014/09/a-rightful-place

Perry, B. & Howard, P. (2008). Mathematics in Indigenous contexts. *Australian Primary Mathematics Classroom*, 13(4), 4–9.

Perso, T. (2012). *Cultural responsiveness and school education: With particular focus on Australia's first peoples; A review & synthesis of the literature.* Menzies School of Health Research, Centre for Child Development and Education, Darwin Northern Territory.

Perso, T. & Hayward, C. (2015). *Teaching Indigenous students: Cultural awareness and classroom strategies for improving learning outcomes.* Sydney: Allen & Unwin.

Praeger, C.E. (2008). The essential elements of mathematics: A personal reflection. *Australian Mathematical Society Gazette*, 35(1), 20–5. Retrieved from https://austms.org.au/wp-content/uploads/Gazette/2008/Mar08/Gazette35(1)WebVersion.pdf

Queensland Curriculum and Assessment Authority (2008). *Relationships to Country: Aboriginal people and Torres Strait Islander people.* Retrieved from https://www.qcaa.qld.edu.au/downloads/approach2/indigenous_res005_0803.pdf

Salter, P. & Maxwell, J. (2016). The inherent vulnerability of the Australian Curriculum's cross-curriculum priorities. *Critical Studies in Education*, 57(3), 296–312.

Sarra, C. (2011). *Strong and smart – Towards a pedagogy for emancipation: Education for first peoples*. Oxford: Routledge.

Sarra, C., Spillman, D., Jackson, C., Davis, J. & Bray, J. (2018). High-expectations relationships: A foundation for enacting high expectations in all Australian schools. *The Australian Journal of Indigenous Education*, 49(1), 32–45.

Sarra, G., Alexander, K.M., Carter, M.G. & Cooper, T.J. (2016). *QUT YuMi Deadly Maths: Making a difference in mathematics learning for Indigenous and non-Indigenous students*. Paper presented at the Aboriginal and Torres Strait Islander Mathematics Alliance Conference.

Smith, D.J. & Armstrong, S. (2013). *If the world were a village of 100 people* (3rd edn). New York: Allen & Unwin Children.

Stronger Smarter Institute (2017). *Implementing the stronger smarter approach*. Stronger Smarter Institute Position Paper. Retrieved from https://rypple.org.au/wp-content/uploads/sites/54/2018/04/K2-Stronger-Smarter-Approach-2017_final-2.pdf

Thornton, S., Statton, J. & Mountzouris, S. (2012). Developing mathematical resilience among Aboriginal students. In J. Dindyal, L.P. Cheng & S.F. Ng (eds), *Proceedings of the 35th Annual Conference of the Mathematics Education Group of Australasia (MERGA)* (250–7). Singapore: MERGA.

Warren, E. & Miller, J. (2016). *Mathematics at the margins*. Singapore: Springer Nature.

YuMi Deadly Centre (2016). *YuMI Deadly Maths: Overview of philosophy, pedagogy, change and culture*. Kelvin Grove: Queensland University of Technology. Retrieved from https://research.qut.edu.au/ydc/wp-content/uploads/sites/181/2019/12/YDM-Book1-Overview-2016.pdf

Yunkaporta, T. (2009). *Aboriginal pedagogies at the cultural interface*. Unpublished professional doctorate (research), James Cook University. Retrieved from https://researchonline.jcu.edu.au/10974/

CHAPTER 8

Assessing mathematics learning

Gregory Hine

LEARNING OBJECTIVES

The learning objectives of this chapter are directly linked to the Australian Professional Standards for Teachers (APST) (Australian Institute for Teaching and School Leadership [AITSL], 2011). After studying this chapter, you should be able to:

- define the term 'assessment' and explain the function of assessment in learning (APST 5.1)
- list, explain and justify assessment types that can be used in mathematics learning (APST 5.1)
- delineate and rationalise key principles for creating assessment tasks in mathematics learning (APST 5.1)
- recall appropriate methods of providing feedback to students (APST 5.2).

Introduction

Consider the following three statements:

> You don't fatten a pig by weighing it. (Anonymous)
>
> Nobody ever got taller by being measured. (Professor Wilfred Cockroft, 1982)
>
> Not everything that can be counted counts. Not everything that counts can be counted. (William Bruce Cameron, sociologist, 1967)

These three statements have not been selected randomly at the beginning of a chapter written principally about assessment in secondary mathematics. Rather, their prominent inclusion affords educators – from preservice to veteran, novice to master – some words of wisdom regarding the central issues of *why*, *how* and *what* as they pertain to assessment. In a similar vein, this chapter addresses assessment through an examination of relevant scholarship and an alignment with principles prescribed by the Australian Curriculum, Assessment and Reporting Authority (ACARA). To begin with, key ideas underpinning the purpose of assessment will be explored. Second, various methods of assessment used currently by secondary mathematics teachers will be outlined and justified. Third, various guidelines for providing effective feedback to students are suggested. Finally, a variety of principles for appropriate assessment practices is offered to guide educators in developing, administering and refining effective tools for measuring student achievement. The theory, research, recommendations and guidelines for best assessment practices offered herein are driven implicitly by the need for mathematical educators to produce and communicate information that is worthwhile and accurate.

Why assess?

Teachers and researchers consistently identify assessment as a critical feature of the instructional process (Bobis, Mulligan & Lowrie, 2013; Killen, 2005; Marsh, 2010). Not surprisingly, those involved in mathematics education view the importance of assessment in a similar light (Van de Walle, Karp & Bay-Williams, 2014; Wiliam, 2011). One question driving the need for assessment is: 'What is the purpose of assessment?' or more specifically 'Why assess?'. Killen (2005) argued that the purpose of assessment is underpinned by two assumptions. First, there must be *something* (e.g. intelligence, aptitude for a particular job, program quality) that can be measured. The second assumption is 'that this factor can be measured in a way that distinguishes between how much of it is possessed by different individuals, by different groups of learners or by different instructional purposes' (Killen, 2005, pp. 101–2). The purpose of assessment is recognised also by national and international authorities who outline reasons for assessment and recommend principles for assessing student work. For instance, ACARA asserts that assessment within the Australian Curriculum takes place in different levels and for different purposes. These levels and purposes include:

- ongoing formative assessment within classrooms for the purposes of monitoring learning and providing feedback, for teachers to inform their teaching. and for students to inform their learning

Assessment: the intentional and unintentional activities undertaken by teachers to gather information about student learning (knowledge, skills, dispositions).

Formative assessment: those planned processes that regularly monitor students' understanding of instructional activities and students during instructional activities.

Summative assessment: those planned processes which demonstrate an accumulated demonstration of learning by a student over a given period of time.

- summative assessment for the purposes of twice-yearly reporting by schools to parents and carers on the progress and achievement of students
- annual testing of Years 3, 5, 7 and 9 students' levels of achievement in aspects of literacy and numeracy, conducted as part of the National Assessment Program – Literacy and Numeracy (NAPLAN)
- periodic sample testing of specific learning areas within the Australian Curriculum as part of the National Assessment Program. (ACARA, 2015)

PAUSE AND THINK

In your preservice education degree or qualification you will be required to develop a *philosophy of teaching and learning* which encompasses significant statements about the profession you are entering. This philosophical statement is comprised usually of the way(s) you view knowledge, the purpose of education, the nature of people and approaches towards best instructional practices. As you gain further classroom experience, this statement can undergo editing and refinement. Ensure that your responses for these statements are kept in an accessible location, so that you can periodically review and revise them.

1.   How prominently do the terms *planning* and *assessment* feature in your philosophy?
2.   What do you see as the connection between the terms *planning* and *assessment*?

Feedback: 'the information gained from formal and informal assessment, provided to students by teachers, about what the student has learnt, needs to learn, mistaken ideas and directions for improvement' (Hosking & Shield, 2001).

In the United States, the National Council of Teachers of Mathematics (NCTM) (1995) posited that teachers assess to monitor student progress, make instructional decisions, evaluate student achievement and evaluate teaching programs. Churchill et al. (2019) asserted that the purpose of assessment included teachers monitoring instructional effectiveness, providing timely feedback to students and parents, and identifying special needs of students. Additionally, Booker et al. (2014) emphasised how assessment helped teachers identify specific areas of mathematical development in students. In turn, identification of such areas is instrumental in subsequent iterations of instructional planning and in communicating mathematical capabilities to students, parents and the wider community.

IN PRACTICE

Why do we assess students?

Clarke and Clarke (2002) asked many mathematics teachers to respond to the question: 'Why do we assess students?' These scholars found that the teachers' responses fell broadly into the following (overlapping) categories:

- To find out what my students know and can do.
- To help me know what to teach next.
- To measure the effectiveness of my teaching.

- To provide feedback to students on their learning.
- To inform parents, other teachers, employers and interested others of academic progress.
- Because my principal/school/community expects it. (2002, p. 1)

Clearly there is much agreement about why teachers engage in the practice of assessment. However, reasons underpinning the selection and implementation of effective assessment options are not always met with the same degree of agreement. For example, Panizzon and Pegg (2007) recorded a self-reported realisation by secondary mathematics educators that their students 'were rarely given the opportunity to explain what they understood conceptually about a mathematics concept' (p. 432). According to these researchers, one surveyed teacher stated that 'while his students could calculate compound interest, only a small proportion were able to articulate what it was (as a concept) in their own words' (Panizzon & Pegg, 2007, p. 432). In addition, Watt (2005) noted that in a contemporary Australian context where there is a 'decreased demand for computational skills, syllabi now emphasise mathematical process as distinct from product' (p. 22). In a similar vein, Clarke (1987) argued that the reconceptualisation of a 'successful mathematics student' was the impetus for widespread changes in assessment practices within Australian schools. Specifically, this change involves a departure from the view that successful mathematics students can 'neatly replicate a learned procedure to a routine task in a familiar context' (Watt, 2005, p. 22). Instead, successful mathematics students can devise problem-solving strategies across various contexts, identify conceptual similarities in different situations, assess the relevance of different procedures to applied contexts, and work productively with others (Clarke, cited in Watt, 2005, pp. 22–3).

SHORT-ANSWER QUESTIONS

1. What is the purpose of assessment?
2. When does assessment in mathematics not work well?
3. How can I ensure that best practices for assessment occur in my secondary mathematics classes?

What is assessment?

Assessment is the term used to describe an individual activity or series of activities undertaken by a teacher in order to gather information about student learning (Marsh, 2010). Earlier in this chapter, some of the reasons that teachers assess student learning were outlined. Some of these reasons included: to make informed decisions about children (Sattler, 2008) and to enhance instructional planning (Ohlsen, 2007; Stiggins, 1994). Within the classroom context – which is one characterised by fairly constant formal and informal assessment over time, and across many dimensions of behaviour – teachers have a wide range of classroom assessment methods to select from (Brualdi, 1998). Some of the more common types of assessment used by secondary mathematics teachers are tests, quizzes, examinations, investigations, extended pieces of work and homework assignments. Irrespective of the type of assessment used, Clarke (1987) argued that it is through selection of assessment that teachers communicate most clearly to students those activities and learning outcomes they value.

Types of assessment

Assessments typically fall into one of three broad categories:

- summative assessments
- formative assessments
- diagnostic assessments.

Each of these categories will be explored below.

Summative assessment

Summative assessment (or assessment *of* learning) is any assessment designed mainly to measure student learning for the purposes of ranking or providing an indication of performance of learning at a fixed point in time (e.g. a score or grade) (Bobis, Mulligan & Lowrie, 2013). According to Van de Walle, Karp and Bay-Williams (2014), 'summative assessments are cumulative evaluations that might generate a single score, such as an end-of-unit test or the standardised test that is used in your state or school districts' (p. 82). These assessments are regarded as those which demonstrate an accumulated demonstration of learning by a student over a given period of time. Van de Walle, Karp and Bay-Williams (2014) note that although summative assessments can be useful for schools, teachers and systems in long-term planning or revising curricula, they are often unhelpful in shaping teaching decisions that require more immediate attention. Such decisions can include the selection of pedagogy regarding particular mathematics topics, or the identification of conceptual misunderstandings that may prevent student growth.

Formative assessment

Formative assessment (or assessment *for* learning) is any assessment for which the main purpose is to enhance learning (Bobis, Mulligan & Lowrie, 2013; Klenowski, 2009; Wiliam, 2011). Formative assessment practices are those methods that assess student learning 'on the go' or 'along the way'. Scholars contend that educators use formative assessments as planned processes to regularly monitor students' understanding of and during instructional activities (Hattie, 2009; Popham, 2008; Wiliam, 2007). Formative assessment practices in secondary mathematics usually take the form of quizzes, chapter tests, student work, standardised exams, as well as questioning or interviewing techniques to inform instruction (Chen et al., 2012).

Diagnostic assessment

Diagnostic assessment: those planned processes of getting in-depth information about an individual student's knowledge and mental strategies about concepts (Van de Walle, Karp & Bay-Williams, 2014).

As students enter classrooms at varying points along their educational journeys, it is considered an efficient exercise to start a new teaching unit by checking their knowledge and understandings (Marsh, 2010). Conducting a diagnostic assessment can assist teachers in ascertaining knowledge about students' prerequisite skills and dispositions towards the learning area. According to Marsh (2010), diagnostic assessments remind teachers to commence their instruction precisely at the level students have reached. A diagnostic assessment is the means of getting in-depth information about an individual student's knowledge and mental strategies about concepts (Van de Walle, Karp & Bay-Williams, 2014). Moreover, diagnostic assessments can help identify the way in which mathematics is understood and used, locate areas of strength and determine underlying weaknesses causing errors or difficulties (Booker, 2011). Although these assessments (which can be conducted through a teacher–student interview) are often labour-intensive, they are rich methods 'that provide evidence of misunderstandings and explore students' ways of thinking' (Booker, 2011, p. 94). During the assessment, a student is given specifically selected problems to complete on a single topic, where the working out or answers supplied can illuminate for teachers a student's misunderstanding or progress in that topic.

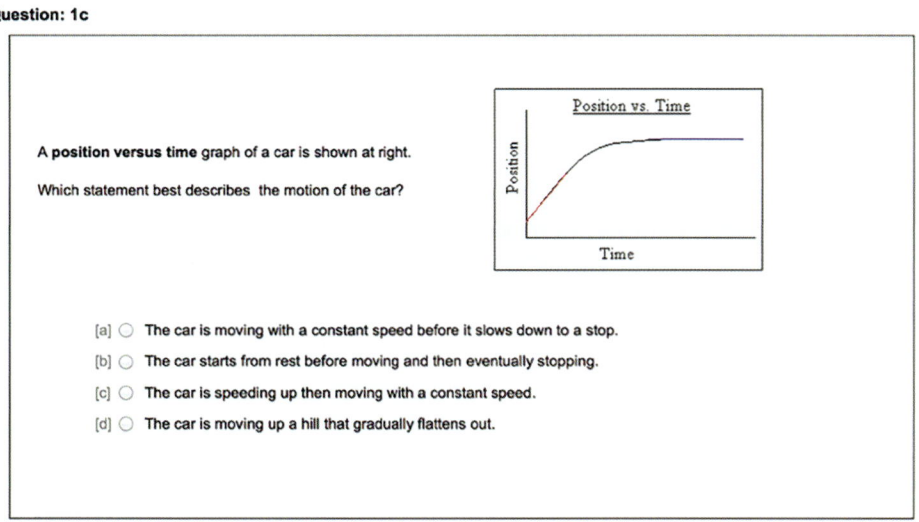

Question: 1c

A **position versus time** graph of a car is shown at right.

Which statement best describes the motion of the car?

Position vs. Time

Position

Time

[a] ○ The car is moving with a constant speed before it slows down to a stop.
[b] ○ The car starts from rest before moving and then eventually stopping.
[c] ○ The car is speeding up then moving with a constant speed.
[d] ○ The car is moving up a hill that gradually flattens out.

Figure 8.1 A diagnostic item in a unit on determining speed
Source: Minstrell, 2001.

Figure 8.1 displays a diagnostic assessment item developed by Minstrell (2015) that can be used in secondary mathematics and science classrooms. This item requires the student to answer a question related to the position–time graph given. Each possible answer is related to a code, known as a facet (Minstrell, 2001), in the diagnostic key illustrated in Figure 8.2. Each facet corresponds to a *facet cluster* (Minstrell, 2001), which acts as a framework for diagnosing how students understand or misunderstand conceptual material.

For instance, if a student writes *(b)* as the correct answer, then the corresponding facet for this response would be *50*. In turn, the facet cluster

Key	Facet	Next Question
a	03	2
b	50	2
c	80	2
d	90	2

Figure 8.2 The diagnosis key for the item in Figure 8.1
Source: Minstrell, 2001.

Facet Cluster - Determining Speed

Facets and facet clusters are a framework for organizing the research on student conceptions so that they are understandable to both discipline experts and teachers. Facet clusters include the explicit learning goals in addition to various sorts of reasoning, conceptual, and procedural difficulties. Each cluster contains the intuitive ideas students have as they move toward scientifically accurate learning targets.

Facets are arranged with the Goal Facets at the top of the page followed by the more problematic facets. Each facet has a two-digit number. The 0X and 1X facets are the learning targets. The facets that begin with the numbers 2X through 9X indicate ideas that have more problematic aspects. In general, the higher facet numbers (e.g., 7X, 8X, 9X) are the more problematic facets. The X0's indicate more general statements of student ideas. Often these are followed by more specific examples, which are coded X1 through X9.

Print Page

Facet Cluster

00   Student correctly describes the motion and interprets the speed of an object for a specific time from the information given. The information may be given in graphs, tables, pictures, or words.

    01   Given speed vs. time data, student correctly identifies an instantaneous speed of an object.

    02   Given position vs. time data, student correctly describes and determines the speed of an object moving uniformly.

    03   Student qualitatively describes the motion of an object from the information given.

---

40   The student misses a portion of the trip during their discussion of the motion.

    41   The student incorrectly describes or omits the middle portion of an object's motion.

    42   The student incorrectly describes or omits the beginning or end of an object's motion.

50   Student incorrectly describes the initial or final conditions of motion of the object.

    51   Initial speed is incorrectly identified as zero, because at the beginning of trips things are not supposed to be moving.

    52   Final speed is incorrectly identified as zero, because at the end of trips things are not supposed to be moving.

70   When asked for the speed at one instant, the student incorrectly reports another quantity or rate.

    71   Student reports the position, change in position or distance traveled.

    72   Student reports the average speed.

    73   Student reports the change in speed.

    74   Student divides the speed by the final time or change in time.

    75   Student divides the change in speed by the final time or change in time.

    76   Student reports zero, the object cannot be moving at an instant in time.

80   Student confuses position vs. time and speed vs. time graphs or data tables.

    81   Student interprets sloping up (or down) on a position graph to mean the object is speeding up (or slowing down).

    82   Student interprets a flat line segment on a position graph to mean the object is moving at constant speed.

    83   Student interprets sloping up (or down) on a speed graph to mean the object is moving with constant speed away from (or toward) the origin.

    84   Student interprets a flat line segment on a speed graph to mean the object is not moving.

90   Student views a position or speed graph as a map of the actual motion.

    91   Student interprets an upward (or downward) sloping graph to mean the object is going up hill (or downhill).

    92   Student interprets a flat line on the graph to mean the object is moving on a flat surface.

Figure 8.3 The facet cluster used to code student responses to assessment items in the determining speed unit
Source: Minstrell, 2001.

Student incorrectly describes the initial or final conditions of the motion of the object describes generally how the student has misunderstood the question. Additionally, the facet numbers *51* and *52* are included to help teachers locate more specifically how the student has applied incorrect reasoning, or experienced conceptual or procedural difficulties. Alternatively, if a student writes *(d)* as the correct answer the facet *90* would be applied, aligning with the facet cluster *Student views a position or speed graph as a map of the actual motion.* Facet numbers *91* and *92* are also available for a more specific diagnosis of conceptual misunderstanding. Minstrell (2001) notes that in relation to facet *50*, facet *80* indicates how a student exhibits greater conceptual or procedural difficulties in completing the given problem (see Figure 8.3).

PAUSE AND THINK

In the previous few pages, the idea of diagnostic assessment has been outlined. For the Diagnostic Test Item – Determining Speed, see if you can determine the Australian Curriculum content descriptions which align with the following courses:

- Year 7
- Year 8
- Year 9
- Year 10
- General Mathematics
- Mathematics Methods.

According to Booker (2011), diagnostic assessment can be expressed as a '3-phase cycle' (Figure 8.4). The cycle is comprised of a series of observations, assumptions and probes that are followed until the underlying causes of difficulties are illuminated.

First, teacher observations focus on individual students and their behaviour during teaching and learning activities as students work individually or cooperatively. Such observations take place to try to determine thinking and understanding of specific concepts and processes. Second, ongoing observations lead to assumptions about a student's understanding and the progress of the class as a whole. These assumptions reflect 'a teacher's knowledge of both the underlying mathematical ideas and experience with a range of children in similar situations' (Booker et al., 2014, p. 32). As a result of such assumptions, teachers form conjectures regarding student mathematical thinking (evidenced in explanations,

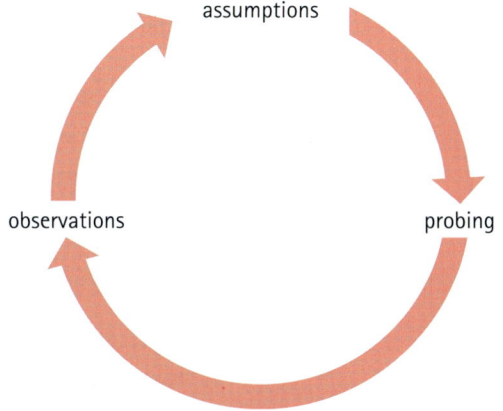

Figure 8.4 3-phase cycle of observations, assumptions, probes
Source: Booker, 2011.

Conceptual knowledge: the knowledge of abstract and general ideas (concepts) and how these ideas can be interconnected.

representations or drawings) and understanding (use of processes, display of conceptual knowledge). Finally, any assumptions made then need to be confirmed through further probing (Booker et al., 2014).

IN PRACTICE

Smart tests

Within Australia a series of diagnostic assessments (known as 'smart tests') have been developed at the University of Melbourne by Stacey et al. (2018). These smart tests (specific mathematics assessments that reveal thinking) are designed to provide teachers with an insightful and informative diagnosis of junior secondary students' conceptual understanding of mathematical topics. After signing up at no cost, teachers gain internet access to a repository of online tests for most mathematical topics. Although tests are completed online, students may use paper, pens and a calculator if required. The teacher receives students' results instantly, where interpretive analysis, further teaching and follow-up testing can occur. Parallel tests are also available for teachers who wish to use a 'pre-test/post-test' approach, or for those who prefer to offer alternative tests for particular students. The smart tests can be accessed online at the Smart Vic website (www.smartvic.com/teacher/).

Common assessment techniques

A variety of assessment techniques are available to secondary mathematics teachers. Commonly used techniques include tests, quizzes, examinations, homework and investigations. Teachers can also measure student learning through the administration of alternative assessments, including student self-assessment, teacher observation, student journalling, oral tasks or practical tasks. When deciding precisely how student progress will be measured, Leder, Brew and Rowley (1999) suggest that teachers avoid an over-reliance on one form of assessment type. These scholars argue that differences in learner characteristics imply that a variety of assessment techniques should be used so that all students may display their mathematical knowledge, skills and abilities effectively (Leder, Brew & Rowley, 1999).

Tests

Tests will always be a part of assessment and evaluation (Van de Walle, Karp & Bay-Williams, 2014). The prevalence of test administration within mathematics classrooms is well documented, with multiple researchers reporting that mathematics teachers rely heavily on traditional tests to assess student learning (Henke, Chen & Goldman, 1999; Senk, Beckmann & Thompson, 1997; Watt, 2005). Senk, Beckmann and Thompson (1997) purport that tests are a popular assessment item because they can often address 'a dozen or more objectives and that they must often be administered in 40–50 minute periods' (p. 210). Moreover, such tests effectively assess aspects of mathematics in an unambiguous and straightforward way, including students' performance on routine skills and algorithms (Clarke & Lovitt, 1987; Grimison, 1992; Stephens, 1988). Research in an Australian secondary school context found that surveyed teachers described how 'the written test was a good measure of mathematical abilities for Years 11–12, less good for Years 9–10, and least good for Years 7–8' (Watt, 2005, p. 37). In some Australian states and territories (e.g. WA), secondary school tests and examinations contain two sections: a section that permits the use of handwritten notes and a calculator and a section that does not permit any resources.

A cautionary note on tests

Although testing is viewed as an efficient and popular assessment technique, multiple commentators advise teachers to exercise caution during the design and administration of tests. For instance, Senk, Beckmann and Thompson (1997) expressed concern regarding the over-reliance on textbook-generated chapter tests for mathematics students. These authors found that such tests were comprised generally of low-level test questions that seldom required students to justify a conclusion, contained few questions set in realistic contexts, and lacked any technological component. To address these research findings, 'greater efforts need to be made to incorporate more reasoning and multistep problem solving, as well as more substantive use of both numerical and graphical technology on tests' (Senk, Beckmann & Thompson, 1997, p. 211). Below are two test items that require learners to demonstrate their knowledge of circular equations. Although both items could be used within a mathematics assessment, Item 2 requires a higher degree of reasoning and demonstration of further multi-step thinking than Item 1.

Item 1

Find the centre and the radius of the circular equation given by

$$(x + 2)^2 + (y + 3)^2 = 9$$

Item 2

AB is the diameter of a circle with equation $x^2 + y^2 = 9$. If A is the point $(-2, -3)$, find the coordinates of B.

In addition, these characteristics suggested that teachers were not assessing according to nationally prescribed guidelines which 'reflect the mathematics that all students need to know and be able to do' (NCTM, 1995, p. 11). Clarke and Clarke (2002) highlighted how for certain students 'the pressure of the test leads to performance that is not representative of their knowledge and understanding' (p. 2). While scholars acknowledge that test administration can be justified on the grounds of increasing reliability and ensuring comparability, validity should not be compromised (Clarke, 1987; Lacey & Lawton, 1981). To enhance validity, Van de Walle, Karp and Bay-Williams (2014) recommend that tests should not be 'a collection of low-level skill exercises that are simple for teachers to mark' but should be designed to 'match the goals of instruction' and to determine 'what concepts students understand and how their ideas are connected' (p. 91). To accomplish this, tests of procedural knowledge should consist of items requiring students to demonstrate a conceptual basis for processes used (Van de Walle, Karp & Bay-Williams, 2014). Some examples of the types of questions teachers can use for enhancing students' mathematical procedural knowledge

Validity: the degree to which an assessment item measures what it has been designed to measure.

Procedural knowledge: the knowledge of processes or steps needed to achieve a particular goal.

are to 'find the odd one out', to 'provide an example of', and to include questions that can be solved in different ways.

Quizzes

Quizzes can serve as an efficient assessment method allowing teachers to evaluate student learning at a specific point in the learning process (Webb, 2001). At a specific point in time, quizzes can provide teachers with important and timely feedback regarding student learning (Ohlsen, 2007). In particular, teachers can use the results of quizzes to 'determine if remediation or researching of key concepts or processes is needed for improved student outcomes' (Ohlsen, 2007, p. 10). In this way, Black and Wiliam (1998) contend that quizzes can be utilised as both a formative assessment method that informs and guides instruction and an evaluation component that can be used in the calculation of student summative grades.

Examinations

Major examinations are often a summative assessment tool that teachers use as a cumulative evaluation of student learning at the end of a chapter or unit (Ohlsen, 2007, p. 10). As such, examinations are designed to last for several hours and may even be divided into sections where resources (e.g. notes, formula sheets, calculators) may or may not be permitted for use.

Homework

A common and expected activity within Australian secondary schools is the allocation of homework tasks. According to Huetinck and Munshin (2008), determining the extent to which homework is used in formative assessment can be a tricky decision for teachers. Following a synthesis of research on homework tasks, Cooper, Robinson and Patall (2006) outlined that homework positively influences both immediate and long-term achievement and learning, and has a variety of non-academic (e.g. greater self-direction, self-discipline) and familial (e.g. awareness of the school–home connection) benefits. On the other hand, potential negative effects of homework can include satiation, denial of access to leisure time and community activities, parental interference, an increase in the onset of cheating, and increased differences between high and low achievers (Cooper, Robinson & Patall, 2006). Furthermore, these authors found that a typical secondary school student who completes homework will outperform classmates who do not complete homework by 69 per cent. For middle school students this figure is 35 per cent.

IN PRACTICE

Guidelines for issuing homework

In their research, Cooper, Robinson and Patall (2006) offer some guidelines regarding the issuing of homework:

- Homework is never given as punishment
- The purpose of homework should be to diagnose individual learning problems

- All students in the same class will be responsible for the same assignment
- Homework assignments should not be used to teach complex skills
- Homework should be a combination of material already covered and simple introduction to material about to be covered
- Homework should include a mixture of both mandatory and voluntary tasks, with the caveat that voluntary should be intrinsically interesting.

Investigations

Alongside quizzes and tests, investigations are another popular method of formative assessment in the secondary mathematics classroom. A mathematical investigation requires students to pose their own problems after an initial exploration of a given mathematical situation (Ronda, 2010; Yeo, 2008). Such exploration – together with the formulation of problems and their corresponding solutions – provides an 'opportunity for the development of independent mathematical thinking' and to engage in 'mathematical processes such as organising and recording data, pattern searching, conjecturing, inferring, justifying and explaining conjectures and generalisations' (Ronda, 2010). Scholars have pointed out that problem-solving tasks and investigations are different activities. Ernest (1991) described problem solving as 'trail-blazing to a desired location' (p. 285) and Pirie (1987) defined investigation as the exploration of an unknown land where 'the journey, not the destination, is the goal' (p. 2). To summarise, problem solving is a convergent activity focused on a well-defined goal and solution, and investigation is a divergent activity with an open goal and solution (Evans, 1987). The open-ended nature of investigative tasks assists students in developing mathematical habits of mind (Lampert, 1990; Ronda, 2010), problem-posing skills (Brown & Walter, 2005), and conjectures and generalisations (Calder et al., 2006). In doing so, students are able to engage in a 'variety of rich mathematical activities that parallel what academic mathematicians do' (Yeo, 2008, p. 613). As Civil (2002) noted, mathematicians investigate and solve mathematical problems.

PAUSE AND THINK

Up until this point in time you would have completed many mathematics assessments. Recall the assessment item that had the most positive influence on you as a learner.

1. What were the key characteristics of this assessment item?
2. When you have compiled a list of characteristics, compare these with a classmate or colleague. What did your comparison reveal?

Alternative assessments

There is a need for teachers to explore alternative assessment methods to assess other instructional goals (Watt, 2005). Some of these methods include:

- student self-assessment
- teacher observation
- student journalling
- oral tasks
- practical tasks.

Student self-assessment

Student self-assessment can provide a greater degree of ownership of the assessment process for students (Clarke & Clarke, 2002). Stenmark (1989) asserted that 'the capacity and willingness to assess their own progress and learning is one of the greatest gifts our students can develop . . . Mathematical power comes with knowing how much we know and what to do to learn more' (p. 26). Van de Walle, Karp and Bay-Williams (2014) argued that student self-assessment should not be a teacher's only measure of student learning or disposition, but rather 'a record of how they perceive their strengths and weaknesses as they begin to take responsibility for their learning' (p. 90). 'As well as assessing their own growth in knowledge and skills', Clarke and Clarke (2002) stated that 'students also have much to offer about their preferred learning styles and ways in which the teaching and learning process can be enhanced' (p. 4). Furthermore, Clarke and Clarke (2002) outlined how some teachers advocated the use of 'student-constructed tests, . . . as part of a review of content prior to formal assessment' (p. 4). According to these commentators, 'students are invited to create assessment tasks that they believe would assess fairly the key ideas in the topic under study. The students then work on each other's tasks' (p. 4). From their research, Clarke and Clarke (2002) concluded that this is considered excellent revision practice, 'and is particularly well received when the teacher makes a commitment to use at least some of the student-created tasks in the final assessment' (p. 4).

Other alternative assessment methods

Other alternative assessment methods are considered valuable in the instructional process. These are tabulated in Table 8.1 with a brief description of the method (Watt, 2005, p. 26).

Table 8.1 Alternative assessment methods

Assessment method	Description of assessment method
Teacher observation	Teachers observe students in structured or unstructured activities and evaluate the quality of student task engagement.
Student journals	Students keep reflective accounts of their mathematics learning and processes of understanding, from which the quality of their task engagement and development may be explored by the assessor.
Oral tasks	Students give short answers, demonstrations, seminar presentations and debates.
Practical tasks	Students use instruments to apply or deduce mathematical principles.

Source: Watt, 2005, p. 26.

National Assessment Program – Literacy and Numeracy (NAPLAN)

The NAPLAN tests are given to students in Years 3, 5, 7 and 9 in all Australian states and territories annually. This standardised testing commenced in 2008 and is overseen by ACARA. NAPLAN tests are comprised of multiple-choice and short-answer items – and in more recent versions of the assessment there are two sections for students to complete: one that requires the use of a calculator and one that does not. Since its inception, NAPLAN testing has been met with healthy scepticism and criticism. It can be argued that high-stakes standardised tests can yield valuable data for individual schools, regions and states, as they 'measure all [mathematical] achievement against a common framework of content and questions' (Booker et al., 2014, p. 35). On the other hand, the broad scope of such testing limits the possibility for teachers to probe appropriately mathematical understanding. Additionally, these tests do not 'allow for the original thinking involved with problem solving, generalisations of ideas, and alternative ways to achieve likely answers' (Booker et al., 2014, p. 35).

Scholars acknowledge that discussing test items with students can be used strategically to enhance mathematical thinking, promote learner confidence and resilience (Anderson, 2009; Martin, 2003). In particular, Anderson (2009) states that using NAPLAN test items within lessons can 'develop students' competence in reading mathematical text, to promote thinking strategies including estimation, and to evaluate alternative solutions for errors and misconceptions' (p. 22). To illustrate, Anderson (2009) outlines how a NAPLAN item can be used with a class of students (see Figure 8.5). To begin with, teachers should mention to students that many errors occur in items about percentages, and that 'multiple-choice items typically include common errors and misconceptions as alternative solutions' (Anderson, 2009, p. 20). In addition to highlighting the need for students to take time to think carefully about the problem, its solution and approaches to verify this solution, Anderson offers some questions to guide student thought:

- How much is the decrease in the electricity bill?
- Is the decrease less or greater than 50% of the first electricity bill?
- What are the fraction equivalents for 25% and 33%? How much would the discount be for each of these? (Anderson, 2009, p. 48)

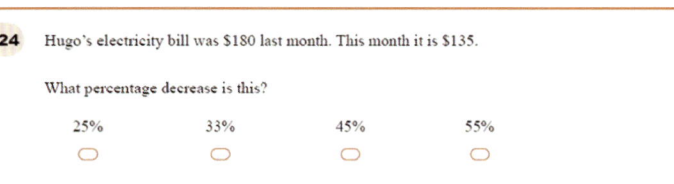

24 Hugo's electricity bill was $180 last month. This month it is $135.

What percentage decrease is this?

25%   33%   45%   55%

Figure 8.5 2008 Year 7 Numeracy calculator allowed test, question 24
Source: MCEETYA, 2008.

PAUSE AND THINK

In Brown and Tang's (2019) paper *Does the NAPLAN match the Australian Mathematics Curriculum?*, the authors explore the extent to which items on NAPLAN mathematics assessments correspond to the Australian Curriculum: Mathematics. Reflect on your own understanding of the question posed by the title of this article, and after reading the article, answer the questions below.

1. What are the two main findings of this research project?
2. How will these findings influence the way you teach mathematics in lower secondary years?

SHORT-ANSWER QUESTIONS

1. From the list of assessments typically used in the secondary mathematics classroom, select one assessment type. Where have you seen this type of assessment work well, and why? What considerations are there in planning for this type of assessment?
2. As part of your own teaching and learning philosophy, outline your stance towards homework.
3. From the alternative assessment types offered in this chapter, which have you experienced as a mathematics learner? Would you use this type of assessment with your students? Why or why not?
4. In your view, how can NAPLAN assessment items be put to good use in the mathematics classroom?

Guidelines for creating assessments in secondary mathematics

Whatever model of instruction is selected to teach secondary mathematics, teachers inevitably must select and create appropriate assessments for students to demonstrate learning. During the first few years of teaching – or even teaching a new course or year level – creating completely original and valuable assessment tasks can be daunting. In addition to meeting the external curricular requirements prescribed by ACARA and Departments of Education, assessments should undergo an internal review process (via mathematics department colleagues) before administration to students. Such reviews can not only help detect simple typographical errors, but also provide useful feedback regarding the procedural complexity, time constraints, feel of comprehension, balance of content – and depending on the class – accessibility for diverse learners. In this section some recommendations and guidelines are provided for secondary mathematics teachers to consider when creating assessment items. Such professional advice is drawn from educational authorities, professional associations and scholarly commentators who commonly underscore the importance of assessment within the mathematics classroom.

Principles for assessment
ACARA

ACARA has recommended assessment practices for teachers in Australian schools to follow. These practices are outlined as a series of steps, using the Australian Curriculum content and achievement standards (ACARA, 2015):

1. Identify current learning and achievement levels.
2. Select appropriate content to teach.

3. Judge the quality of learning as demonstrated by students.
4. Review teaching programs for further instructional delivery.

Specifically, these practices are outlined as:

Teachers use the Australian Curriculum achievement standards and content to identify current levels of learning and achievement, and then to select the most appropriate content (possibly from across several year levels) to teach individual students and/or groups of students. Teachers develop teaching programs designed to build on current learning. In each class, there may be students with a range of prior achievement (below, at, and above the year level expectations).

Teachers also use the achievement standards, at the end of a period of teaching, to make on-balance judgments about the quality of learning demonstrated by the students – that is whether they have achieved below, at, or above the standard. To make judgments, teachers draw on assessment data that they have collected as evidence during the course of the teaching period.

If an individual student's achievement is below the expected standard, this suggests that the teaching programs and practice should be reviewed to better assist the student in their learning in the future. It also suggests that additional support and teaching that targets the student's specific needs is necessary to ensure that the student does not fall behind. (ACARA, 2015)

These practices align with the standard Plan – Teach – Evaluate model espoused by Barry and King (1998, p. 328) and with those practices offered in many teacher education programs.

The Australian Association of Mathematics Teachers

The Australian Association of Mathematics Teachers (AAMT) (2017) described how highly accomplished teachers use a range of assessment strategies that are appropriate, fair and inclusive, and which inform learning and action. First, the AAMT recommended that teachers should assess appropriately by matching the purpose to the information required, and by assessing the full range of learning goals via a range of strategies. Second, teachers should assess student learning in a fair and inclusive manner by:

- involving students in the processes for assessing their learning
- using assessment strategies and tasks that are as fair as possible to students, and inclusive of those from a variety of backgrounds
- assessing in ways that are clear and transparent
- making fair and inclusive judgements
- assessing 'through planned means and opportunities that arise in their work with students'
- ensuring 'students are familiar with the genres of items used in their own assessment programs and in those of education authorities' (AAMT, 2017).

Third, the AAMT recommends that teachers use assessment of students' mathematics to inform learning and action. This can be achieved by reflecting on assessment information and subsequently planning student learning experiences, providing purposeful feedback to students about learning, and sharing assessment information with relevant key stakeholders.

Rich assessment tasks

According to Wiliam (2007), assessment should become a means of fostering growth in mathematics. Teachers must exercise care when designing assessments so that 'there is a balance in the questions that are asked, and that the range of problems reflects due importance given to each aspect of the topic under review' (Booker et al., 2014, p. 31). These authors offer broad recommendations for teachers designing assessments, including ensuring that students can develop concepts in depth, and emphasising a wide range of problem-solving processes in reaching a solution. Additionally, practical activities (e.g. specific topics within measurement, geometry, statistics and probability) need 'to be seen to have an equal weighting with the written tasks associated with computation and problem solving' (Booker et al., 2014, p. 31). It is with the above comments in mind that the topic of *rich tasks* in mathematics can be discussed.

Rich tasks encourage students to think creatively, work logically, and communicate their ideas within a mathematical context (Piggott, 2015). Such tasks also provide opportunities for students to 'synthesise their results, analyse different viewpoints, look for commonalities, and evaluate findings' (Piggott, 2015). Various scholars suggest that rich assessment tasks contain certain characteristics (Ahmed, 1987; Clarke & Clarke, 2002; Department for Education and Skills, 2007; Piggott, 2015) including:

> **Rich assessment tasks:** planned tasks that ensure students can develop concepts in depth and emphasise a wide range of problem-solving processes in reaching a solution.

- be accessible to all learners
- engage the learner through different levels of challenge
- allow the learner to make decisions about which methods or approaches to use
- 'encourage students to disclose their own understanding of what they have learned'
- 'allow students to show connections they are able to make between the concepts they have learned'
- provide a range of student responses, including a chance for students to show all they know about the relevant content
- provide an opportunity for students to transfer knowledge from a known context to a less familiar one.

Figure 8.6 is an example of a rich mathematical task (Clarke & Clarke, 2002) that could be used with secondary students.

2. Area = Perimeter

```
                6
┌──────────────────────────────┐
│                              │
│3                             │
│                              │
└──────────────────────────────┘
```

Consider the rectangle with dimensions 6 units by 3 units. We can calculate easily that the area is 18 square units and the perimeter is 18 units. So, if we ignore the units, *the magnitude of the area and perimeter are the same* for this shape.

Investigate the following questions, reporting what you find:

- Are there other rectangles that have this property? Please explain.

- Are there any circles for which area = perimeter? (ignoring units) Please explain.

- Are there any squares for which area = perimeter? (ignoring units) Please explain.

- (Extension task) Select another kind of shape (e.g., triangle, hexagon, etc.), and explore the situations in which the area is the same as the perimeter, ignoring units.

Figure 8.6 Rich assessment task
Source: Clarke & Clarke, 2002, p. 7.

Designing mathematics assessments: guidelines for teachers

IN PRACTICE

Cangelosi (2003) offers some guidelines for teachers designing mathematics assessments. His systematic approach is offered in the way of four phases:

- clarifying the learning goal
- designing relevant mini-experiments and storing and accessing them via a computerised folder structure
- developing a test blueprint
- synthesising the test.

Clarifying the learning goal

Before designing an achievement test, teachers should answer the question 'achievement of what?' (Cangelosi, 2003). In doing so, the relevant learning goals for which the test is designed and developed must be clarified. After the learning goals are developed, allocate a weight for each learning goal according to its relative importance to the goal of the unit, chapter or syllabus.

Designing relevant mini-experiments and storing and accessing them via a computerised folder structure

Cangelosi (2003) recommends that for each unit or chapter, teachers create an electronic folder containing mini-experiments. These mini-experiments are questions and activities that link specific mathematical content to the unit learning objectives. Over time, this 'bank' of mini-experiments will become larger and will not only assist teachers in creating valid and useable tests, but save them valuable time. For instance, such an electronic folder may take the hierarchical format of: Year 10 Mathematics > Number & Algebra > Chapter 1 > Learning Objectives > Mini-experiments. Additional folders would be created for the other two mathematical strands, chapters, learning objectives and mini-experiments.

Developing a test blueprint

According to Cangelosi (2003), a test blueprint is 'an outline specifying the features you want to build into a unit test you plan to develop from the prompts stored in your mini-experiments files' (p. 310). Such a test blueprint contains the following features:

- the title of the unit
- anticipated administration dates and times
- provisions for accommodating students with special needs
- approximate number and types of mini-experiments to be included
- an approximation of the maximum possible score for the measurement
- how points should be distributed among the objectives that define the goal (based on the weights of the objectives)
- the overall structure of the measurement
- for summative evaluations, the method for converting scores to grades.

Synthesising the test

The final stage of creating a test is to synthesise the measurement according to the variety of question formats and overall question complexity. Cangelosi (2003) recommends that teachers organise tests so that 'students have less-consuming formats (e.g. multiple choice) before respond-ing to those that are more time consuming (e.g. essay)'. To counter this, grouping prompts together with the same format 'simplifies the directions and prevents students from having to reorient their thinking frequently due to changes in format' (Cangelosi, 2003, p. 314). Additionally, prompts using the same format should be sequenced from less difficult to more difficult. Sequencing prompts this way enables students to attempt a greater proportion of the test, rather than becoming 'stuck' on a hard prompt at any early point of the assessment.

Rubrics

A rubric (or scoring framework) is an assessment tool designed to assist planning, learning and evaluation processes. Rubrics provide students and teachers with established criteria about the intentions of a learning task which are linked to identified curriculum outcomes

(Churchill et al., 2019; Siemon et al., 2015). Specifically, rubrics are comprised of two elements: the criteria about the learning task, and a scale or gradations of quality for each criterion (Lang & Evans, 2006). These elements are illustrated in the example below. Following the idea that rubrics communicate explicit pathways for developmental learning or graded performance, Churchill et al. (2019) argue that the effectiveness of rubrics rests in the specified 'content', or the development of performance statements. For instance, if a rubric is well written it 'can inform planning, strategy selection, communication and reporting, and many other aspects of the assessment process' (Churchill et al., 2019, p. 436). When designing performance assessments, Huetinck and Munshin (2008) recommend that teachers develop holistic rubrics immediately after writing the task, and well before giving the assessment to students. During the rubric-drafting phase, these authors stress the importance of using an even number of assessment points (e.g. 4 or 6). According to relevant literature of rubric development (Danielson, 1997; Stenmark, 1991), scorers are less discriminatory using a rubric with an uneven number of assessment points. Acknowledging that the use of an odd-numbered (e.g. 5-point) rubric may appear a more logical process to inevitably standardise grades, Huetinck and Munshin (2008) herein amplify:

> The natural tendency is to place many students in the middle scoring category (a 3 on a 5-point rubric) without examining the nuances of differences that indicate when more work is needed (1 or 2 on a 4-point rubric) versus communicating to students that they have shown satisfactory achievement of the learning goal (3 or 4 on a 4-point rubric). (2008, p. 367)

Other authors advise against the development of rubrics becoming a solitary task for teachers; rather, these should be created by teacher teams and in partnership with students (Lang & Evans, 2006).

Using a rubric: the lightbulb problem   CONNECTION

This Statistics and Probability learning task can be used for Year 10 students. The task is accompanied by a rubric that has been developed in accordance with the Australian Curriculum: Mathematics for the Year 10/Year 10A Data Representation and Interpretation content description. The task requires students to use their knowledge of statistical calculations (mean, median, mode, range, standard deviation), representations (stem and leaf diagrams, boxplots) and interpretations of data before making an informed judgement. The accompanying scoring rubric uses a four-point assessment scale with the performance criteria.

The lightbulb problem: The maintenance staff of Happee High School wish to make a bulk purchase of 1000 lightbulbs. To assist with this decision, the maintenance staff have obtained the results of longevity tests of three lightbulb brands, Brand A, Brand B and Brand C. The number of hours each brand of lightbulb will last (before blowing out) are displayed in Table 8.2.

It is also known that the prices per lightbulb are $0.38, $0.32 and $0.34 for Brands A, B and C respectively. Using the information from the table, determine which brand of lightbulb should be purchased, justifying your answer with appropriate statistical calculations and representations.

► ► **Table 8.2** Longevity tests of three lightbulb brands

Brand A	Brand B	Brand C
100	112	120
121	104	115
107	118	110
132	98	118
124	140	122
105	112	115
111	92	113
95	109	114
135	111	116
126	96	112
112	115	114
120	145	115

Table 8.3 Mathematical task: The lightbulb problem scoring rubric

Category	Score	Description
No response	0	The problem has not been attempted, or it has been restated without attempting any calculations.
Minimal	1	The response demonstrates a minimal knowledge of univariate statistics (mean, median, mode, range, standard deviation) with few of these measures of central tendency calculated correctly for each brand. The response indicates a graphical representation (e.g. a boxplot) containing several errors for each brand. A recommendation to purchase Brand C is justified as the cheapest option (by approximately $20) solely using calculations (no comparison of data from the three brands).
Emerging	2	The response demonstrates an emerging knowledge of univariate statistics (mean, median, mode, range, standard deviation) with some of these measures of central tendency calculated correctly for each brand. The response indicates a graphical representation (e.g. a boxplot) containing few errors for each brand. A recommendation to purchase Brand C is justified using a comparison of data from one or two brands, with calculations showing that it is approximately $20 cheaper to do so.
Satisfactory	3	The response demonstrates a satisfactory knowledge of univariate statistics (mean, median, mode, range, standard deviation) with most of these measures of central tendency calculated correctly for each brand. The response indicates a graphical representation (e.g. a boxplot) for each brand. A recommendation to purchase Brand C is justified using a comparison of data from one or two brands (standard deviation and range), with calculations showing that it is approximately $20 cheaper to do so.

Table 8.3 (*cont.*)

Category	Score	Description
Excellent	4	The response demonstrates a thorough knowledge of univariate statistics (mean, median, mode, range, standard deviation) with all these measures of central tendency calculated correctly for each brand. The response indicates a graphical representation (e.g. a boxplot) for each brand. A recommendation to purchase Brand C is justified using a comparison of data from two or more brands (standard deviation and range), with calculations showing that it is approximately $20 cheaper to do so.

PAUSE AND THINK

In this chapter you have looked at guidelines and principles for creating assessments for the secondary mathematics classroom. From what has been offered, together with your own professional learning and experience in education, devise a set of guidelines that will guide the way you plan, implement and evaluate assessments.

SHORT-ANSWER QUESTIONS

1. According to ACARA, what are some of the key assessment practices for mathematics learning?
2. The AAMT describes how highly accomplished teachers use a range of assessment strategies that are appropriate, fair and inclusive. List and explain three strategies which you plan to use in your own classroom.
3. Using your own words, describe what is meant by the term 'rich assessment tasks'. Provide an example of a rich assessment task you could use in your teaching.
4. Describe what a rubric is when assessing mathematical learning. Outline the key components of a rubric, and provide some recommendations for developing a rubric for the assessment of learning.

Providing feedback to students

Clarke (2001) argued that wherever possible, forms of assessment should be used that raise students' self-esteem, enable students to create success criteria and assist students to organise their individual targets. As such, the provision of carefully considered and delivered feedback can be instrumental in this regard. Feedback refers to 'the information gained from formal and informal assessment, provided to students, by teachers, about what the student has learnt, needs to learn, mistaken ideas and directions for improvement' (Hosking & Shield, 2001, p. 289). The key aim of feedback should be to bridge the gap between an actual level of performance and a desired learning goal (Ramprasad, 1983; Sadler, 1989). According to

Latham et al. (2011), feedback on paper, spoken or conveyed through facial expressions can have a 'branding' effect. These authors suggest that verbal, gestural or written feedback has to be thoughtfully provided and it should be specific to the individual. Before providing feedback to a student, Latham et al. (2011) recommend that teachers ask themselves a number of questions about the comments that could be made (Latham et al., 2011). A list of hypothetical questions is included below, with corresponding examples of teacher-generated feedback. For instance:

- What in particular can I comment on? *Be careful to multiply first and then add in number sentences.*
- What in particular do I hope to clarify? *Between line three and line four of working out I had difficulty in following the steps you took.*
- What do I hope to extend? *Well done – could you have arrived at the same answer another way?*
- How would I feel about receiving that comment? *Your use of algebra here is great! With continued work in rearranging equations this result will continue to improve.*
- How can I help students improve what they have done? *Remember to multiply the indices of all bases (numbers and letters) within the parentheses by the index outside the parentheses.*

Figure 8.7 Providing student feedback

PAUSE AND THINK

The theory included in this chapter indicated that feedback must be specifically targeted to a student's learning – for instance, what the student has achieved, areas for improvement, and ways in which this improvement can be reached. Additionally, this feedback must raise students' self-esteem, which can initially seem difficult if mathematics teachers rely on written feedback. Think back to when you were a student in a secondary mathematics classroom. Try to remember an assessment item where the teacher provided you with some constructive, specific and useful feedback. In recalling this feedback, respond to the following questions and prompts:

1. How was the feedback given to you (e.g. written or oral)?
2. For what assessment task was the feedback given?
3. Describe the teacher's feedback in terms of content.
4. How did this feedback make you feel about your efforts in secondary mathematics?
5. How did you respond to this feedback?
6. What advice would you give this teacher, based upon the feedback received?

Conclusion

Assessment comprises the intentional and unintentional activities undertaken by teachers to gather information about student learning (knowledge, skills, dispositions). Assessment practices assist teachers with monitoring student learning and making instructional decisions, provide valuable insight to teachers for evaluating learning programs and generate valuable information for feedback and reporting purposes. Common assessment techniques used in mathematical learning include tests, quizzes, examinations, homework and investigations. Teachers can also measure student learning through the administration of alternative assessments, such as student self-assessment, teacher observation, student journalling, oral tasks or practical tasks. When deciding precisely how student progress will be measured, Leder, Brew and Rowley (1999) suggest teachers avoid an over-reliance on one form of assessment type.

ACARA has recommended assessment practices for teachers in Australian schools to follow. These practices are outlined as a series of steps, using the Australian Curriculum content and achievement standards. In a position paper, the AAMT described how highly accomplished teachers use a range of assessment strategies that are appropriate, fair and inclusive, and which inform learning and action. Various authors provide key principles for the development of rich mathematical tasks, which encourage students to think creatively, work logically and communicate their ideas within a mathematical context. These tasks are accessible and engaging to all learners who are involved through different levels of challenge and allow the learners to make decisions about which methods or approaches to use. Cangelosi (2003) offers some guidelines for teachers designing mathematics assessments: clarifying the learning goal, designing relevant mini-experiments and storing and accessing them via a computerised folder structure, developing a test blueprint, and synthesising the test.

Wherever possible, forms of assessment should be used that raise students' self-esteem, enable students to create success criteria and assist students to organise their individual targets. Feedback refers to 'the information gained from formal and informal assessment, provided to students by teachers, about what the student has learnt, needs to learn, mistaken ideas and directions for improvement' (Hosking & Shield, 2001). To provide effective feedback to students, teachers should consider the following questions: What in particular can I comment on? What in particular do I hope to clarify? What do I hope to extend? How would I feel about receiving that comment? How can I help students improve what they have done?

BRINGING IT TOGETHER

1. What are the main reasons for assessing mathematical learning?
2. Outline and describe some assessment practices teachers can use in a mathematics classroom. Alongside each practice, note one strength and one limitation which teachers must be mindful of when using that practice to measure student learning.
3. During your next practicum experience, select a secondary mathematics textbook. List its strengths and weaknesses as a determining factor in the assessment plan for that class.
4. Outline some recommendations to bear in mind when providing feedback to students about their mathematical learning.
5. NAPLAN is a topic of continued heated debate across Australia. In this chapter, NAPLAN has been described and amplified with the inclusion of some pedagogical considerations. Develop an argument (for or against) as to whether this statewide, standardised test should be taken and passed by all students, and then defend your position with evidence.
6. Select a Lower Secondary Mathematics strand and a year level. For a particular topic, develop both a traditional and formative assessment (e.g. an end of chapter test) and an alternative assessment (e.g. a performance or interview). How can the development of different assessments as a first step in planning (i.e. constructive alignment or 'backwards planning') help teachers to plan a conceptually coherent unit of work for students?

REFERENCES

Further resources

Ahmed, A. (1987). *Better mathematics: A curriculum development study based on the low attainers in mathematics project*. Indiana: HMSO.

Anderson, J. (2009). *Using NAPLAN items to develop students' thinking skills and build confidence. Proceedings of the 22nd Biennial Conference of the Australian Association of Mathematics Teachers (AAMT) Inc.* (45–52). Adelaide: AAMT.

Australian Association of Mathematics Teachers (AAMT) (2017). *The practice of assessing mathematics learning*. Retrieved from https://www.aamt.edu.au/About-AAMT/Position-statements/Assessment/(language)/eng-AU

Australian Curriculum, Assessment and Reporting Authority (ACARA) (2015). *Implications for teaching, assessment and reporting*. Retrieved from https://www.australiancurriculum.edu.au/f-10-curriculum/implications-for-teaching-assessing-and-reporting/

Australian Institute for Teaching and School Leadership (AITSL) (2011). *Australian Professional Standards for Teachers*. Melbourne: AITSL.

Barry, K. & King, L. (1998). *Beginning teaching and beyond* (3rd edn). Katoomba: Social Science Press.

Black, P. & Wiliam, D. (1998). Inside the black box: Raising standards through classroom assessment. *Phi Delta Kappan*, 80(2), 139–48.

Bobis, J.M., Mulligan, J.T. & Lowrie, T. (2013). *Mathematics for children: Challenging children to think mathematically* (4th edn). Frenchs Forest: Pearson.

Booker, G. (2011). *Building numeracy: Moving from diagnosis to intervention*. Melbourne: Oxford University Press.

Booker, G., Bond, D., Sparrow, L. & Swan, P. (2014). *Teaching primary mathematics* (5th edn). Frenchs Forest: Pearson.

Brown, P. & Tang, K.-S. (2019). *Does the NAPLAN match the Australian Mathematics Curriculum?* Mathematics Education Research Group of Australasia. Paper presented at the Annual Meeting of the Mathematics Education Research Group of Australasia (MERGA) (42nd, Perth, Western Australia, Jun 30–Jul 4, 2019). Retrieved from https://merga.net.au/common/Uploaded%20files/Annual%20Conference%20Proceedings/2019%20Annual%20Conference%20Proceedings/Brown%20Tang_RP120.pdf

Brown, S.I. & Walter, M.I. (2005). *The art of problem posing* (3rd edn). Mahwah, NJ: Erlbaum.

Brualdi, A. (1998). Implementing performance assessment in the classroom. *Practical Assessment, Research & Evaluation*, 6(2). Retrieved from http://pareonline.net/getvn.asp?v=6&n=2

Calder, N., Brown, T., Hanley, U. & Darby, S. (2006). Forming conjectures within a spreadsheet environment. *Mathematics Education Research Journal*, 18(3), 100–16.

Cameron, W.B. (1967). *Informal sociology: A casual introduction to sociological thinking*. New York: Random House.

Cangelosi, J.S. (2003). *Teaching mathematics in secondary and middle school: An interactive approach* (3rd edn). Upper Saddle River, NJ: Pearson Education.

Chen, C., Crockett, M.D., Namikawa, T., Zilimu, J. & Lee, S.H. (2012). Eighth grade mathematics teachers' formative assessment practices in SES-different classrooms: A Taiwan study. *International Journal of Science and Mathematics Education*, 10, 553–79.

Churchill, R., Godinho, S., Johnson, N.F., Keddie, A., Letts, W., Lowe, K., Mackay, J., McGill, M., Moss, J., Nagel, M.C. & Shaw, K. (2019). *Teaching: Making a difference*. Wiley.

Civil, M. (2002). Everyday mathematics, mathematicians' mathematics, and school mathematics: Can we bring them together? In E. Yackel (series ed.), M.E. Brenner & J.N. Moschkovich (monograph eds), *Everyday and academic mathematics in the classroom* (40–62). Reston, VA: National Council of Teachers of Mathematics.

Clarke, C. & Lovitt, C. (1987). MCTP assessment alternatives in mathematics. *Australian Mathematics Teacher*, 43(3), 11–12.

Clarke, D. (1987). A rationale for assessment alternatives in mathematics. *Australian Mathematics Teacher*, 43(3), 8–10.

Clarke, D. (1998). *Securing the future: Subject based assessment materials for the School Certificate-Mathematics*. Ryde: Department of Education.

Clarke, D. (2007). *Constructive assessment in mathematics*. Berkeley, CA: Key Curriculum Press.

Clarke, D. & Clarke, B. (2002). *Using rich assessment tasks in mathematics to engage students and inform teaching*. Background paper for Seminar for Upper Secondary Teachers, Stockholm.

Clarke, S. (2001). *Unlocking formative assessment*. London: Hodder & Stoughton.

Cockcroft, W.H. (1982). *The Cockcroft report: Mathematics counts*. Report of the Committee of Inquiry into the Teaching of Mathematics in Schools under the Chairmanship of Dr W.H. Cockcroft. Retrieved from http://www.educationengland.org.uk/documents/cockcroft/cockcroft1982.html

Cooper, H.M., Robinson, J.C. & Patall, E.A. (2006). *Does homework improve academic achievement? A synthesis of research, 1987–2004*. Review of Educational Research, American Education Research Association.

Danielson, C. (1997). *A collection of performance tasks and rubrics: Middle school mathematics*. Larchmont, NY: Eye on Education Inc.

Department for Education and Skills (2007). *Mathematics at key stage 4: Developing your scheme of work*. Norwich: Crown.

Ernest, P. (1991). *The philosophy of mathematics education*. London: Falmer Press.

Evans, J. (1987). Investigations: The state of the art. *Mathematics in School*, 16(1), 27–30.

Grimison, L. (1992). *Assessment in mathematics: Some alternatives*. Mathematics Education Research Group of Australasia 15th Annual Conference. University of Western Sydney.

Hattie, J. (2009). *Visible learning: A synthesis of over 800 meta-analyses relating to achievement*. New York: Routledge.

Henke, R.R., Chen, X. & Goldman, G. (1999). What happens in classrooms? Instructional practices in elementary and secondary schools: 1994–1995. *Educational Statistics Quarterly*, 1(2), 7–13.

Hosking, P. & Shield, M. (2001). *Feedback practices and the classroom culture*. Paper presented at the 24th Annual Mathematics Education Research Group of Australasia Conference, Sydney.

Huetinck, L. & Munshin, S.N. (2008). *Teaching mathematics in the 21st century: Methods and activities for grades 6–12* (3rd edn). Upper Saddle River, NJ: Pearson Education.

Killen, R. (2005). *Programming and assessment for quality teaching and learning*. South Melbourne: Thomson Social Science Press.

Klenowski, V. (2009). Assessment for learning revisited: An Asia-Pacific perspective. *Assessment in Education: Principles, Policy & Practice*, 16(3), 263–8. https://doi:10.1080/09695940903319646

Lacey, C. & Lawton, D. (1981). *Issues in evaluation and accountability*. London and New York: Metheun.

Lampert, M. (1990). When the problem is not the question and the solution is not the answer: Mathematical knowing and teaching. *American Educational Research Journal*, 27, 29–63.

Lang, H.R. & Evans, D.N. (2006). *Models, strategies, and methods for effective teaching*. Boston, MA: Pearson.

Latham, G., Blaise, G., Dole, S., Faulkner, J. & Malone, K. (2011). *Learning to teach: New times, new practices* (2nd edn). South Melbourne: Oxford University Press.

Leder, G., Brew, C. & Rowley, G. (1999). Gender differences in mathematics achievement here today and gone tomorrow? In G. Kaiser, E. Luna & I. Huntley (eds), *International comparisons in mathematics education. Studies in mathematics education series 11* (213–24). London: Falmer Press.

Marsh, C.A. (2010). *Becoming a teacher: Knowledge, skills and practices* (5th edn). Frenchs Forest: Pearson.

Martin, A.J. (2003). *How to motivate your child for school and beyond*. Sydney: Bantam.

MCEETYA (2008). *NAPLAN – Year 7 Numeracy (calculator allowed)*. Retrieved from https://docs.acara.edu .au/resources/200807_NAPLAN_2008_Final_Test_Numeracy_Year_7_calculator.pdf

Minstrell, J. (2001). Facets of students' thinking: Designing to cross the gap from research to standards-based practice. In K. Crowley, C. Schunn, & T. Okada (eds), *Designing for science*. Mahwah, NJ: Lawrence Erlbaum Associates.

Minstrell, J. (2015). *Diagnostic assessment*. Retrieved from http://www.facetinnovations.com/daisy-public-website/fihome/services/diagnosticassessment

National Council of Teachers of Mathematics (1995). *Assessment standards for school mathematics*. Reston, VA: Author.

Ohlsen, M.T. (2007). Classroom assessment practices of secondary school members of NCTM. *American Secondary Education*, 36(1), 4–14.

Panizzon, D. & Pegg, J. (2007). Assessment practices: Empowering mathematics and science teachers in rural secondary schools to enhance student learning. *International Journal of Science and Mathematics Education*, 6, 417–36.

Piggott, J. (2015). *Rich tasks and contexts*. Retrieved from http://nrich.maths.org/5662

Pirie, S. (1987). *Mathematical investigations in your classroom: A guide for teachers*. Basingstoke: Macmillan.

Popham, W.J. (2008). *Transformative assessment*. Alexandria, VA: Association for Supervision and Curriculum Development.

Ramprasad, A. (1983). On the definition of feedback. *Behavioral Science*, 28(1), 4–13.

Ronda, E. (2010). *What is mathematical investigation?* Retrieved from http://math4teaching.com/2010/03/09/what-is-mathematical-investigation/

Sadler, D.R. (1989). Formative assessment and the design of instructional systems. *Instructional Science*, 18, 119–44.

Sattler, J. (2008). *Assessment of children: Cognitive foundations*. La Mesa, CA: Jerome's Sattler.

Senk, S.L., Beckmann, C.E. & Thompson, D.R. (1997). Assessment and grading in high school mathematics classrooms. *Journal for Research in Mathematics Education*, 28(2), 187–215.

Siemon, D., Beswick, K., Brady, K., Clark, J., Faragher, R. & Warren, E. (2015). *Teaching mathematics: Foundations to middle years* (2nd edn). South Melbourne: Oxford University Press.

Stacey, K., Steinle, V., Price, B. & Gvozdenko, E. (2018). Specific mathematics assessments that reveal thinking: An online tool to build teachers' diagnostic competence and support teaching. In T. Leuders, K. Philipp & J. Leuders (eds), *Mathematics teacher education: Diagnostic competence of mathematics teachers*, 11(241–61). Springer.

Stenmark, J.K. (1989). *Assessment alternatives in mathematics*. Berkeley, CA: University of California.

Stenmark, J.K. (1991). *Mathematics assessment: Myths, models, good questions, and practical suggestions*. Reston, VA: National Council of Teachers of Mathematics.

Stephens, M. (1988). AAMT discussion on assessment and reporting in school mathematics. *The Australian Mathematics Teacher*, 44(1), 16a–16c.

Stiggins, R.J. (1994). *Student-centred classroom assessment*. New York: Macmillan Publishing Company.

Van de Walle, J.A., Karp, K.S. & Bay-Williams, J.M. (2014). *Elementary and middle school mathematics: Teaching developmentally* (8th international edn). Essex: Pearson Education.

Watt, H.M.G. (2005). Attitudes to the use of alternative assessment methods in mathematics: A study with secondary mathematics teachers in Sydney, Australia. *Educational Studies in Mathematics*, 58, 21–44.

Webb, D.C. (2001). *Instructionally embedded assessment practices of two middle grade mathematics teachers*. Doctoral dissertation, University of Wisconsin-Madison, 2001.

Wiliam, D. (2007). Keeping learning on track: Classroom assessment and the regulation of learning. In F. Lester (ed.), *Second handbook of research on mathematics teaching and learning* (1053–98). Charlotte, NC: Information Age Publishing.

Wiliam, D. (2011). What is assessment for learning? *Studies in Educational Evaluation*, 37, 3–14.

Yeo, J.B.W. (2008). Secondary school students investigating mathematics. In G. Merrilyn, B. Ray & M. Katie (eds), *Navigating currents and charting directions: Proceedings of the 31st Annual Conference of the Mathematics Education Research Group of Australasia*, (613–619). Adelaide: MERGA.

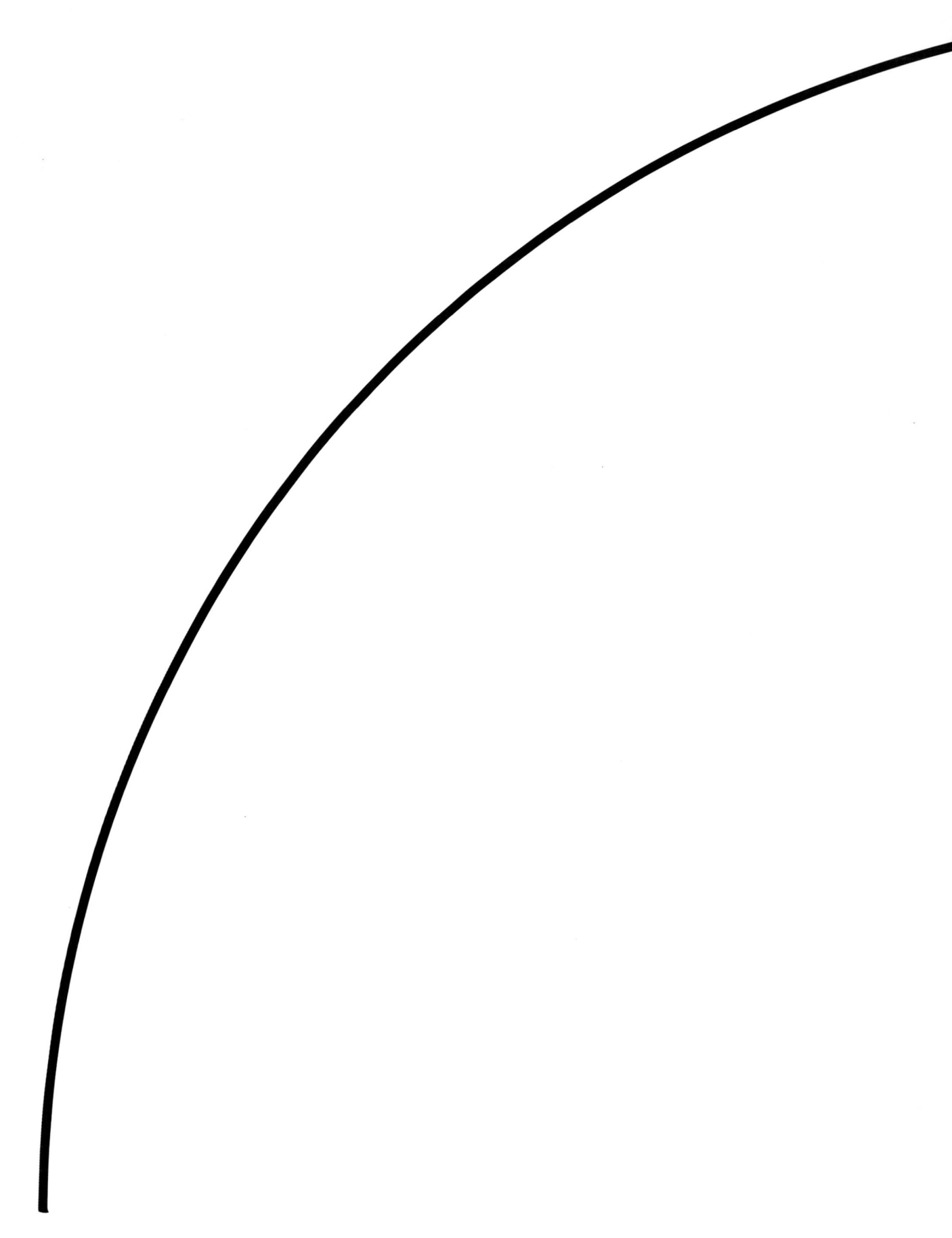

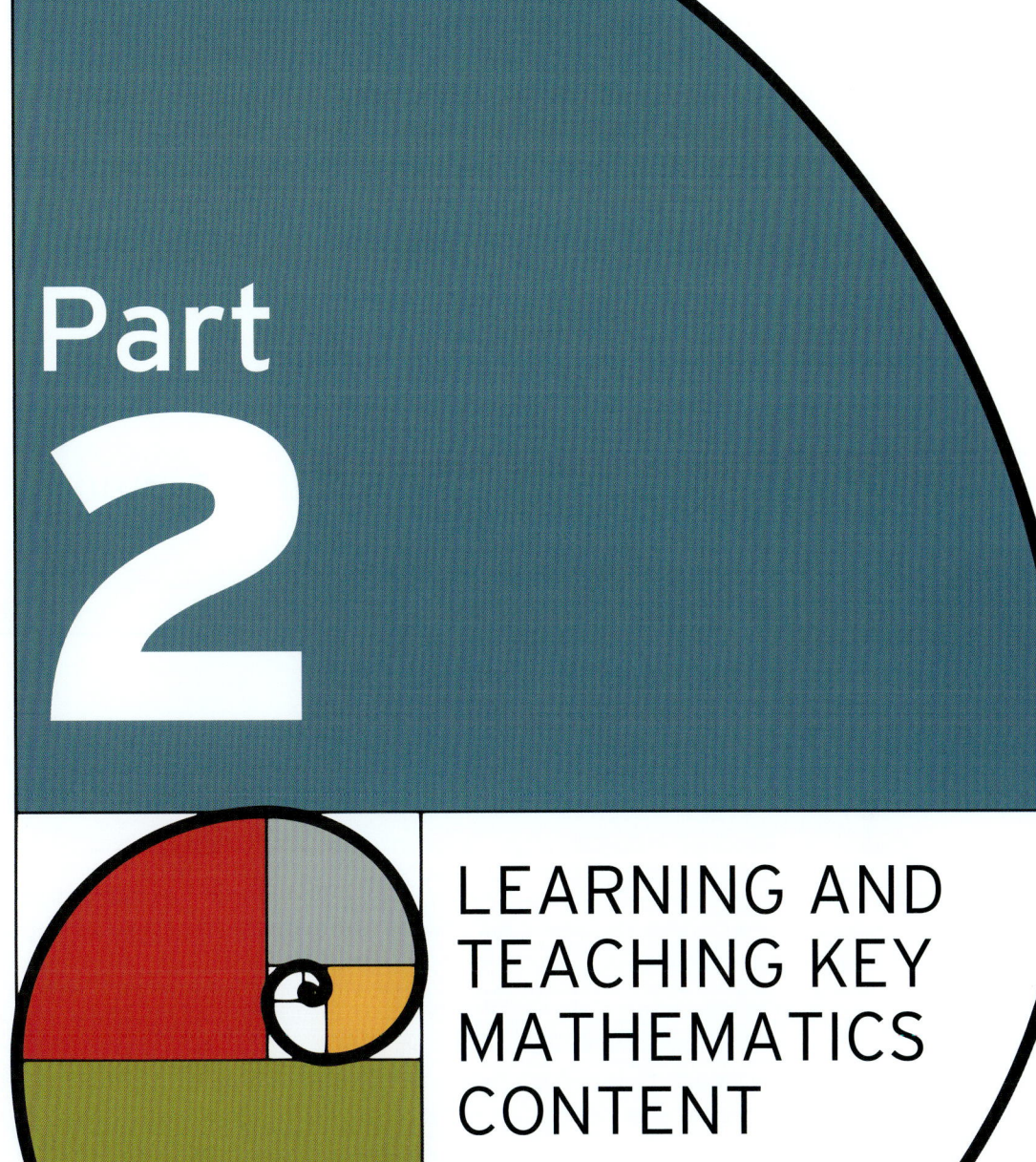

Part

2

LEARNING AND
TEACHING KEY
MATHEMATICS
CONTENT

The Mathematics Curriculum: a guide for teaching and learning

Judy Anderson

LEARNING OBJECTIVES

The learning objectives of this chapter are directly linked to the Australian Professional Standards for Teachers (APST) (Australian Institute for Teaching and School Leadership [AITSL], 2011). After studying this chapter, you should be able to:

- describe the structure of the Australian curriculum for mathematics and present arguments about the role of mathematics in the school curriculum (APST 2.1)
- identify the differences between the four proficiency strands (APST 2.1)
- present examples of how the general capabilities and cross-curriculum priorities can be taught through mathematics lessons (APST 2.4, 2.5, 2.6)
- identify the key components of a good mathematics lesson plan and explain why each is important for teaching and learning (APST 3.1, 3.4, 4.1)
- plan different types of mathematics lessons including concept or skills development focused, problem-solving focused or integrated and connected to other learning areas (APST 3.2, 3.3).

Introduction

The school curriculum in Australia has three dimensions – subjects or learning areas, general capabilities (e.g. critical thinking) and cross-curriculum priorities (e.g. Sustainability) (Australian Curriculum, Assessment and Reporting Authority [ACARA], 2012). Each learning area has a separate curriculum document which describes the essential skills and knowledge for that subject and provides advice about how to embed the general capabilities and cross-curriculum priorities into learning opportunities for students.

The Australian Curriculum: Mathematics F–10 is comprised of four proficiency strands (Understanding, Fluency, Problem Solving and Reasoning), and three content strands (Number and Algebra, Measurement and Geometry, and Statistics and Probability). The curriculum document presents content appropriate for each year of schooling, thus enabling teachers to plan lessons based on the content descriptions while at the same time ensuring they pay attention to the proficiencies. Teachers are also able to plan problem-solving lessons based on rich tasks or unfamiliar problems. In some schools, an integrated curriculum is developed as teachers from different subject areas design learning tasks or projects which connect knowledge and understanding for their students. This approach frequently requires the planning process of backward mapping whereby the anticipated product or outcome is determined before identifying suitable evidence of success and then designing learning goals. Regardless of the planning approach adopted, teachers are responsible for documenting student learning against the curriculum, providing quality feedback to students, and reporting learning outcomes to students and parents.

According to legislation, each state and territory in Australia is responsible for developing and implementing its own school curriculum. While attempts were made in the 1990s to develop a more national approach to school curriculum, no agreement was reached until earlier this century when state and territory governments agreed to collaborate on the development of the first national curriculum for Australian schools. The rationale for one curriculum for all Australian students was to improve quality, equity and accessibility, with the suggestion that 'a national curriculum would play a key role in delivering quality education' and that it would be 'world class' (ACARA, 2010). Further, one curriculum would mean:

- a united focus on how student learning can be improved to achieve national goals
- greater attention devoted to equipping students with skills, knowledge and capabilities necessary to enable them to effectively engage with and prosper in society
- more efficient development of high-quality resources
- greater consistency for mobile student and teacher populations (ACARA, 2010).

The newly formed Australian Curriculum, Assessment and Reporting Authority (ACARA) developed the first Australian curriculum from 2007 in consultation with representatives from each of the states and territories. However, when and how to implement the Australian Curriculum is the responsibility of state and territory curriculum authorities. Several papers informed and guided the development of the Australian Curriculum including the *Shape of the Australian Curriculum* paper (ACARA, 2012). A key goal was to develop successful learners, confident individuals, and active and informed citizens. To achieve this goal, the curriculum was developed in consultation with a range of stakeholders including teachers, school leaders, school system personnel, parents and students. Initially developed for English, Mathematics, the Sciences and History, the proposed curriculum was endorsed by education ministers in 2012 for implementation in Foundation, the first year of compulsory formal school education, to Year 10.

The curriculum for the senior years of schooling (Years 11 and 12) was also developed by ACARA – four differentiated courses were identified to cater for appropriate pathways beyond schooling. Because of the diversity of approaches to curriculum organisation and end-of-school examinations, states and territories agreed to embed the content from these courses into their existing programs in 2014 – for further information see ACARA's webpage on the Senior Secondary Curriculum.

While the term 'curriculum' can be used as a broad description of the 'decisions and choices by teachers and students that together construct their experience in classrooms in a multitude of formal and informal contexts' (Smith & Ewing, 2002, p. 28), in this chapter, the term 'curriculum' is used to describe the plan of learning about content and skills to be taught in schools. The school curriculum is traditionally presented in separate subject areas and in some jurisdictions the separate subject area documents are each referred to as a syllabus.

Three representations of curriculum have been described – the intended curriculum, the implemented curriculum and the attained curriculum, although a more nuanced set of representations might include:

- Visionary: the ideas, ideals and intentions underpinning choices in the curriculum.
- Written: how the intentions are elaborated and specified in a written format.
- Perceived: how teachers interpret the intended curriculum.
- Operational: how the curriculum is enacted in classroom practice.
- Experiential: how students experience the curriculum.
- Attained: the learner outcomes of the enacted curriculum (van den Akker, 2018).

The vision for the Australian Curriculum is presented in the goals of schooling, recently updated (Council of Australian Governments Education Council, 2019), and in the rationale and aims of written curriculum documents, which also include essential curriculum components such as content, learning activities, resources and assessment advice.

Scope and sequence chart: a document which connects 'concepts that refer to the overall organisation of the curriculum to ensure its coherence and continuity. Scope refers to the breadth and depth of content and skills to be covered. Sequence refers to how these skills and content are ordered and presented to learners over time' (UNESCO-IBE, 2013).

ACARA describes the Australian Curriculum as three-dimensional since it includes learning areas, general capabilities and cross-curriculum priorities (see Figure 9.1). It states: 'These all contribute to a well-rounded education of all Australian students, providing the knowledge, understanding and skills needed for life and work in the 21st century' (ACARA, 2020a). In the latest online version of the Australian Curriculum, disciplinary knowledge, skills and understanding are described in eight learning areas: English, Mathematics, Science, Health and Physical Education, Humanities and Social Sciences, The Arts, Technologies and Languages, with the last four of these written to include multiple subjects.

In schools, leaders are responsible for ensuring the curriculum documents are used to plan the teaching and learning program for students. The perceived curriculum might be represented through whole school scope and sequence charts. This planning approach helps to ensure all teachers address the whole curriculum across all years. Based on the scope and sequence charts, teachers develop programs and units of work for their students.

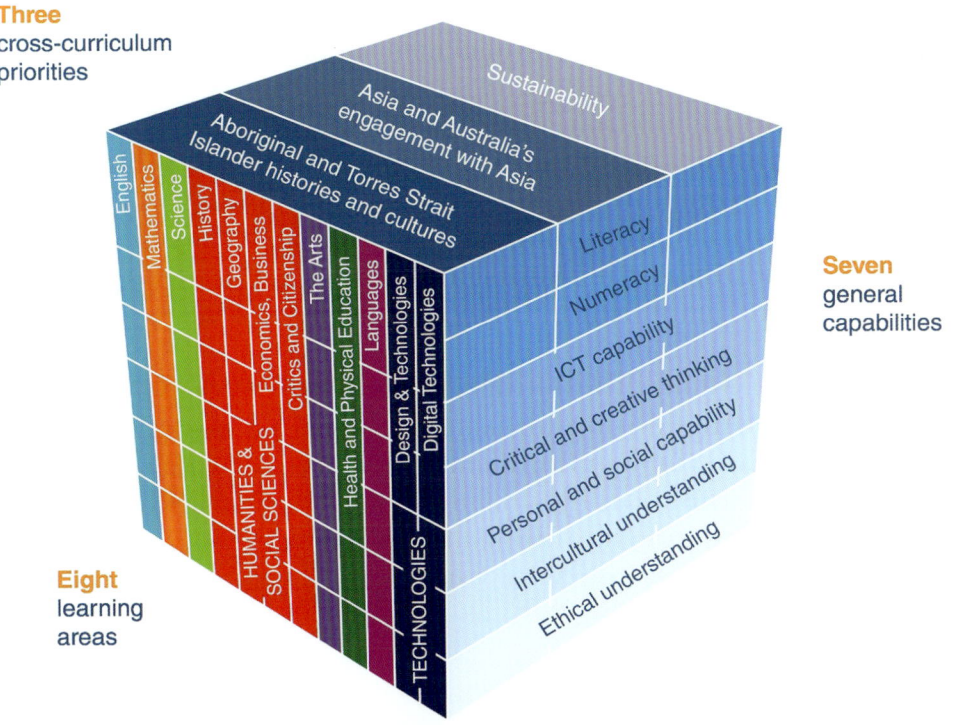

Figure 9.1 A representation of the three dimensions of the Australian Curriculum
Source: ACARA, 2020k.

This chapter presents a description of the structure of the Australian Curriculum: Mathematics F–10, including an overview of the content strands, followed by descriptions of the proficiency strands. Information about each of the general capabilities and the cross-curriculum priorities follows with connections made to how they might be addressed within mathematics lessons. To assist with planning for teaching and learning, key components of good mathematics lessons are presented, and followed by advice about planning different types of mathematics lessons.

The structure of the Australian Curriculum: Mathematics F–10

The Australian Curriculum: Mathematics has three content strands – Number and Algebra, Measurement and Geometry, and Statistics and Probability – and four proficiency strands – Understanding, Fluency, Problem solving and Reasoning. Compared to many earlier curriculum documents in Australia, the content strands are grouped in pairs, allowing opportunities to teach mathematical concepts together and highlight the connections between them.

Within each of the three content strands, content descriptions specify what is to be learned, and 'achievement standards describe the depth of understanding and the sophistication of knowledge and skill expected of students at the end of each year level or band of years' (ACARA, 2020b). While there is only one curriculum for mathematics and the content

Content descriptions: include knowledge, understanding and skills, described at a year level or band of years in the Australian Curriculum.

is specified at each year level, a Year 10A is included to supplement the content at the Year 10 level for students aiming to study the more advanced mathematics subjects in Years 11 and 12.

While the mathematics learning area is a component of one dimension of the Australian Curriculum, the other two dimensions are equally important – general capabilities and cross-curriculum priorities. The remainder of this section of the chapter examines the rationale and aims of the mathematics curriculum and then focuses on describing the structure of the three content strands.

Rationale and aims for mathematics in the curriculum

Before considering the structure of the content strands in the mathematics curriculum, it is important to understand the role of mathematics in society and its purpose in the school curriculum. These are usually expressed in curriculum documents as the rationale and the aims. While many of us take the importance of mathematics for granted, that is not the case for all students, particularly for those who find the subject difficult and not necessarily meaningful.

PAUSE AND THINK

If a student asked you why they needed to learn mathematics, what would you say?

Most people see a practical role for mathematics in the school curriculum which helps citizens navigate and manage their daily lives – this would involve managing finances, reading timetables, interpreting information and making judgements about claims based on data, and planning and organising events. This practical mathematics is frequently described as 'numeracy' and is encompassed in the Australian Curriculum general capability, Numeracy, which is described later in this chapter. However, others see a specialised role which enables learners to develop a deep foundation of mathematical principles that facilitates the further study of more sophisticated mathematics. Sullivan (2011) argues school mathematics should provide opportunities for both perspectives, although part of the teacher's role is to adapt the learning experiences to the needs of the learner.

The rationale for the Australian Curriculum: Mathematics provides powerful information which helps to justify its role in the development of essential mathematics skills in each of the content strands, its usefulness for life and work, its cultural and historical purpose, and its role in helping to solve problems and issues facing our world. While mathematics is described as a separate subject with three content strands, connections can be made within the content strands of the mathematics curriculum as well as with other subjects such as science, health, geography and technology. These connections need to be emphasised by teachers at every opportunity so that learning does not become too atomised, painting a picture of mathematics as disconnected facts and procedures. By presenting mathematics to students as having a key role in solving authentic problems, as being able to provide important information about current affairs, and as helping to analyse data and predict future scenarios in medical and scientific fields, teachers can bring mathematics to life for students who might find the subject uninspiring.

The aims for the Australian Curriculum: Mathematics are succinct, useful statements that can guide our overall approach to designing learning experiences that engage and enthuse our students. They indicate we need to ensure students:

- are confident, creative users and communicators of mathematics, . . .
- develop an increasingly sophisticated understanding of mathematical concepts and fluency with processes, and are able to pose and solve problems and reason . . .
- recognise connections between the areas of mathematics and other disciplines and appreciate mathematics . . . (ACARA, 2020c)

For these aims to be realised, we need to plan tasks, lessons and units of work that provide students with success, that allow them to develop deep and connected understandings of mathematics content, and that encourage them to be curious, pose their own questions and use their knowledge in meaningful ways both within mathematics as well as across the whole school curriculum. The following sections describe each of the content strands of the mathematics curriculum and highlight key concepts and issues to be addressed – further detail about the teaching and learning of each of these strands can be found in Chapters 10 to 13 in this volume.

Lynn Arthur Steen

CONNECTION

Lynn Arthur Steen (1961–2015) was a mathematician who sought to develop accessible and meaningful curriculum for his undergraduate students as well as advocating for inclusive school mathematics curriculum. He was influential in mathematics education and edited a well-received volume in 1990 called *On the Shoulders of Giants: New Approaches to Numeracy* which suggested 'the key issue for mathematics education is not whether to teach fundamentals but which fundamentals to teach and how to teach them' (p. 2). Further, he suggested traditional school mathematics picks a few strands and 'arranges them horizontally to form the curriculum: first arithmetic, then simple algebra, then geometry, then more algebra, and finally — as if it were the epitome of mathematical knowledge — calculus. This layer-cake approach to mathematics education effectively prevents informal development of intuition along the multiple roots of mathematics ... To help students see clearly into their own mathematical futures, we need to construct curricula with greater vertical continuity, to connect the roots of mathematics to the branches of mathematics in the educational experience of children' (p. 4). The chapters in his book present five examples to exemplify the developmental power of deep mathematical ideas which could be used as a different way to organise curriculum – dimension, quantity, uncertainty, shape and change.

Number and Algebra

The Number and Algebra content strand is presented as six sub-strands to be taught within designated year levels as indicated in brackets: number and place value (F–8); fractions and decimals (1–6); real numbers (7–10); money and financial mathematics (1–10); patterns and algebra (F–10); and linear and non-linear relationships (7–10). In this strand, students gradually develop understanding of the number system, including the magnitude and properties of number; they develop computational strategies, recognise patterns, and learn about variable and function; and they describe relationships, determine generalisations, recognise equivalence and solve equations.

When you look at the Australian Curriculum: Mathematics online, you will notice that for each content strand, and at each year level, there are lists of content descriptions with elaborations. Content descriptions include knowledge, understanding and skills, described at a year level or band of years in the Australian Curriculum. For example, for Year 7 Number and Algebra, the first content description under the sub-strand Number and Place Value is:

> Investigate index notation and represent whole numbers as products of powers of prime numbers (ACMNA149) – notice the code which stands for 'Australian Curriculum Mathematics Number and Algebra description 149'

This content description is accompanied by three elaborations with one example as follows:

> defining and comparing prime and composite numbers and explaining the difference between them. (ACARA, 2020d)

Listing content as topics does not highlight the important overarching concepts or 'big ideas' that are critical for developing students' deep knowledge and understanding of number and algebra. Siemon, Bleckly and Neal (2012) describe a big idea as:

- an idea, strategy, or way of thinking about some key aspect of mathematics without which, students' progress in mathematics will be seriously impacted
- encompasses and connects many other ideas and strategies
- serves as an idealised cognitive model, that is, it provides an organising structure or a frame of reference that supports further learning and generalisations . . . (p. 22)

Not all researchers agree on a definitive list of big ideas, but for number, most would include trusting the count, place-value, multiplicative thinking, partitioning, proportional reasoning, and finally, reasoning (Siemon, 2006).

One important big idea, proportional reasoning, underpins much of the learning in number and algebra, particularly at the secondary school level. Proportional reasoning involves using relational rather than absolute thinking to analyse change over time or compare relationships. A deep understanding of proportional reasoning is necessary to fully understand rational number, ratio, rates, problems in financial mathematics, and change in relationships, and these concepts are themselves the basis for trigonometry, probability and calculus, and hence required in other strands such as scale and similar figures in measurement and geometry. Finally, being able to use proportional reasoning understanding to solve problems in context, such as percentage change in financial mathematics, is crucial.

Designing learning experiences for students that support the development of these big ideas requires a careful consideration of the content descriptions within a particular year level of schooling but also across year levels. It is critical when planning lessons to read the content descriptions for the year level you are teaching and also for the year level below and above since not all students will have developed the requisite knowledge and be ready for your lesson while others may already know and understand the main ideas. A resource

Elaborations: are optional and are provided to give teachers ideas about how they might teach the content.

which helps this process is the Sequence of Content published on the Australian Curriculum website (ACARA, 2020e) for F–6 and 7–10 for each strand.

The introduction of algebra into the mathematics curriculum from the first years of schooling occurred early this century, with a '. . . recent emphasis on patterning and early algebraic reasoning in the preschool and early grades. Early algebraic thinking develops through an awareness of the structural relationships of patterns and later in the structure of arithmetic' (Mulligan, Cavanagh & Keanan-Brown, 2012, p. 47). Connecting arithmetic thinking with algebraic thinking needs to occur from the early years so that students can identify and represent patterns and make generalisations using inductive reasoning. Formal algebra is not introduced until Year 7 when students learn about variables, and then beyond that, they learn about linear and non-linear relationships.

Measurement and Geometry

The Measurement and Geometry content strand is presented as five sub-strands to be taught within designated year levels as indicated in brackets: using units of measurement (F–10); shape (F–7); geometric reasoning (3–10); location and transformation (F–7); and Pythagoras and trigonometry (9–10). In this strand, students develop understanding of size, shape, relative position and movement of two-dimensional figures in the plane and three-dimensional objects in space; they define, compare and construct figures and objects; and they learn to develop geometric arguments. Students develop knowledge and understanding of measurements of quantities and learn how to choose appropriate metric units as well as build understanding of the connections between units and calculate derived measures such as area, speed and density. While Lowrie, Logan and Scriven (2012) suggest this strand assists teachers to make connections between measurement and geometry, they note a lack of reference to spatial and visual reasoning – a shortcoming given the frequent use of technology to teach and support learning of the content in this strand.

There are several important aspects to learning measurement and geometry, which need to be emphasised by teachers. Developing the correct terminology is critical even though students are encouraged to use 'everyday' language in the early years of schooling. Students need to be provided with 'hands on' measuring experiences and able to manipulate shapes and objects to identify different views in a range of orientations. For example, students need to be able to recognise shapes where the base is horizontal to the bottom of a page or screen as well as when it is vertical or at a different angle. This knowledge building connects with transformational geometry. Students must also develop deep understandings about the relationships between length, area and volume, so that they appreciate what occurs to the area and volume, for example, when the side length of an object is doubled. Measurement provides a practical context for a lot of number work (e.g. rounding and approximation) – an example of integration across the content strands.

Statistics and Probability

The Statistics and Probability content strand is presented as two sub-strands to be taught within designated year levels as indicated in brackets: chance (1–10); and data

Sequence of Content: the term used for the sequence of concepts in the Australian Curriculum and is available on the Australian Curriculum website.

Spatial and visual reasoning: spatial reasoning is the ability to imagine or visualise in one's mind the positions of objects, their shapes, their spatial relations to one another and the movement they make to form new spatial relations. Visual reasoning is the process of manipulating one's mental image of an object to reach a certain conclusion.

representation and interpretation (F–10). Statistics and probability concepts build and are developed together, beginning with recognising and analysing data, drawing inferences, summarising and interpreting data, assessing likelihoods and assigning probabilities, building more sophisticated approaches to evaluation, and developing intuitions about data.

According to Watson and Neal (2012), 'doing statistics' is about carrying out investigations, so this strand offers unique opportunities for teachers to also address problem solving and reasoning from the proficiency strands. As the authors appropriately note, statistics is not just about drawing a graph or calculating a mean but rather, it is about answering a question which requires data collection, data representation, data reduction, chance and inference while also considering variation. Across the year levels, students are introduced to more and more sophisticated techniques, which can be used in their investigations with the added benefit that these strategies can be used in many contexts and across the curriculum.

SHORT-ANSWER QUESTION

What types of investigations could students conduct in Year 9 to support the development of knowledge and understanding in statistics and probability?

The proficiency strands of the Australian Curriculum: Mathematics

The three content strands in the mathematics curriculum are accompanied by four proficiency strands – Understanding, Fluency, Problem Solving and Reasoning. While problem solving has been explicitly included in mathematics curriculum documents since the 1980s (Anderson, 2014), the other three proficiencies have been implied as essential elements of learning, but for the first time they are named as strands of the curriculum and were adapted from the five proficiencies described by Kilpatrick, Swafford and Findell (2001) (see Table 9.1).

Table 9.1 The proficiencies of Kilpatrick, Swafford and Findell (2001) aligned with the Australian Curriculum: Mathematics

Proficiencies from Kilpatrick, Swafford and Findell (2001)	The Australian Curriculum
Conceptual Understanding – comprehension of mathematical concepts, operations and relations	Understanding
Procedural Fluency – skill in carrying out procedures, flexibly, accurately, efficiently and appropriately	Fluency
Strategic Competence – ability to formulate, represent and solve mathematical problems	Problem Solving
Adaptive Reasoning – capacity for logical thought, reflection, explanation and justification	Reasoning
Productive Disposition – inclination to see mathematics as sensible, useful and worthwhile, with a belief in diligence and self-efficacy	NA

Kilpatrick, Swafford and Findell (2001) explained that 'proficiency' is used to describe what 'it means for anyone to learn mathematics successfully' (p. 5). In addition, they noted 'these [proficiencies] are not independent; they represent different aspects of a complex whole ... the five strands are interwoven and interdependent in the development of proficiency in mathematics' (p. 116). The authors used an image of a twisted rope with five 'strands' to represent the linking and connectedness of the proficiencies. Note that Kilpatrick, Swafford and Findell (2001) described a fifth proficiency, productive disposition, which has not been included in the Australian Curriculum: Mathematics. Table 9.2 presents the definitions of each of the proficiencies from the Australian Curriculum: Mathematics.

Table 9.2 Brief descriptions for each of the proficiencies

Proficiency	Description
Understanding	Students build a robust knowledge of adaptable and transferable mathematical concepts. They make connections ... They develop an understanding of the relationship between the 'why' and the 'how' of mathematics ...
Fluency	Students develop skills in choosing appropriate procedures, carrying out procedures flexibly, accurately, efficiently and appropriately, and recalling factual knowledge and concepts readily ...
Problem Solving	Students develop the ability to make choices, interpret, formulate, model and investigate problem situations, and communicate solutions effectively ...
Reasoning	Students develop an increasingly sophisticated capacity for logical thought and actions, such as analysing, proving, evaluating, explaining, inferring, justifying and generalising ...

Source: ACARA, 2020f.

PAUSE AND THINK

If you were planning to teach Year 7 students to add fractions, how could you provide opportunities within the lesson to develop understanding, fluency, problem solving and reasoning?

To highlight the embedding of the proficiencies in the content descriptions, a Year Level Description is presented at the beginning of each year of the Australian Curriculum: Mathematics, followed by statements referring to each of the proficiencies appropriate for that year level. These statements are rather general and perhaps not explicit enough to help new teachers plan to include the proficiencies in mathematics lessons. Other resources are available providing examples of tasks for each of the proficiencies – see, for example, the advice and work samples on the ACARA website (ACARA, 2020g), Sullivan (2011), and Chapter 5 in this volume.

SCENARIO

How do I include problem solving in a lesson on finding the surface area of objects?

During professional experience, a student teacher received feedback for her Year 9 lesson on finding the surface area of objects that suggested she was only providing formulae and not allowing the students to problem solve and investigate the ideas for themselves. Her mentor teacher suggested she visit the Australian Curriculum website to read the Level Description for Year 9 which describes how the proficiencies are embedded into the content descriptions. By focusing on the Measurement and Geometry strand, she also recommended the student read the content descriptions and the elaborations related to finding the surface area of objects.

Question

Based on the advice provided in the curriculum descriptions, how could problem solving be introduced into this lesson?

While the statements provided on the Australian Curriculum website communicate how teachers might include the proficiencies in lessons, many teachers worry about preparing students for assessment tasks which may not include assessment of the proficiencies. For example, authentic problem solving takes time so it can be difficult to include problem-solving questions in short, timed tests. Further advice about assessment is included in Chapter 8 of this volume. As well as focusing on the strands in the curriculum, all the Australian Curriculum learning areas are required to include general capabilities and cross-curriculum priorities where appropriate.

SHORT-ANSWER QUESTIONS

1. Compare the description of 'conceptual understanding' in Table 9.1 with the description of 'understanding' in Table 9.2. By reflecting on how you know when you understand something new, how would you define 'understanding'?
2. What strategies would you use to build 'productive disposition' (Table 9.1) in a mathematics lesson?

General capabilities and cross-curriculum priorities

While the learning areas form one dimension of the Australian Curriculum, the general capabilities and cross-curriculum priorities form the other two, as presented in the three-dimensional curriculum model in Figure 9.1. There are seven general capabilities and three cross-curriculum priorities. These are discussed in the following sections.

General capabilities

The seven general capabilities are literacy, numeracy, information and communication technology capability, critical and creative thinking, personal and social capability, ethical understanding, and intercultural understanding (see Figure 9.2). In the Australian Curriculum, capability encompasses knowledge, skills, behaviours and dispositions that are essential for students to participate fully in their learning at school and beyond.

Currently there is a clear agenda to prepare students for the workplace and to ensure they have developed a suite of twenty-first-century skills (Care, Griffin & Wilson, 2018). The general capabilities in the Australian Curriculum align with many lists of twenty-first-century skills. To support the implementation of general capabilities in lessons, each is included within the content of the mathematics curriculum and is identified on the website beside relevant content descriptions with a distinguishing icon. When accessing the Australian Curriculum online, it is possible to apply a filter to the content descriptions to identify where the capabilities have been embedded in the content descriptions. However, teachers are encouraged to also develop their own approaches to addressing the general capabilities in mathematics lessons.

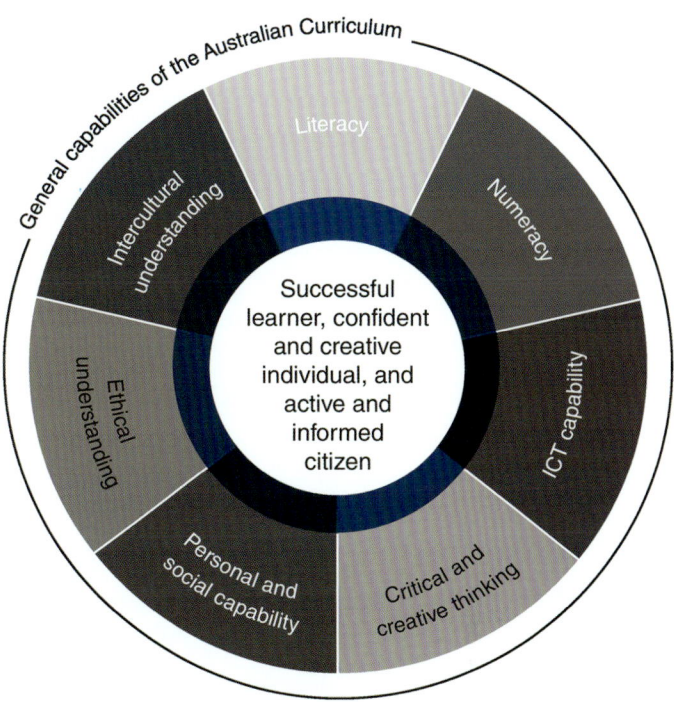

Figure 9.2 The general capabilities in the Australian Curriculum
Source: ACARA, 2020h.

In addition to identifying content descriptions in the mathematics curriculum where a capability can add richness to student learning, each capability is presented as a learning continuum to describe the relevant knowledge, skills, behaviours and dispositions at particular points of schooling. One important consideration is the relationship between mathematics and numeracy.

Mathematics and numeracy

In the Australian Curriculum: Mathematics, it is noted 'mathematics makes a special contribution to the development of numeracy in a manner that is more explicit and fore grounded than is the case in other learning areas' (ACARA, 2012). The definition of numeracy on the Australian Curriculum website (ACARA, 2020h) suggests becoming numerate means developing the disposition to use and apply mathematics confidently and purposefully across subject areas, and in many situations.

So, what is the difference between mathematics and numeracy? Goos et al. (2019, p. 18) investigated this question, suggesting that 'numerate people engage with the world and its diverse contexts and situations' while mathematicians 'look for patterns and formulate conjectures'. Mathematics is more abstract whereas numeracy requires a focus on the setting and context to solve problems which might be confronted in everyday experiences, including school or work.

To become numerate, school students need to use mathematics in different contexts and to solve many different types of problems. While they can learn mathematics within mathematics lessons, they need to then have a range of experiences to use their mathematics to solve problems in all school subjects. Goos (2018, p. 62) is critical of the treatment of numeracy in the Australian Curriculum because it lacks reference to the 'critical orientation to the use of mathematics to make decisions and judgements, and support or challenge an argument or position'. Another way to develop students' capabilities is through the cross-curriculum priorities, the final dimension in the Australian Curriculum.

Cross-curriculum priorities

There are three cross-curriculum priorities in the Australian Curriculum which provide national, regional and global dimensions to enrich the curriculum:

- Aboriginal and Torres Strait Islander Histories and Cultures
- Asia and Australia's Engagement with Asia
- Sustainability.

Sustainability: one of three cross-curriculum priorities in the Australian Curriculum. It addresses the ongoing capacity of Earth to maintain all life. Sustainable patterns of living meet the needs of the present without compromising the ability of future generations to meet their needs (ACARA, 2020j).

The priorities are embedded in content descriptions as relevant. For mathematics, some reference is made to Indigenous mathematics, including representations of number, space and pattern. The contribution of Asian mathematicians is noted as well as the mathematics in games, art and architecture. Sustainability connects with problem solving and modelling in real-world contexts through the impact of humans on the environment.

The cross-curriculum priorities provide unique opportunities for students to engage in investigations which foster the problem solving and reasoning proficiencies as well as many of the general capabilities, including critical and creative thinking, personal and social capability, ethical understanding and intercultural understanding. Through the application and evaluation of statistical data, students can deepen their understanding of the lives of Aboriginal and Torres Strait Islander peoples. Investigations involving data collection, representation and analysis can be used to examine issues pertinent to the Asian region. For Sustainability, students can investigate the health of the local ecosystem; propose a budget for local changes to the environment; measure, monitor and quantify changes in ecological systems over time; and use statistical analysis to enable the prediction of probable futures which informs decision making (ACARA, 2020i).

Teachers have used sustainability as a theme for cross-curriculum project work, thus connecting mathematics with the other STEM subjects of science, technology and engineering (Anderson et al., 2019). Sustainability also connects with geography, history, and health and physical education. Further information about planning investigations and cross-curriculum projects is included later in this chapter and in Chapter 5.

SHORT-ANSWER QUESTIONS

1. What is the difference between mathematics and numeracy?
2. What types of experiences in mathematics would support the development of the critical and creative thinking capability in mathematics lessons?
3. How could an open-ended investigation such as 'Is our school environmentally friendly?' promote the general capabilities as well as the Sustainability cross-curriculum priority?

Planning for teaching and learning

To teach mathematics effectively, teachers require mathematical 'content knowledge, general pedagogical knowledge, curriculum knowledge, pedagogical content knowledge, knowledge of learners, knowledge of education contexts, and knowledge of educational ends, purposes and values' (Shulman, 1987, p. 8). These knowledges are informed by your own experiences of learning mathematics but then developed and refined during your preservice education program and field experience placements. Curriculum knowledge is clearly important as this guides what you teach, to whom and when, so this section of the chapter includes consideration of the curriculum as perceived by teachers, as operationalised in classrooms and as experienced by students (van den Akker, 2018).

Planning for teaching and learning can take many forms since there needs to be planning at the system level, the whole school level and the year level, as well as at the classroom level. There will be policies and procedures from the school system or jurisdiction and from the school principal. The head or coordinator of the mathematics department has responsibility for developing a whole school approach to the teaching and learning of mathematics in collaboration with the mathematics teachers. Policies and procedures would include:

- a scope and sequence and associated learning programs for each course at each year level, including teaching activities mapped against curriculum content, and strategies for registration and evaluation
- teaching and learning time allocated to each year level
- accommodations for students with additional needs
- allocation of textbooks, resources and technological tools
- assessment approaches and timetables
- homework expectations for each year level
- roles and responsibilities of mathematics teachers in the team.

Individual mathematics teachers are responsible for:

- following the year level scope and sequence plan and program and registering content addressed
- planning tasks, lessons and units of work for each class they teach
- designing assessment tasks
- assessing and providing feedback to students

- gathering evidence of learning and reporting to parents and students
- reflecting on practice and compiling evidence that they meet the AITSL standards.

Beginning teachers will work with a mentor to support them in this work. As a mathematics teacher, you will spend time planning, and for most beginning teachers planning good lessons is critical to engaging and enthusing your students. This section of the chapter includes advice about the features of a good mathematics lesson plan, students' views on mathematics lessons, and advice about planning a range of different types of mathematics lessons.

Features of a good mathematics lesson plan

Teachers are expected to plan tasks, lessons and units of work for each class they teach. To plan lessons, some teachers begin with a task they believe will help the students learn the concept they need to teach and then create a lesson involving the task. Others create lessons based on a curriculum content description and then look for tasks or activities which will support student learning of that concept. Both approaches will work if you consider the overall structure of the lesson, plan within the time allowed for the lesson, and consider the resources needed, the learning needs of the students, and how you will manage and organise the classroom and seating arrangements.

Before discussing the components of a good mathematics lesson, it is worth reflecting on the types of lessons you enjoyed most when you were learning mathematics at school.

PAUSE AND THINK

What types of mathematics lessons did you prefer most in secondary school? What did you do during the lesson? What did the teacher do? How were the lessons structured?

You will continue to develop your knowledge and understanding of teaching mathematics as you work with your own students during your career, but being prepared is critical to enhancing your efficacy to teach and to instil confidence in your students that you do know and understand how to teach. Preparation for teaching includes reading the curriculum, reviewing advice about teaching the topic, examining available resources and working with a mentor. Lesson planning requires time and effort, and it means you think carefully about several features of the lesson beforehand, you have written down what you plan to do, and you have a backup plan if things go awry, particularly when using technology. Many student teachers initially plan lessons which take much longer to implement than anticipated. This is not surprising as it is difficult to plan for students you do not know, and you are probably drawing on your experiences as a learner. Working with your professional experience supervisor and mentor assists with adjusting expectations and designing lessons which better meet the needs of the students in each class.

To support the development of good lesson planning, three components are necessary:

1. specify learning objectives, including knowledge and skills
2. select and sequence learning experiences
3. evaluate the effectiveness of the learning experiences.

The three components of good lesson planning need elaboration because each component involves multiple decisions, as follows:

1. specify learning objectives including knowledge and skills – how does this lesson fit into the overall plan for learning this topic? How did students cope with the last lesson? What issues arose and need to be addressed in this lesson?
2. select and sequence learning experiences – what teaching strategies will best support learning this mathematics? How will I begin the lesson? Which tasks will I use? Will students work alone, in pairs, or in some other arrangement? How long will they need to do each task? What will I do if students struggle to begin? What will I do if students finish quickly? How will I conclude the lesson? What homework will I set?
3. evaluate the effectiveness of the learning experiences – how will I get feedback from the students on the lesson? What happened during the lesson that I need to address in the next lesson? Who struggled and needs more support? Who needs more challenging tasks? How effective were my teaching strategies?

To support lesson planning, many sources have examples of lesson planning templates and most teacher education providers present student teachers with their preferred templates – see, for example, the University of Notre Dame's School of Education Lesson Plan template (www.notredame.edu.au/__data/assets/word_doc/0009/13122/Lesson-Plan-Template.doc).

This template takes planning lessons to the next level and includes a list of important features of a well-planned lesson that also connect to the Australian Curriculum:

1. Lesson organisation including year level, date, learning area (in our case, mathematics), curriculum topic
2. Students' prior knowledge – what mathematics do students need to know to be able to engage with this lesson?
3. General capabilities – which ones are incorporated into the lesson and in what ways?
4. Cross-curriculum priorities (when appropriate)
5. Mathematical proficiencies – briefly describe how the lesson incorporates understanding, fluency, problem solving and reasoning
6. Lesson (or learning) objectives – two to four main learning outcomes – see additional advice below
7. Teachers' preparation and organisation for the lesson – include seating plan, resources and materials for students to use, technology requirements
8. Provision for students at risk – what supports/accommodations are needed?
9. Lesson evaluation – to be completed after the delivery of the lesson
10. Lesson delivery – introduction, lessons steps and lesson closure – include content, processes and pedagogical approaches used by the teacher as well as actions/responses required of the students

Learning objectives: brief statements that describe what students will be expected to learn by the end of the lesson. Learning objectives are designed for most students to achieve within the lesson; they should be limited to 2 to 4; and they should be observable.

Learning experiences: include all tasks and activities students participate in during lessons. Learning experiences include listening to teacher explanations and directions as well as completing exercises, discussing ideas with peers, completing assessment tasks and investigating problems alone or in a group.

11. Transition to the next lesson – what will you need to do before the next lesson?
12. Assessment – how will you judge the learning objectives have been met and by whom? Allow opportunities for students to provide feedback on their learning.

The lesson delivery on p. 2 of the template is divided into the three key phases of the lesson, usually referred to as introduction, body and conclusion. Within each of these phases, it is useful to describe the mathematical content and processes to be addressed as well as the teachers' pedagogical practices, including such things as explanation, demonstration, calling for and answering questions, organising students into groups, distributing materials or resources, and scaffolding the learning – see Chapter 2 for a list of scaffolding practices. Some templates divide the three key phases of the lesson into two columns headed 'Teacher's actions' and 'Students' actions'. This might help you to keep in mind what you will be doing and also what your students will be doing. Students need to do more than just listen to the teacher. Consider how you can make learning more active.

> **Teachers'**
> **pedagogical**
> **practices:** include the
> learning activities
> adopted by the teacher
> to help students learn
> the content or concepts
> for the lesson.
> Such practices might
> include connecting
> new content to
> past learning,
> demonstrating what
> students are to do,
> setting high
> expectations, providing
> regular and timely
> feedback and creating
> inclusive classrooms.

PAUSE AND THINK

Use the template to think about how you would begin a Year 9 lesson on back-to-back stem-and-leaf plots and histograms that describe data, using terms including 'skewed', 'symmetric' and 'bi-modal' (ACMSP282). How could you introduce the lesson so that students are motivated to learn more about this way of representing data?

Aspects of the lesson planning process require careful consideration. For example, to complete the lesson (or learning) objectives, it is critical to use verbs which allow you to easily assess whether students have achieved the objectives. For example:

- interpret a back-to-back stem-and-leaf plot to compare distributions of the two data sets
- construct a back-to-back stem-and-leaf plot from two sets of data
- describe similarities and differences of the two data sets in the back-to-back stem-and-leaf plot.

The verbs 'interpret', 'compare', 'construct' and 'describe' are easily understood and can be observed in students' responses to questions during the lesson. Avoid using more abstract terms such as 'understand' or 'know' as these are difficult to define and hence difficult to assess. It is advisable to limit the number of objectives so that most students in the class can achieve them during class time, although it is likely you may use similar objectives in subsequent lessons.

Students' prior knowledge needs to include the mathematical ideas which are necessary for students to engage with the new content. Regarding the stem-and-leaf plot lesson, students will need to understand how to record data in a simple table, possibly with tally marks, how to interpret distribution of data including range and outliers, and how to draw a simple stem-and-leaf plot, although this could certainly be included within this lesson on back-to-back stem-and-leaf plots.

Several researchers have sought students' views about good mathematics lessons and, unsurprisingly, their responses can be culturally varied. In a large comparative international study, *The Learner's Perspective Study* conducted by Clarke, Keitel and Shimizu (2006), students' reactions to secondary mathematics lessons were investigated in several countries (Anthony et al., 2013). Students from Singapore and Hong Kong valued clear teacher explanations and demonstration of procedures followed by individual seatwork for practice. Sullivan, Clarke and O'Shea (2010) surveyed more than 900 Year 5 to 8 Australian students and concluded they valued clarity of explanations, learning experiences which connected to their everyday lives, a variety of activities and lesson types, inclusion of challenging tasks and working in groups. Surveying your students about their mathematics learning experiences is a valuable way to gain feedback on your lessons and to support your reflection on practice.

Approaches to teaching mathematics: types of lessons

Mathematics lessons can take many forms depending on their purpose. Such purposes could include: introducing new concepts and ideas; focusing on consolidation; applying knowledge to new contexts; revising several topics for a forthcoming examination; investigating an unfamiliar idea; or visiting a local business or listening to a guest speaker to see how mathematics is used in the workplace. Regardless of the purpose of the lesson, planning is critical, with the teacher considering each of the features noted earlier and ensuring the learning objectives are clear. This section examines some approaches to teaching mathematics and the structure of lessons, including a more traditional mathematics lesson, a problem-solving mathematics lesson, a rich and balanced mathematics lesson, and a connected or integrated lesson.

Building on the work of the Trends in International Mathematics and Science Study video study to compare Year 8 mathematics lessons from several countries, an international comparative study of mathematics classrooms by Clarke, Keitel and Shimizu (2006) confirmed country level differences in lesson structure. For example, the traditional mathematics lesson structure was more prevalent in the United States and Australia while the problem-solving lesson structure was more prevalent in Japan. However, these scholars also found many variations in these patterns and argued that there is a wide variety of ways to introduce a lesson, to introduce new content and to organise practice.

Traditional mathematics lessons for concept or skill development

When selecting and sequencing learning experiences, traditional mathematics lessons in secondary classrooms in the United States follow a familiar pattern, which Stigler and Hiebert (1999) referred to as a culturally based teacher script:

- reviewing previous material
- demonstrating how to solve the problems for today
- practising
- correcting seatwork and assigning homework.

In their review of Australian Year 8 lessons, Hollingsworth, Lokan and McCrae (2003) noted that because of the prevalence of this lesson structure, there was limited evidence of the use of complex problem solving. This is of concern given 'problem solving is one of the most fundamental goals of teaching mathematics, but also one of the most elusive' (Stacey, 2005, p. 341). This traditional approach to planning lessons may not support the embedding of all the proficiency strands, particularly problem solving and reasoning. However, it is useful for concept or skill development and frequently used to teach mathematics.

Problem–solving mathematics lessons

With increased use of higher-order, more open-ended mathematics problems, Japanese lessons followed a slightly different pattern (Stigler & Hiebert, 1999):

- reviewing the previous lesson
- presenting the problem for the day
- students working individually or in groups
- discussing solution methods
- highlighting and summarising the major points.

Based on Japanese lesson structure, Sullivan (2011) suggested teachers plan using the terms launch, explore, summarise and review. This recommendation is informed by four of the elements of Japanese mathematics lessons which typically begin with a problem. The second phase involves individual students working on the problem before sharing with peers – as this is taking place, the teacher observes and listens to the discussion, giving feedback and deciding which students, and in what order, they will share their ideas with the whole class in the third phase of the lesson. Finally, the teacher summarises the key ideas for the lesson. Having students share their strategies for solving the problem enables them to hear the different approaches to problem solving and supports and values the development of a shared learning community within the class where all contributions are acknowledged and valued.

Rich and balanced mathematics lessons

Traditional mathematics lessons focus on students listening to an explanation before practising examples of similar questions, while problem-solving lessons focus on using a challenging or unfamiliar problem as the main task for the whole lesson. Rich and balanced lessons include additional features such as:

- discussion of social issues
- use of videos or other technology
- alternative representations, including concrete materials
- mathematical modelling (see Lovitt & Clarke, 2011).

While not all features would appear in every mathematics lesson, over the course of a unit of work or throughout the year, Lovitt and Clarke recommended teachers reflect on their current practice and endeavour to incorporate some of these features in lessons when appropriate. Beginning teachers should try some of these ideas with a class which is open to providing feedback on new approaches and willing to play an active role in the evaluation process. After learning how to use these features and refining your practice, you can plan to include a different type of lesson in each unit plan throughout the year.

What are the features of rich and balanced mathematics lessons?

Lovitt and Clarke (2011) frequently work with teachers to help them reflect on current practice to expand their repertoire of lesson features. In their paper, they describe three rich, balanced lessons for different grade levels which have a variety of lesson design features. Read their article 'A Designer Speaks' in *Educational Designer* and the teachers' reactions to each of the three lessons.

Questions

1. During your learning of mathematics, did you ever experience a lesson like these? If so, how did you feel about this approach to learning mathematics?
2. Do you agree with the teachers' reactions and the features identified in each lesson?

Integrated lessons: science, technology, engineering and mathematics (STEM)

In some schools, connecting knowledge by working with teachers in other subject areas to design tasks, lessons and units of work can help to engage students, particularly if the connection is based around a theme or local issue. Secondary school teachers of science, technology, engineering and mathematics have been designing lessons and units of work based on the Sustainability cross-curriculum priority from the Australian Curriculum. Using project-based learning, students are given a driving question to guide their inquiry. In teams, they use knowledge and skills from any of the STEM subjects to design an inquiry approach, conduct the investigation, collect and analyse data or design solutions so that they can present findings to an audience. Typically, the whole project may take several weeks or a whole term. Individual lessons may need to focus on content as students begin to ask questions involving the use of new STEM concepts and ideas. Teachers who have participated in such connected projects and lessons have found them to be time-consuming but rewarding since they frequently witnessed increased student engagement (Anderson et al., 2019).

When using project-based learning, it can be difficult to match the learning objectives to the curriculum before the project begins. The more open-ended the project is, teachers may be unsure what students will decide to investigate. For example, if the driving question posed to the students is as follows, how might students respond?

Is our school environmentally friendly?

In groups, students need to decide what the question means, and what data or information they need to collect to answer the question. Some may decide to focus on consumption of utilities such as electricity and water and, based on data, recommend ways to reduce consumption to the principal. Others might decide to explore how resources are recycled or repurposed within the school and report to the principal or parent body that strategies should be put in place to reduce printing and paper usage. Further, groups could investigate the sustainability of food consumed from the canteen and recommend planting a school garden or designing a chicken enclosure to sell eggs at the local market.

In a similar project with Year 1 to 2 students, the children decided to investigate the rubbish found in the school bins, and with the support of their teachers, approached the

Project-based learning: a teaching method in which students learn by actively engaging in real-world and personally meaningful projects. Such projects usually take several lessons, if not weeks, to complete and most projects require students to work in small collaborative groups.

local council to have recycling bins placed in the school. They then went about teaching children from other year groups to put the correct things in the correct bins – they have featured on the ABC television program *War on Waste*. At the secondary school level, schools have used many ways to investigate sustainability issues including designing an energy efficient house, designing watering systems to water plants when they have dried out, designing and building gardens with special features, investigating utility consumption in students' households, and designing energy saving devices.

The approach to planning for these projects is often quite different to the usual process of planning lessons and units of work. While you may have an overall outcome that you want the students to achieve, identifying specific learning objectives for lessons may be drafted for the unit but finalised after the project – this approach has been referred to as backward design or backward mapping (Wiggins & McTighe, 2005). Backward design is a method of educational curriculum design that sets goals prior to selecting instructional methods and types of assessment. Backward design curriculum usually involves three stages. First, desired results or project outcomes are identified. This stage focuses on the larger ideas and skills that students should learn, considering both goals and curriculum expectations. The second step focuses on determining suitable evidence levels that confirm that the desired results identified in the first stage have occurred. The third step focuses on designing activities to achieve learning goals.

SHORT-ANSWER QUESTIONS

1. Planning can begin with a topic. Choose a topic from the Australian Curriculum: Mathematics and develop a lesson plan on the lesson planning template. Which features of the lesson plan do you find most difficult to develop?
2. Planning can begin with a rich task. Take the *Licorice Factory* task presented in Chapter 4 of this volume and use the lesson planning template to write up the lesson plan using this task.

Conclusion

The Australian Curriculum has three dimensions including learning areas (subjects), general capabilities and cross-curriculum priorities. All dimensions are considered important to prepare students with the necessary twenty-first-century skills for active participation in civic life, in work, and in future study. At the school level, curriculum documents are typically used to design school-based scope and sequence charts, and teaching and learning programs for each course in each year of schooling. Teachers are responsible for following these programs, designing teaching and learning activities for each of their classes, assessing student learning, and reporting to students and parents.

The Australian Curriculum: Mathematics F–10 has three content strands and four proficiency strands. All strands are important and require careful attention when planning lessons and learning experiences for students. Planning mathematics lessons which engage and enthuse students is a critical component of teachers' work. While we may want to teach the way we were taught, not all of our students enjoy mathematics or understand its utility. It is through thorough and thoughtful planning that we can begin to enthuse our students, particularly if we consider ways to connect to their lived experiences and the issues they are concerned about.

BRINGING IT TOGETHER

1. What are the three dimensions of the Australian Curriculum and what is the purpose of each dimension?
2. What is the difference between mathematics and numeracy?
3. What are the components of a good mathematics lesson and what is the purpose of each component?
4. When you begin teaching, what questions will you need to ask the mathematics coordinator or head teacher to help you start planning for teaching and learning? Include the documents you think you will need and the types of resources you believe you will need to access for you to plan effectively.
5. During your first two years of teaching, you will need to collect evidence of successfully meeting the AITSL standards at the 'proficient' level. Regarding implementing the curriculum and planning for teaching and learning, what types of evidence do you think you should collect?

REFERENCES

Further resources

Anderson, J. (2014). Forging new opportunities for problem solving in Australian mathematics classrooms through the first national mathematics curriculum. In Y. Li & G. Lappan (eds), *Mathematics curriculum in school education* (209–29). Dordrecht: Springer.

Anderson, J., Wilson, K., Tully, D. & Way, J. (2019). 'Can we build the wind powered car again?' Students' and teachers' responses to a new integrated STEM curriculum. *Journal of Research in STEM Education*, 5(1), 20–39.

Anthony, G., Kaur, B., Ohtani, M. & Clarke, D. (2013). The learner's perspective study: Attending to student voice. In B. Kaur, G. Anthony, M. Ohtani & D. Clarke (eds), *Student voice in mathematics classrooms around the world* (1–12). Rotterdam: Sense Publishers.

Australian Curriculum, Assessment and Reporting Authority (ACARA) (2010). *Australian curriculum information sheet: Why have an Australian Curriculum?* Sydney: ACARA.

Australian Curriculum, Assessment and Reporting Authority (ACARA) (2012). *The shape of the Australian Curriculum, version 4.0.* Sydney: ACARA.

Australian Curriculum, Assessment and Reporting Authority (ACARA) (2020a). *Foundation: Year 10 Curriculum.* Retrieved from https://www.acara.edu.au/curriculum/foundation-year-10

Australian Curriculum, Assessment and Reporting Authority (ACARA) (2020b). *Structure.* Retrieved from https://www.australiancurriculum.edu.au/f-10-curriculum/structure/

Australian Curriculum, Assessment and Reporting Authority (ACARA) (2020c). *Understand how mathematics works: Rationale and aims.* Retrieved from https://australiancurriculum.edu.au/f-10-curriculum/mathematics/

Australian Curriculum, Assessment and Reporting Authority (ACARA) (2020d). *Mathematics.* Retrieved from https://australiancurriculum.edu.au/f-10-curriculum/mathematics/

Australian Curriculum, Assessment and Reporting Authority (ACARA) (2020e). *Sequence of content F–6 and 7–10*. Retrieved from https://australiancurriculum.edu.au/media/3680/mathematics_-_sequence_of_content.pdf

Australian Curriculum, Assessment and Reporting Authority (ACARA) (2020f). *Key ideas*. Retrieved from https://australiancurriculum.edu.au/f-10-curriculum/mathematics/key-ideas/

Australian Curriculum, Assessment and Reporting Authority (ACARA) (2020g). *Mathematics proficiencies, work samples, portfolios and illustrations of practice*. Retrieved from https://www.australiancurriculum.edu.au/resources/mathematics-proficiencies/

Australian Curriculum, Assessment and Reporting Authority (ACARA) (2020h). *General capabilities*. Retrieved from https://australiancurriculum.edu.au/f-10-curriculum/general-capabilities/

Australian Curriculum, Assessment and Reporting Authority (ACARA) (2020i). *Cross-curriculum priorities*. Retrieved from https://australiancurriculum.edu.au/f-10-curriculum/cross-curriculum-priorities/

Australian Curriculum, Assessment and Reporting Authority (ACARA) (2020j). *Sustainability*. Retrieved from https://www.australiancurriculum.edu.au/f-10-curriculum/cross-curriculum-priorities/sustainability/

Australian Curriculum, Assessment and Reporting Authority (ACARA) (2020k). *Using the three dimensions of the Australian Curriculum*. Retrieved from https://www.australiancurriculum.edu.au/resources/student-diversity/planning-for-student-diversity/using-the-three-dimensions-of-the-australian-curriculum/

Australian Institute for Teaching and School Leadership (AITSL) (2011). *Australian Professional Standards for Teachers*. Melbourne: AITSL.

Care, E., Griffin, P. & Wilson, M. (2018). *Assessment and teaching of 21st century skills*. Cham: Springer Nature.

Clarke, D., Keitel, C. & Shimizu, Y. (2006). *Mathematics classrooms in twelve countries: The insider's perspective*. Rotterdam: Sense Publishers.

Council of Australian Governments Education Council (2019). *Alice Springs (Mparntwe) Education Declaration*. Retrieved from https://docs.education.gov.au/documents/alice-springs-mparntwe-education-declaration

Goos, M. (2018). The mathematics learning area: Conforming, reforming or transforming? In A. Reid & D. Price (eds), *The Australian Curriculum: Promises, problems and possibilities* (55–64). Canberra: The Australian Curriculum Studies Association.

Goos, M., Geiger, V., Dole, S., Forgasz, H. & Bennison, A. (2019). *Numeracy across the curriculum: Research-based strategies for enhancing teaching and learning*. Sydney: Allen & Unwin.

Hollingsworth, H., Lokan, J. & McCrae, B. (2003). *Teaching mathematics in Australia: Results from the TIMSS 1999 video study*. Camberwell: Australian Council of Educational Research.

Kilpatrick, J., Swafford, J. & Findell, B. (2001). *Adding it up: Helping children learn mathematics*. Washington, DC: National Academies Press.

Lovitt, C. & Clarke, D. (2011). The features of a rich and balanced mathematics lesson: Teacher as designer. *Journal of the International Society for Design and Development in Education*, 4, 1–25. Retrieved from https://www.educationaldesigner.org/ed/volume1/issue4/article15/

Lowrie, T., Logan, T. & Scriven, B. (2012). Perspectives in geometry and measurement in the Australian Curriculum: Mathematics. In B. Atweh, M. Goos, R. Jorgensen, & D. Siemon (eds), *Engaging the Australian National Curriculum: Mathematics – Perspectives from the field* (71–88). Retrieved from https://www.merga.net.au/Public/Public/Publications/Engaging_the_Australian_curriculum_mathematics_book.aspx

Mulligan, J., Cavanagh, M. & Keanan-Brown, D. (2012). The role of algebra and early algebraic reasoning in the Australian Curriculum: Mathematics. In B. Atweh, M. Goos, R. Jorgensen & D. Siemon (eds), *Engaging the Australian National Curriculum: Mathematics – Perspectives from the field* (47–70).

Retrieved from https://www.merga.net.au/Public/Public/Publications/Engaging_the_Australian_curriculum_mathematics_book.aspx

Shulman, L. (1987). Knowledge and teaching: Foundations of the new reform. *Harvard Educational Review*, 57(1), 1–22.

Siemon, D. (2006). *Assessment for common misunderstandings materials*. Prepared for and published electronically by the Victorian Department of Education and Early Childhood Development. Retrieved from http://www.education.vic.gov.au/studentlearning/teachingresources/maths/common/default.htm

Siemon, D., Bleckly, J. & Neal, D. (2012). Working with the big ideas in number and the Australian Curriculum: Mathematics. In B. Atweh, M. Goos, R. Jorgensen & D. Siemon (eds), *Engaging the Australian National Curriculum: Mathematics – Perspectives from the field* (19–46). Retrieved from https://www.merga.net.au/Public/Public/Publications/Engaging_the_Australian_curriculum_mathematics_book.aspx

Smith, D. & Ewing, R. (2002). Curriculum studies: Storylines. *Change: Transformations in Education*, 5(1), 26–45.

Stacey, K. (2005). The place of problem solving in contemporary mathematics curriculum documents. *Journal of Mathematical Behaviour*, 24, 341–50.

Steen, L.A. (ed.) (1990). *On the shoulders of giants: New approaches to numeracy*. Washington, DC: National Academy of Sciences.

Stigler, J. & Hiebert, J. (1999). *The teaching gap*. New York: Free Press.

Sullivan, P. (2011). *Teaching mathematics: Using research-informed strategies*. Camberwell: Australian Council for Educational Research. Retrieved from https://research.acer.edu.au/cgi/viewcontent.cgi?article=1022&context=aer

Sullivan, P., Clarke, D. & O'Shea, H. (2010). Students' opinions about characteristics of their desired mathematics lesson. In L. Sparrow, B. Kissane & C. Hurst (eds), *Shaping the future of mathematics education. Proceedings of the 33rd Annual Conference of the Mathematics Education Research Group of Australasia (MERGA)* (531–8). Fremantle: MERGA.

UNESCO-IBE (2013). *Glossary of curriculum terminology*. UNESCO International Bureau of Education. Retrieved from http://www.ibe.unesco.org/fileadmin/user_upload/Publications/IBE_GlossaryCurriculumTerminology2013_eng.pdf

van den Akker, J. (2018). *Bridging curriculum design and implementation*. Working paper for the Organisation for Economic Co-operation and Development. Retrieved from https://www.oecd.org/education/2030-project/contact/Bridging_curriculum_redesign_and_implementation.pdf

Watson, J. & Neal, D. (2012). Preparing students for decision making in the 21st century – Statistics and probability in the Australian Curriculum: Mathematics. In B. Atweh, M. Goos, R. Jorgensen & D. Siemon (eds), *Engaging the Australian National Curriculum: Mathematics – Perspectives from the field* (89–114). Retrieved from https://www.merga.net.au/Public/Public/Publications/Engaging_the_Australian_curriculum_mathematics_book.aspx

Wiggins, G. & McTighe, J. (2005). *Understanding by design* (2nd edn). Alexandria, VA: Association for Supervision and Curriculum Development.

CHAPTER
10

Number and algebra

Bing H. Ngu

LEARNING OBJECTIVES

The learning objectives of this chapter are directly linked to the Australian Professional Standards for Teachers (APST) (Australian Institute for Teaching and School Leadership [AITSL], 2011). After studying this chapter, you should be able to:

- understand the algebraic concepts of the unknown, pronumeral and variable (APST 2.1, 2.3)
- unpack linear equations in a hierarchical level of complexity (APST 2.1, 2.3)
- appreciate algebraic problem-solving in real-life contexts (APST 2.1, 2.3, 3.3)
- transfer algebraic problem-solving skills to the science curriculum (APST 2.1, 2.3, 3.3).

Introduction

Many children and adults have reservations about algebra. Why? The abstractness of algebra partly lies in its use of variables and pronumerals. Nonetheless, algebra expressed in a mathematical principle to solve a range of problems is what makes it powerful (Kieran, 1992). Algebra skills are essential to engage successfully with higher-order mathematical thinking skills in advanced mathematics. Algebra skills are useful not only for solving real-life problems (e.g. 'If your father wants to increase your weekly allowance of \$20 by 5%, what is your new allowance?') (Ngu, Yeung & Tobias, 2014), but are also transferrable to other curriculum domains such as physics and chemistry (e.g. 'A solution contains 1.1 g of sodium nitrate $NaNO_3$ in 250 ml, what is the molarity of this solution?') (Ngu & Yeung, 2013; Ngu, Yeung & Phan, 2015). Success in algebra will assist students to pursue a STEM-related career.

Despite the prominent role of algebra in the secondary mathematics curriculum, there is limited evidence of an efficient use of algebra, particularly in middle school students (Stacey & MacGregor, 1999). Such findings suggest that secondary students perceive algebra, which requires abstract thinking and reasoning, a challenging topic to learn and master. Algebra requires proficiency in using variables (symbols) to represent and express mathematical relationships.

To assist secondary students in building a foundation for algebra knowledge, this chapter highlights several aspects of teaching and learning algebra based on the Australian Curriculum: Mathematics. In particular, this chapter shares instructional practices that are supported by research. Most of the instructional practices are part of the authors' ongoing mathematics education research with collaborators. Unpacking the complexity of linear equations hierarchically has provided a much-needed framework for teaching and learning linear equations. Highlighting the link between fractions, percentages and decimals enhances the learning of structurally similar linear equations. Similarly, the development of prior knowledge (e.g. collect like terms, expand the bracket) impacts upon the learning of complex linear equations involving multiple solution steps. Proficiency in equation-solving skills can pave the way for adopting an algebraic approach in solving real-life problems. The power of algebraic problem solving is also demonstrated in its transferability to the science, technology, engineering and mathematics (STEM) curriculum, such as in the science domain.

The unknown, pronumeral and variable

Central to arithmetic is the concept of number sense, whereas central to algebra is an understanding of the concept of variable. A good starting point to learn algebra is to learn the meaning of the unknown, pronumeral and variable.

In algebraic thinking and learning, the unknown is represented by a symbol such as a letter (e.g. x). In the context of solving linear equations, x stands for a number. For example, $x - 4 = 9$, where x stands for 13 so that the left side of the equation equals to its right side. Nonetheless, mathematics education researchers and educators tend to use the pronumeral instead of the unknown in linear equations. The word 'pronumeral' is derived from a Latin

Unknown: something whose value needs to be found.

Pronumeral: use of a letter to stand for a number in the context of equation solving.

Variable: use of a letter to stand for different numbers.

word in which 'pro' means 'for' and 'numeral' means 'number'. In solving a simple linear equation such as $n + 2 = 8$, mathematics educators would regard n as a pronumeral because it stands for a number. More specifically, n stands for a specific number, which in this case is 6 and not any other number.

What about a variable? A variable is a letter that can stand for different numbers. For example, if y is a variable, then it can be any number (e.g. 1, 0, 5 or 7). Let us consider a formula such as $s = \frac{d}{t}$, where s stands for speed, d stands for distance, and t stands for time. Each of these letters, s, d and t, is a variable. The value of s will vary depending on the values of d and t. The concept of a variable whereby a letter stands for any number is central to algebraic thinking and learning.

Teaching the concept of variable

How shall we teach the concept of variable? Consider $2 + ? = 6$ in which '?' represents an unknown number. The sum of 2 and the unknown number will give a value of 6. Having learned the concept of an unknown number, the next step is to introduce a letter such as n to present the unknown number, $2 + n = 6$. In the context of solving linear equations, n is called a pronumeral. It is important to emphasise that we can use any letter to present the unknown number or the pronumeral, and the letter stands for a number or quantity. To further expand the idea of variable, we can assign students to work on activities involving growing patterns. The purpose is to highlight how a mathematical rule based on the concept of variable represents a powerful tool to solve problems in real-life contexts.

Consider the following problem situated in a real-life context (Cai, 2005, p. 340):

> Sally is having a party.
> The first time the doorbell rings, 1 guest enters.
> The second time the doorbell rings, 3 guests enter.
> The third time the doorbell rings, 5 guests enter.
> The fourth time the doorbell rings, 7 guests enter.
> The guests continue to arrive, with two more people than the previous time.
>
>   (i) How many guests will enter on the 10th ring? Explain how you found
>   your answer.
> (ii) Assuming the pattern continues, describe in words or write a rule to determine
>   which doorbell ring had 99 guests arrive.
> (iii) What is the benefit of having this rule?

Using g to represent the number of guests, and n to indicate when the doorbell rings (i.e. which doorbell), we can draw a table to show the growing pattern (Table 10.1). The number of guests that corresponds to the 10th ring is 19. Obviously, it will be a laborious process and therefore an inefficient strategy to continue the growing pattern in Table 10.1 in order to *determine which doorbell ring had 99 guests arrive*. In contrast, based on the relation portrayed in the growing pattern, we can generate a mathematical rule such as $g = 2n - 1$ to determine the value of g when the value of n varies or vice versa. To generate a mathematical relationship such as $g = 2n - 1$ may pose a challenge. One strategy is to use trial and error to derive an expression such as $2n - 1$ so that it produces the number of guests (g) in the growing pattern. The number of guests (g) and doorbell rings (n) represents two variables. The number of guests (g) depends on which doorbell (n) rings. Thus, the doorbell (n) is an independent variable and the guest (g) is a dependable variable. This mathematical rule would be far more superior than the use of a table to solve the problem. For example, by substituting $g = 99$ in the rule $g = 2n - 1$, we can easily obtain the value for n, which is 50. The benefit of the mathematical rule is obvious. It highlights the power of generalised information expressed as a mathematical rule to solve real-life problems. The number of guests (g) will vary depending on when the doorbell rings (n).

Table 10.1 The growing pattern depicting the relation between doorbell (n) and guest (g)

n (doorbell)	1st	2nd	3rd	4th	5th	6th	7th	8th	9th	10th
g (guest)	1	3	5	7	9	11	13	15	17	19

Misconception about variables

It is not uncommon for students to regard a variable as an object rather than a number or quantity. To assist students in differentiating between a variable and an object, we can ask them to formulate an equation to solve the following problem:

> The cost of an avocado is $3, and the cost of a pear is $1. If I buy 2 avocados and 4 pears, what is the total cost?

Solution:

Let a = number of avocados, and p = number of pears

$$\text{Total cost} = \$3a + \$1p$$
$$= \$3 \times 2 + \$1 \times 4$$
$$= \$10$$

It is important to emphasise that a and p stand for variables and not for objects such as avocado and pear. Obviously, students can choose to use other methods to solve the problem instead of the algebra approach. Nonetheless, there are two reasons why it is important to emphasise algebraic problem solving. First, the algebra approach allows the calculation of the total cost easily where a and p can stand for any number. Second, it is important to expose middle school students to algebraic problem solving wherever appropriate so as to pave the way for them to pursue senior mathematics, in which algebraic problem solving becomes a dominant component of the curriculum.

Mathematical representations in the mathematics classroom   SCENARIO

In a Year 9 class, you are teaching the concept of mathematical representations. Consider the following activity as one you would pose to your students.

> In my garden one day I saw a tiny plant with 4 leaves (Day 1). The following day, it had grown even more leaves. Each day I noticed it continued to grow in the same way. (Wilkie & Clarke, 2016)

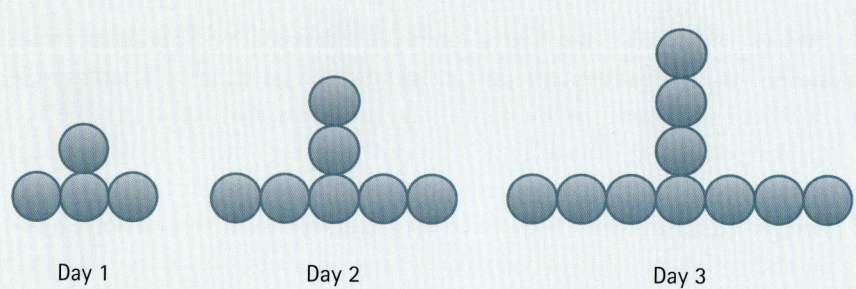

Figure 10.1 The growing pattern of leaves across Day 1, Day 2 and Day 3
Source: Wilkie & Clarke, 2016.

1. Draw a picture of what the plant will look like on Day 4.
2. Complete the table below.

Day	1	2	3	4
Number of leaves				

3. How many leaves will the plant have on Day 10?
4. Write a mathematical rule (equation) to show the relation between the Day (D) and Number of leaves (n) (i.e. what is the formula for this pattern?)
5. How would you explain question 4 to a student who was struggling to understand creating an equation?
6. What is the benefit of having a mathematical rule?

PAUSE AND THINK

There are six times as many students as professors at this university. Use S for the number of students and P for the number of professors. Write the equation for this problem (Clement, 1982). Most students wrote $6S = P$. Is this correct? Why?

Algebraic expression problems

After learning the concept of a variable, we can proceed to learn algebraic expressions, which represents an essential component of linear equations. Consider a linear equation such as $3x + 4 = 16$. The component $3x + 4$ constitutes an algebra expression. More specifically, an algebraic expression comprises terms that can be added or subtracted. A term may consist of the following:

1. A number (e.g. 8)
2. A pronumeral (e.g. x)

3. Two different pronumerals multiplied (e.g. xy)
4. A pronumeral and a number multiplied or divided (e.g. $3x$, $\frac{1}{2}y$)
5. A pronumeral with an index (e.g. x^2, x^3)
6. A combination of pronumerals and numbers (e.g. $4ab^2$)

When the term is a number (e.g. 8), it is called a constant because the value does not change. The number in front of a pronumeral is called a coefficient. For example, 3 and $\frac{1}{2}$ are coefficients of $3x$ and $\frac{1}{2}y$, respectively. Let us examine an example of an algebraic expression:

$$6x^2 - 8xy + \frac{y}{7} - 11$$

- the number of terms: 4 terms (i.e $6x^2$, $-8xy$, $+\frac{y}{7}$ and -11)
- the coefficient of the third term: $\frac{1}{7}$
- the constant term: -11

Prior to introducing simple linear equations, it is important to strengthen students' knowledge of algebraic expressions. One of these is collecting like terms whereby we can add or substract like terms. Like terms contain the same pronumeral (e.g. $3x$ and $9x$), whereas unlike terms contain different pronumerals (e.g. $4x$ and $6y$). It should be noted that xy and yx are like terms because $xy = yx$. We can simplify an algebraic expression such as $2x - 7y + 10x + 21$ in which we add the like terms (i.e. $2x + 10x$), resulting in $12x - 7y + 21$.

PAUSE AND THINK

Assume a correct algebraic equation is represented as $5x + 12x = 17x$. However, there may be a tendency for students to incorrectly represent the equation as $5x + 12 = 17x$ instead. Why? How can teachers help students to overcome such an error?

One way to gain insights into algebraic thinking and reasoning is to compare how mathematical operations are applied to arithmetic situations and algebraic situations via analogical reasoning. Structure mapping theory (Gentner, 1983) 'predicts that successful mapping of the structural elements of a new problem (target) with a learned problem (source) is likely to result in analogical transfer'. Learning by analogy, underpinned by structure mapping theory, has provided a strong theoretical framework for research on mathematics education (Richland, Zur & Holyoak, 2007).

Arithmetic operation and algebraic operation

By drawing on prior knowledge of number sense in the context of arithmetic operation, the learning of algebraic expression problems can be facilitated through algebraic operation (Table 10.2). In essence, analogical comparison between arithmetic operation and algebraic operation enables the students to transfer their prior knowledge of number sense to learn algebraic expression problems.

Learning by analogy: mapping of the structural elements of a new problem (e.g. $10\%x = 16$) with a learned problem (e.g. $2x = 6$) is likely to result in analogical transfer. Both $10\%x = 16$ and $2x = 6$ share a similar problem structure and thus can be solved using the same method.

Table 10.2 Analogical comparison between arithmetic operation and algebraic operation

Arithmetic operation	Algebraic operation
Addition and subtraction	
$2 + 0 = 2$	$a + 0 = a$
$2 - 0 = 2$	$a - 0 = a$
$2 + 5 = 5 + 2$	$a + b = b + a$
Multiplication and division	
$3 \times 1 = 3$	$a \times 1 = a$
$3 \div 1 = 3$	$a \div 1 = a$
$3 \times 4 = 4 \times 3$	$a \times b = b \times a$

Putting arithmetic operation and algebraic operation side-by-side reveals that algebraic operation is analogous to the arithmetic operation. For example, as indicated in Table 10.2, there is one-to-one mapping of elements between $2 + 0 = 2$ and $a + 0 = a$ in terms of the underlying relational structure. The idea of analogical comparison can also facilitate the teaching and learning of the expansion of algebraic expression problems. It should be noted that knowing how to expand the bracket of algebraic expressions represents another important prerequisite knowledge of solving linear equations.

PAUSE AND THINK

How would you teach $0 \div x$?

Arithmetic manipulation and algebraic expression manipulation

As shown in Table 10.3, there are two ways to calculate $2(5 + 1)$, and one of these is based on the expansion of the bracket. However, there is only one way to expand the algebraic expression problem, $3(a + 1)$. It is important to highlight the one-to-one mapping of elements between the expansion of $2(5 + 1)$ and $3(a + 1)$. This will help students to draw upon their prior knowledge of $2(5 + 1)$ to make sense of $3(a + 1)$. For example, $3 \times a$ is $3a$.

Table 10.3 Analogical comparison between arithmetic manipulation and algebraic expression manipulation

Arithmetic manipulation		Algebraic expression manipulation
(i) $\begin{aligned} & 2(5+1) \\ & = 2 \times 6 \\ & = 12 \end{aligned}$	(ii) $\begin{aligned} & 2(5+1) \\ & = 2 \times 5 + 2 \times 1 \\ & = 12 \end{aligned}$	$\begin{aligned} & 3(a+1) \\ & = 3 \times a + 3 \times 1 \\ & = 3a + 3 \end{aligned}$

PAUSE AND THINK

How would you help students to differentiate between $x + 3x$ and $x \times 3x$? Provide reasons to support your teaching ideas.

SHORT-ANSWER QUESTION

We can use a diagram to scaffold the expansion of $3(a + 1)$. Draw two diagrams to show the expansion of $2(5 + 1)$ and $3(a + 1)$, respectively. (*Hint: regarding* $2(5 + 1)$, *for example, draw two rectangles where they share the same dimension of 2.*)

Highlight the similarity and difference between the two diagrams and provide reasons to support your interpretation.

Important points when teaching number and algebra

IN PRACTICE

1. Hands-on activities connect the concrete growing patterns and the mathematical rule that illustrates the relationship between the independent variable and the dependent variable.
2. It is important to distinguish between a variable and an object in algebraic problem solving.
3. The prior knowledge of arithmetic manipulation could facilitate the learning of algebraic manipulation through reasoning by analogy.

Linear equations

Equation solving acts as a bridge between pre-algebra and algebra learning. The transition from arithmetic to algebraic reasoning is notoriously difficult for secondary school students to achieve (National Research Council, 2001). Therefore, a renewed focus on classifying linear equations provides a much-needed framework for teaching and learning linear equations. Specifically, within the framework of cognitive load theory, we will examine an innovative way to classify the complexity of linear equations hierarchically using the number of the relational line and the operational line. Moreover, we will examine different instructional methods to facilitate learning of linear equations from a cognitive load theory perspective. For example, is the inverse method better than the balance method for learning linear equations? Why?

Relational line: the quantitative relation between elements where the left side of the equation equals to its right side.

Operational line: the application of a mathematical operation to change the state of the equation, but at the same time to maintain the equality of the equation.

CONNECTION   John Sweller: cognitive load theory

Working memory: limited capacity to process information.

Long-term memory: unlimited capacity that stores a huge number of schemata.

Schema: 'a cognitive construct that permits us to treat multiple elements of information as a single element categorised according to the way in which it will be used' (Pawley et al., 2005, p. 76).

John Sweller, an Emeritus Professor at the University of New South Wales, pioneered cognitive load theory in the 1980s. Cognitive load theory is an instructional theory that addresses the instructional design issues. Instructional designers need to consider the limitation of working memory and the long-term memory that stores a huge number of schemas in varying degree of specification. The working memory is restricted when processing novel information but not schemas, which can be retrieved from the long-term memory. Many years later, empirical studies conducted by cognitive load researchers around the globe continue to advance cognitive load theory (Sweller, van Merrienboer & Paas, 2019).

Previous work on classifying linear equations

Mathematics education researchers have classified linear equations as one-step equations (e.g. $2x = 8$), two-step equations (e.g. $-5x + 2 = 17$), and equations with variables on both sides (e.g. $9 - x = 2x - 6$) (Matsuda et al., 2013). This manner of classifying linear equations is based on the number of operations required to solve these equations. For example, performing a single operation such as $\div 2$ on both sides is required to solve the one-step equation such as $2x = 8$. Other mathematics education researchers have classified linear equations as multi-step equations when they either involve a bracket (e.g. $4(x + 1) = 20$) or variables on both sides plus brackets (e.g. $5(x - 3) = 2(x - 3) + 16$) (Rittle-Johnson & Star, 2007).

Simple linear equations have been classified as one-step (e.g. $x + 2 = 6$) and two-step equations ($2x + 4 = 10$), whereas more complex linear equations have been classified as equations with grouping symbols (e.g. $4(x + 1) = 12$) and equations with pronumerals on both sides (e.g. $5a + 7 = 2a - 1$) (McSeveny, Conway & Wilkes, 2004). Mathematics educators have extended the classification of linear equations to include three-step and four-step equations (Vincent et al., 2011). Interestingly, linear equations with a negative pronumeral (e.g. $6 - x = 15$) have been classified as 'equations with negative numbers' (Vincent et al., 2011, p. 411) or 'one-step equations with negative integers' (McSeveny, Conway & Wilkes, 2004, p. 245).

Clearly, up to this point, neither mathematics education researchers nor mathematics educators have provided a systematic way of classifying linear equations in a hierarchical level of complexity. Based on cognitive load theory, we argue that the use of operational and relational lines to classify linear equations has captured the complexity of linear equations hierarchically (Ngu & Phan, 2016b).

Relational line and operational line

Cognitive load theory (Sweller, Ayres & Kalyuga, 2011) provides guidelines to design effective instructions across various disciplines. Central to cognitive load theory is how people process information and learn. The interaction between elements within learning material constitutes element interactivity, and therefore the intrinsic cognitive load imposed on learning. An element refers to anything that needs to be learned (e.g. a number, a symbol, a procedure) (Chen, Kalyuga & Sweller, 2017). Let us consider $x + 3$. The learner needs to know individual elements (i.e. x, $+3$) separately as a term, and the interaction between these two elements to form an algebraic expression such as $x + 3$. The complexity of learning material is proportionate to the degree of element interactivity.

Element interactivity: the interaction between elements in the learning material.

To illustrate the complexity of linear equations, we use relational and operational lines to describe the solution procedure of linear equations. Figure 10.2 depicts the solution procedure of $x + 6 = 11$ using the balance method. The balance method emphasises the concept of 'balance' (i.e. perform -6 on both sides) when solving linear equations. A relational line (Lines 1 and 3) indicates the relation between elements so that the left side of the equation equals to its right side. An operational line (Line 2) refers to the performance of the same operation (e.g. -6 on both sides) to alter the problem state of the equation and yet at the same time to maintain the equality of the equation.

As depicted in Figure 10.2 (see also Table 10.4), Line 1 refers to a relational line. It comprises three elements (x, $+6$, 11) and three concepts:

1. $x + 6 = 11$ represents an algebraic sentence in which x (pronumeral) can be replaced by a number.
2. The equal sign (=) describes a relationship between the left and right sides so that they are equal.
3. To find x, the learner needs to perform the same operation on both sides (what is done on the left side should also be done on the right side) to balance the equation.

Equal sign: mathematical equivalence in the context of equation solving.

The learner needs to manipulate the interaction between these three elements and concepts simultaneously. Line 2 refers to an operational line. It has one element (-6). The learner needs to perform an operation (-6) on both sides (cancel -6 with $+6$ on the left side, the same -6 must be done on the right side) in order to maintain the equivalence of the equation. Line 3 refers to a relational line and it has two elements (x, 5). If the learner can process the preceding Lines 1 and 2 successfully, then x equals 5 being the solution would seem obvious.

Line 1	$x + 6 = 11$	(-6 on both sides)
Line 2	$-6 \quad -6$	
Line 3	$x = 5$	

Figure 10.2 Solution procedure of a simple equation

Naturally, the higher the number of operational and relational lines, the higher the degree of element interactivity, and thus the complexity of the linear equations (Table 10.4). For example, as shown in Table 10.4, the equation $\frac{6}{a} = 2$, involving two operational and three relational lines, is more complex than $x + 6 = 11$ (Figure 10.2), which involves one operational and two relational lines. While the number of operational and relational lines acts as a point of reference to capture the complexity of the linear equations, the complexity of linear equations is also affected by the nature of the element (simple or complex). 'In mathematics learning, the presence of a complex element inevitably increases its complexity' (Ngu & Phan, 2016b). It is a known fact that operating negative numbers (Ayres, 2001) and fractions (Cramer & Wyberg, 2009) poses a challenge to students. Moreover, students often fail to mathematically reason and make connections between fractions, decimals and percentages

(Parker & Leinhardt, 1995). Therefore, when the number of operational and relational lines is kept constant, the presence of a complex element will increase the complexity of the linear equations.

Table 10.4 Solution procedure: one operational line and two relational lines, two operational lines and three relational lines

One operational line and two relational lines, two operational lines and three relational lines				
Solution procedure	Number of operational line	Number of relational line	Type of element	Comment
$x + 2 = 11$ $-2 \quad -2$ $x = 9$	1	2	simple	• positive number: 2, 11
$a - 3 = -7$ $+3 \quad +3$ $a = -4$	1	2	complex	• negative number: $-3, -7$
$\frac{p}{8} = 4$ $\times 8 \times 8$ $p = 32$	1	2	complex	• fraction • positive number
$\frac{1}{3}x = 5$ $\times 3 \quad \times 3$ $x = 15$	1	2	complex	• fraction • positive number
$\frac{x}{0.5} = 7$ $\times 0.5 \times 0.5$ $x = 3.5$	1	2	complex	• fraction • decimal number: 0.5
$\frac{6}{a} = 3$ $\times a \times a$ $6 = 3a$ $\div 3 \div 3$ $2 = a$	2	3	complex	• pronumeral as a denominator: $\frac{6}{a}$
$8 - n = 0$ $-8 \quad -8$ $-n = -8$ $\div(-1) \div(-1)$ $n = 8$	2	3	complex	• negative pronumeral: $-n$

Note: when the number of operational lines and relational lines is kept constant, the presence of a complex element increases its complexity. For example, $\frac{x}{0.5}$ is more complex than $\frac{p}{8} = 4$ owing to the presence of a decimal number in the former.

Source: Ngu & Phan, 2016b.

PAUSE AND THINK

1. Solve this linear equation, $x - 12\%x = 72$. What is the number of operational and relational lines in the solution procedure? How would you teach this type of linear equation?
2. Would this linear equation $5m = 2m + 1$ pose a challenge to students? Why or why not?

SHORT-ANSWER QUESTIONS

1. Solve the three linear equations below. Having done that, we want you to: (1) identify the number of operational and relational lines, and (2) identify the nature of the element (simple or complex). Can you see a pattern here? What is the significance of the pattern in terms of teaching and learning linear equations?
   a. $4(2 - w) = 4$
   b. $2(r + 5) - 7 = 11$
   c. $8(p + 2) - 6p - 4 = 18$
2. How would you teach $15\%x = 30$? Is it important to teach students this type of linear equation? Why or why not?

Teaching and learning linear equations in a hierarchical level of complexity

According to cognitive load theory, our human cognitive architecture comprises a working memory and a long-term memory. The working memory can only process about four elements of information at any given time (Cowan, 2001); moreover, information readily disappears without rehearsal. In contrast, the long-term memory stores a huge number of schemata (learned information) in varying degrees of specification. In the mathematics domain, a schema is defined as a cognitive construct that permits us to treat multiple elements of information as a single element categorised according to the way in which it will be used (Pawley et al., 2005, p. 76). The constraint of working memory decreases when processing a schema retrieved from the long-term memory. Accordingly, the benefit of sequencing linear equations is in line with the impact of prior knowledge of a simpler linear equation upon the teaching and learning of a more complex equation (van Merrienboer, Kirschner & Kester, 2003).

A more complex linear equation requires an extra solution step (or steps) than the preceding simpler linear equation. In effect, there is a hierarchical level of schemas spanning from a lower-level schema of simple linear equations to a higher-level schema of complex linear equations. The lower-level schema of simple linear equations is embedded within the higher-level schema of complex linear equations (Ngu & Phan, 2016b).

(a)

Line 1	$\frac{x}{2} = 7$ ($\times 2$ on both sides)
Line 2	$\frac{x}{2} \times 2 = 7 \times 2$
Line 3	$x = 14$

(b)

Line 1	$\frac{n}{3} + 2 = 8$ ($- 2$ on both sides)
Line 2	$\frac{n}{3} + 2 - 2 = 8 - 2$
Line 3	$\frac{n}{3} = 6$ ($\times 3$ on both sides)
Line 4	$\frac{n}{3} \times 3 = 6 \times 3$
Line 5	$n = 18$

Figure 10.3 Use of the balance method to solve (a) and (b). Lines 1, 2, and 3 in (a) are similar to Lines 3, 4 and 5 in (b)

Figure 10.3 depicts the solution procedure of a more complex equation (b) preceded by a simpler equation (a). For example, the equation $\frac{n}{3} + 2 = 8$ (two operational and three relational lines) is more complex than the equation $\frac{x}{2} = 7$ (one operational and two relational lines). Once the students have acquired the schema for the equation $\frac{x}{2} = 7$, this schema will be stored in the long-term memory. Subsequently, when presented with the equation $\frac{n}{3} + 2 = 8$, the students can retrieve the schema for the equation $\frac{x}{2} = 7$, which is the same as a subset of the complex equation, $\frac{n}{3} = 6$. Mathematics educators can draw students' attention to map Lines 1, 2 and 3 in the equation $\frac{x}{2} = 7$, which are similar to Lines 3, 4 and 5 in the equation $\frac{n}{3} = 6$. The learning of a more complex equation $\frac{n}{3} + 2 = 8$, becomes the learning of Lines 1 and 2 only. Therefore, this manner of teaching and learning the complex equation $\frac{n}{3} + 2 = 8$ is beneficial because it builds upon the prior knowledge of a simpler equation, thus reducing working memory load.

In summary, sequencing linear equations in a hierarchical level of complexity based on the number of operational and relational lines allows mathematics educators to map structurally similar operational and relational lines across linear equations. Ultimately, the learning of a more complex equation is preceded by a simpler equation. This would be an efficient way of teaching and learning linear equations because it will impose relatively little cognitive load.

PAUSE AND THINK

The teaching and learning of a more complex equation $5x - 8 = 3x + 4$ (three operational and four relational lines) can be built upon the prior knowledge of the simpler equation $3x - 4 = 5$ (two operational and three relational lines). How would you scaffold this teaching idea?

Structurally similar linear equations

The nature of the element will affect the complexity of linear equations when the number of operational and relational lines is kept constant. Consider two equations such as $\frac{y}{3} = 2$ and $\frac{x}{0.5} = 9$. Both equations have one operational and two relational lines; and, they share a similar structure and therefore the same degree of element interactivity. Nonetheless, the presence of a complex element such as a decimal in the equation $\frac{x}{0.5} = 9$ renders it more complex than the equation $\frac{y}{3} = 2$ (Ngu & Phan, 2017). Because both equations are structurally similar, the equation $\frac{x}{0.5} = 9$ is analogous to the equation $\frac{y}{3} = 2$. That is, there is one-to-one correspondence in elements (e.g. 9 matches with 2). The teaching and learning of structurally similar equations should begin with a simpler equation. In this case, it should begin with the equation $\frac{y}{3} = 2$. Once students have acquired schema for the equation $\frac{y}{3} = 2$, they can retrieve the same schema to solve a structurally similar equation such as $\frac{x}{0.5} = 9$ via reasoning by analogy. Figure 10.4 shows pairs of structurally similar equations in which the presence of a complex element in the complex equation poses a challenge to students. To overcome this, mathematics educators could engage students in analogical reasoning to learn structurally similar equations.

Simpler equation (source)	Complex equation (target)	Complex element in the complex equation
1. $6x = 36$	$10\%x = 20$	percentage
2. $m + 8 = 12$	$p + \frac{1}{2} = 2$	fraction
3. $2b + 1 = 3$	$0.2x + 0.1 = 0.3$	decimal

Figure 10.4 Pairs of structurally similar equations

PAUSE AND THINK

Consider the advantage of using a structurally simpler linear equation to teach a complex linear equation. How would you teach a complex linear equation involving a fraction, $4x + \frac{3}{5} = 1$?

Equation-solving methods

The balance method (Figure 10.5) is a popular method for teaching equation-solving in Western countries. In contrast, some Asian countries prefer the inverse method (Figure 10.5) for equation-solving. A review by Cai et al. (2005) indicates that some Asian countries (e.g. Singapore, China, South Korea) have introduced the inverse method in their primary mathematics curriculum. What is the difference between the balance method and the inverse method? As shown in Figure 10.5, the main difference between the balance and inverse methods lies in the operational line (-5 on both sides vs. $+5$ becomes -5). Regarding the balance method, interaction between elements occurs on both sides of the equation; however, for the inverse method, the interaction between elements occurs on one side of the equation. Thus, for each operational line, the inverse method only incurs half as many interactive elements as the balance method. In the study conducted by Ngu, Chung and Yeung (2015), the inverse method was better than the balance method for equations involving more than one operational and two relational lines. In particular, as shown in Figure 10.6, the inverse method was better than the balance method for equations involving a negative pronumeral. The inverse method incurs fewer operational lines (1 vs. 2) and relational lines (2 vs. 3) than the balance method. Therefore, the inverse method is more efficient than the balance method because it involves a lower degree of element interactivity.

Balance method			Inverse method	
Line 1	$x + 5 = 9$	(-5) on both sides	Line 1	$x + 5 = 9$ $(+5$ becomes $-5)$
Line 2	-5 -5		Line 2	$x = 7 - 5$
Line 3	$x = 4$		Line 3	$x = 4$

Figure 10.5 The balance and inverse methods to solve a simple linear equation involving one operational and two relational lines

Balance method		Inverse method	
$6 - p = 0$	$(-6$ on both sides$)$	$6 - p = 0$	$(-p$ becomes $+p)$
$-p = -6$	[note: $-p$ means $(-1 \times p)$]	$6 = p$	
$\div(-1)$ $\div(-1)$	$(\div(-1)$ on both sides$)$		
$p = 6$			

Figure 10.6 The use of the balance method and the inverse method to solve a linear equation involving a negative pronumeral

Balance concept: the operational line depicting a mathematical operation applied to both sides of the equation to preserve its equality (e.g. $+2$ on both sides).

Inverse concept: the operational line depicting an inverse application of a mathematical operation to preserve its equality (e.g. -2 becomes $+2$).

Nonetheless, mathematics educators in Western countries tend to advocate the balance method because the balance concept (e.g. -5 on both sides) highlights the 'balance' in which the same must be done on both sides of the equation to maintain the equality of the equation. They tend to view the inverse method as 'change side, change sign', which may not adequately address the equality of the equation after performing the inverse concept (e.g. move $+5$ on the left side of the equation to become -5 on the right side). Nonetheless, the inverse concept is linked to primary numeracy; for example, $3 + 5 = 8$ is the same as $3 = 8 - 5$ and $5 = 8 - 3$ (Warren & Cooper, 2005). Prior knowledge of the inverse concept is likely to help students learn the inverse concept in the context of linear equations. A study by Ngu and Phan (2016a) confirms such a prediction. Students were able to judge pairs of equations (e.g. $x + 2 = 8$, $x = 8 - 2$) as equivalent after being exposed to either the balance method or the inverse method. The ability to judge such pairs of equations as equivalent reflects the grasp of the conceptual knowledge involved in equation solving. The findings indicate that students would not be disadvantaged if they were exposed to the inverse method. Therefore, the use of the inverse method as an additional method to the balance method will likely benefit student learning of linear equations.

PAUSE AND THINK

The '=' sign concept in the context of solving linear equations presents a challenge to secondary school students. Are the equation pairs below equivalent? Why?

a. $p - 7 = 11$ $p = 11 + 7$

b. $2x + 6 = 5$ $2x + 6 - 6 = 5 - 6$

Researchers have provided empirical evidence in developing flexibility in solving equations involving a bracket (Rittle-Johnson & Star, 2007). For example, there are two ways to solve the equation $3(x + 1) = 6$ (Figure 10.7). The first step of the flexible method is to divide both sides of the equation by 3; whereas the first step of the distributive method is to expand the bracket. The researchers have found that the flexible method is more efficient than the distributive method. Such findings are in line with the impact of the number of operational and relational lines upon the complexity of linear equations. The flexible method incurs fewer operational lines (1 vs. 2) and relational lines (3 vs. 4) than the distributive method. Therefore, mathematics educators should endeavour to promote the development of flexibility in equation-solving.

Flexible method		Distributive method	
Line 1	$3(x + 1) = 6$ ($\div 3$ on both sides)	Line 1	$3(x + 1) = 6$ (expand the bracket)
Line 2	$(x + 1) = 2$ (-1 on both sides)	Line 2	$3x + 3 = 6$ (-3 on both sides)
Line 3	$-1 \quad -1$	Line 3	$-3 \quad -3$
Line 4	$x = 1$	Line 4	$3x = 3$ ($\div 3$ on both sides)
		Line 5	$\div 3 \quad \div 3$
		Line 6	$x = 1$

Figure 10.7 The flexible method and the distributive method in solving a complex linear equation involving multiple steps

PAUSE AND THINK

How would you teach a linear equation such as $12\% x = 40$? Justify your answer.

SHORT-ANSWER QUESTIONS

1. Solve the following linear equations using both the balance method and the inverse method. Which of these methods is easier in terms of the number of operational and relational lines in the solution procedure? Why?
   a. $19 - 8x = 3$
   b. $4(n + 2) = 3n + 11$
2. The linear equations involving fractions often pose a challenge to students. Solve the following linear equations with fractions in two ways. Record the number of operational and relational lines for each way. Which way is more flexible and efficient? Why?
   a. $\frac{x}{2} + \frac{x}{5} = 14$
   b. $\frac{x}{6} = \frac{2}{3}$

IN PRACTICE

Teaching ideas

1. Use the number of relational and operational lines as a point of reference for classifying linear equations.
2. Identify the types of elements (simple vs. complex) in the linear equations (e.g. 6 vs. 14%).
3. An effective approach to teach structurally similar equations is via analogical reasoning.
4. Identify instructional methods that would reduce the level of element interactivity for learning linear equations.

Algebraic problem solving in real-life contexts

The use of a mathematical rule to solve growing pattern problems reveals the power of algebra in solving word problems in real-life contexts. Indeed, in a cross-cultural comparative study, Cai (2005) found that Chinese students outperformed US students on certain types of word problems because Chinese students relied on algebra strategies. In view of the merit of algebra in problem solving, how can mathematics educators help students to acquire skills in algebraic problem solving? Mayer (1985) proposed five types of knowledge to solve word problems:

- linguistic
- factual
- schematic
- strategic
- algorithmic.

According to Mayer (1985), the main hurdle to solve word problems lies in learners' ability to engage in problem translation and integration in order to generate a solution. Problem translation requires the use of linguistic and factual knowledge to translate each proposition from the problem text in light of its context. Schematic knowledge is needed to set up an equation integrating relevant information (values, variables and their relationship). The computation of a solution requires strategic and algorithmic knowledge (or equation-solving skills). Successful and unsuccessful problem solvers undertaking word problems generally differed in their ability to represent word problems as a mathematics-specific abstraction (or an equation) that could generate a problem solution (Hegarty, Mayer & Monk, 1995). In light of the findings, one of the challenges to solve word problems is the translation of the problem situation for subsequent formulation of an equation to solve for the unknown variable. Because solving the equation to find the unknown variable represents part of the algebraic problem-solving process, gaining proficiency in equation-solving skills presents another challenge to learners.

There are different types of algebra word problems in real-life contexts. Mayer (1985) classified 1097 algebra word problems based on: (1) families that represent the source of formula (e.g. *distance = rate × time*), (2) categories that represent the story line (e.g. *motion*) and (3) templates that represent a specific propositional structure (e.g. *round trip*). In this chapter, we highlight the teaching and learning of percentage change problems in real-life contexts.

Percentage change problems: real–life problems involving an increase or decrease in percentage.

Teaching and learning percentage change problems

Our human cognitive architecture comprises a long-term memory which stores a large number of schemas, and a working memory which is restricted in processing novel information but not schemas that can be retrieved from the long-term memory (Sweller, Ayres & Kalyuga, 2011). In light of the respective roles of the working memory and long-term memory, the design of instructional material needs to minimise the number of interacting elements in the working memory, particularly for complex material imposing high element interactivity. One strategy to reduce the degree of element interactivity is by building on the learners' existing schemas (prior knowledge) that are already in the long-term memory (Ngu, Yueng & Tobias, 2014).

Consider a word problem in the real-life context of a junior secondary school: *Last semester John scored 70 marks for a mathematics test. He has improved his mathematics marks by 10% this semester. Find John's mathematics marks for this semester.* The flexibility of algebraic problem solving allows mathematics educators to use two algebraic approaches to solve the word problem.

The equation approach 1

The teaching and learning of the equation approach 1 emphasises a two-part process. The first part focuses on revising the prior knowledge of the percentage quantity (e.g. $72 \times 12\%$). The second part focuses on the translation of the problem situation and the formulation of an equation to generate a problem solution.

Percentage quantity: a specific percentage of a quantity.

Part 1: review of the percentage quantity

$\frac{1}{2} \times 20 = 10$	$\frac{1}{4} \times 12 = 3$
$50\% \times 20 = 10$	$25\% \times 12 = 3$
$30\% \times 20 = 6$	

Students can use a calculator to verify the answers.

The revision of the percentage quantity capitalises on the learning by analogy theory. It draws students' attention to the relation between percentage and quantity (e.g. $25\% \times 12 = 3$) and their prior knowledge of the relation between fraction and quantity (e.g. $\frac{1}{4} \times 12 = 3$).

Part 2: problem translation and the formulation of an equation

Figure 10.8 The equation approach 1 for the percentage change problem

The horizontal line in Figure 10.8 provides aid in translating the problem situation. It depicts the underlying problem structure as comprising the original mark plus the increased mark. Based on the information in the diagram, the next step is to formulate an equation for John's new mark.

Let the new mark be x

Step 1: x = original mark + increased mark
Step 2: $x = 70 + (70 \times 10\%)$
Step 3: $x = 77$

Step 1 expresses the information in the horizontal line in an equation. Step 2 replaces the original mark as well as the increased mark (percentage quantity) with numerical values. Essentially, it consists of two concepts: (1) the multiplicative relation between 70 and 10% to form $(70 \times 10\%)$, and (2) the sum of 70 and $(70 \times 10\%)$. The computation of Step 2 gives rise to the problem solution (Step 3).

The revision of the percentage quantity (Part 1) would help students to reinforce their schema for percentage quantity so that they can treat $(70 \times 10\%)$ as a single element rather than multiple interactive elements, hence reducing the burden on working memory. Processing the relation between the two elements in Part 2, 70 and $(70 \times 10\%)$ would constitute relatively low cognitive load, and therefore facilitate the acquisition of schema for the percentage change problem.

We could argue that it may be easier for students to use a new mark instead of x because of the challenge involved in using a variable. Nonetheless, it is important to provide opportunity for middle school students to engage in algebraic thinking skills wherever appropriate so as to pave the way for them to pursue algebraic problem-solving skills in solving real-life problems.

The equation approach 2

The flexibility of algebraic problem solving allows mathematics educators to explore an alternative instructional approach for percentage change problems. The equation approach 2 represents such an alternative. The main feature of the equation approach 2 is the use of a diagram to align the original mark (70) with 100%, and the new mark (x) with (110%), in which 110% represents the percentage after an increase of 10% in the mark (Figure 10.9). Essentially, the diagram would help students to visualise the relation between the quantity and its corresponding percentage. Based on proportional reasoning, we can form an equation such as $\frac{70}{100} = \frac{x}{110}$, and solve for x. Hence, given appropriate preparation in establishing a schema for how to

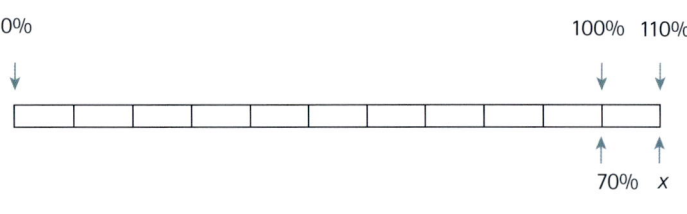

Figure 10.9 The equation approach 2 for the percentage change problem

solve equations involving two fractions, the degree of element interactivity and therefore the cognitive load involved in the equation approach 2 would have been lowered. Accordingly, equation approach 2 would facilitate the teaching and learning of percentage change problems.

SHORT-ANSWER QUESTIONS

Solve the two percentage change problems using both the equation approach 1 and the equation approach 2. Which approach would you prefer? Justify your rationale from a cognitive load perspective.

1. John's annual car insurance costs $950 plus 10% GST. How much does John need to pay?
2. Which is the better deal on a mobile phone marked at $2400: (a) a discount of 40%, or (b) a discount of 20% and then a further discount of 20%?

Equation approach vs. unitary approach

SCENARIO

A study by Ngu et al. (2016) revealed that the equation approach was better than the unitary approach for learning percentage change problems. The equation approach is similar to the equation approach 1 as described under the section 'Teaching and learning percentage change problems'. The equation approach integrates relevant information and expresses it in an equation to solve for the unknown variable. The unitary approach emphasises a unit percentage. The solution procedure of the unitary approach involves calculating 1% of the given quantity, and then a multiple of this quantity, which is the solution.

Question
Which is the better deal on a digital camera marked at $2100?
    a. a discount of 20%
    b. a discount of 10% and then a further discount of 10%.

Question

Examine the solution strategies provided by two students who were exposed to the unitary approach and the equation approach, respectively (Figure 10.10). Did both students solve (a) and (b) correctly? Which of these two approaches would you adopt for teaching percentage change problems? Why?

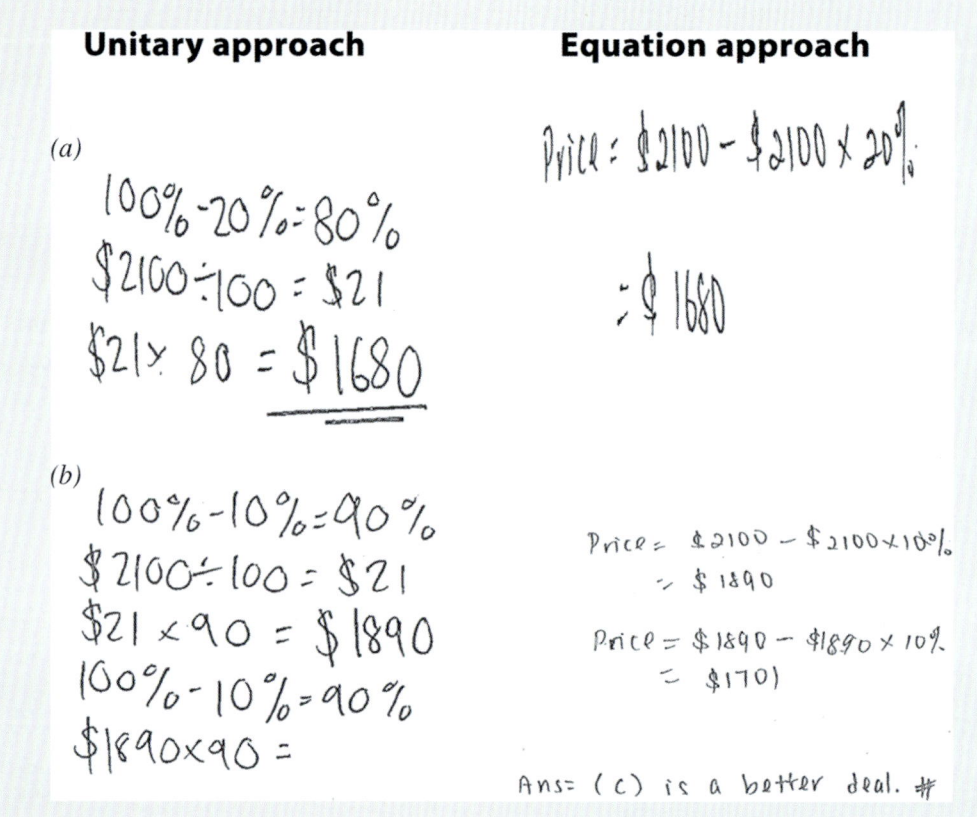

Figure 10.10 Solution strategies of two students who were exposed to the unitary approach and the equation approach

Source: Ngu et al., 2016.

Teaching and learning challenging percentage change problems

Students tend to make errors on challenging percentage change problems such as 'After a 12% markup, the shoes now cost $34. How much did they originally cost?' (Parker & Leinhardt, 1995, p. 448). They tend to retrieve prior knowledge of percentage quantity $34 \times 12\%$ (wrong in this case) and rely on their intuition to subtract ($34 \times 12\%$) from $34 as they may have perceived that the original cost is less than $34. Clearly, this type of percentage change problem poses a challenge to students. Notwithstanding that there are other instructional approaches to teach and learn challenging percentage change problems (e.g. the unitary approach), two algebraic approaches for teaching and learning challenging percentage change problems will now be discussed.

The equation approach 1

The teaching and learning of the above challenging percentage change problem requires the learners to identify the relevant information from the problem text (12%, $34), specify key words such as 'markup' and 'original price', and construct a relation between values and a variable (the original cost) in an equation. Then, the learners need to solve for the unknown variable, which is the original cost.

Let x be the original cost.

Figure 10.11 The equation approach 1 for the challenging percentage change problem

Step 1	Markup price = original cost + increased amount
Step 2	$\$34 = x + x \times 12\%$
Step 3	$\$34 = x(1 + 12\%)$
Step 4	$x = \$34 \div (1 + 12\%)$
Step 5	$x = \$30.35$

Answer: The original cost is $\$30.35$.

For novice learners, the solution procedure would constitute high element interactivity because of the interaction between elements on each solution step and across the solution steps. Nonetheless, the degree of element interactivity depends on the knowledge base of the learners in the algebra domain. For example, if the learners have prior knowledge in algebraic problem solving (e.g. $x \times 25\% = \$200$, solve for x), then it is likely that they can treat $x \times 12\%$ as a single unit (Kalyuga, 2007), thus reducing the degree of element interactivity. Furthermore, prior knowledge of the factorisation skill would help learners to factorise the algebraic expression, $x + x \times 12\%$. More importantly, familiarity with how to solve $4n = 20$ may assist the learners in transferring this skill to solve an equivalent equation, $\$34 = x(1 + 12\%)$. Therefore, learners' prior knowledge of basic algebra skills would affect the degree of element interactivity associated with equation approach 1. An important message for mathematics educators is to strengthen students' algebra skills (e.g. the variable, x, how to set up an equation, factorisation and equation-solving skills) prior to introducing equation approach 1 for challenging percentage change problems.

The equation approach 2

Figure 10.12 shows the underlying problem structure. It clearly shows that the markup price of $\$34$ aligns with 112% (i.e. after 12% increase), and the original cost (x), which is yet to be found, aligns with 100%. In other words, the diagram will help students to visualise the relation between quantity and its corresponding percentage. Once again, the main idea is to capitalise on the proportional reasoning concept to form an equation such as $\frac{x}{100} = \frac{34}{112}$ and solve for x. Thus, given appropriate schema reinforcement regarding how to solve equations involving two fractions, the equation approach 2 would likely impose low cognitive load, thus facilitating the teaching and learning of challenging percentage change problems.

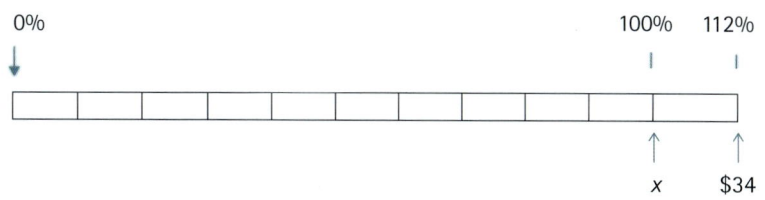

Figure 10.12 The equation approach 2 for the challenging percentage change problem

Comparing the two approaches, the equation approach 2 may be easier than the equation approach 1 for challenging percentage problems because the variable (x) appears only once in equation approach 2 but twice in equation approach 1. Therefore, it is important to assess the prior knowledge of students before exposing them to either the equation approach 1 or equation approach 2, or both.

In summary, when teaching percentage change problems, it is important to organise the material in a hierarchical level of complexity. Evidently, the prerequisite knowledge is the percentage quantity concept. Students not only need to have a schema for finding the part percentage quantity (e.g. $55 \times 15\%$), they also need to have a schema for finding the whole percentage quantity (e.g. $x \times 25\% = 56$, solve for x). Empirical studies have indicated that students have great difficulty in finding whole percentage quantity (Baratta et al., 2010). Accordingly, greater attention is required to help students acquire the schemata for the percentage quantity concept before teaching them percentage change problems. More importantly, students need to have schemas for solving a range of linear equations, as equation-solving skills are another prerequisite for success in algebraic problem solving.

PAUSE AND THINK

Solve the two challenging percentage change problems using both the equation approach 1 and equation approach 2. Which approach would you use to teach your students? Justify your rationale from a cognitive load perspective.

1. The enrolment of Year 11 at a school this year is 42, which includes a 2% increase from the previous enrolment. What was the previous enrolment?
2. Kikal experiences a steady decline in population growth of 2% every year. If the current population is 560 000, what was the original population?

SHORT-ANSWER QUESTIONS

Tina paid $10.95 for a pizza which includes a 38% delivery charge on the price paid. What was the price of the pizza without the delivery charge? Would you buy a pizza that includes a delivery charge? Why?

IN PRACTICE

Keep this in mind when teaching number and algebra

1. The flexibility of algebraic problem solving enables us to generate two types of equation approaches.

2. The diagram of both the equation approach 1 and equation approach 2 scaffolds the problem situation.

3. Proficiency in linear equations is critical to ensure the success of using both the equation approach 1 and equation approach 2 for learning percentage problems.

Transferability of algebraic problem solving to the science curriculum

Algebraic word problems represent a significant part of the secondary school curriculum. The Australian Curriculum: Mathematics highlights the need to connect mathematics across various discipline domains. Secondary school students are expected to learn how to solve both *within-domain* algebra word problems in the mathematics curriculum and *between-domain* algebra word problems, particularly in science subjects (such as in a chemistry or physics context). For example, regarding Newton's second law of motion, $F = ma$ (where F = force, m = mass, and a = acceleration), students need to understand the relation between F, m and a so that they can solve for the unknown variable (e.g. m), if the values of the other two variables (F and a) are given. Several researchers have provided empirical evidence of the transferability of prior knowledge in algebraic word problems to successfully solve physics problems requiring mathematical manipulation (Bassok & Holyoak, 1989). More recent research has investigated how to facilitate the teaching and learning of algebra word problems in a chemistry context (Ngu & Yeung, 2013; Ngu, Yeung & Phan, 2015). Specifically, we will examine learning of molarity chemistry problems. Unless you have studied chemistry as part of the senior secondary science curriculum, you may not be familiar with the molarity chemistry problems.

Molarity chemistry problems

A molarity problem such as the one below is an example of an algebraic word problem in a chemistry context.

Molarity: number of moles of solute/volume in one litre of solution.

   A sample of vinegar is found to contain 15 grams of acetic acid, CH_3COOH, in 316 mL of the solution. Calculate the molarity of the acetic acid in vinegar (H = 1, C = 12, O = 16). Note that H = 1 means the atomic mass of hydrogen (H) is 1. The molar mass (MM) of the acetic acid is the sum of the atomic mass of all elements (Ngu & Yeung, 2013).

   How can mathematics educators teach molarity chemistry problems? More specifically, how can mathematics educators construct an equation to solve the molarity problem?

   Central to the molarity problem is the concept of the molarity. The molarity of a solution is defined as the moles of solute per litre of solution. Thus, the concept of molarity is related to a chemical substance dissolving in water to form a specific concentration. Several chemical concepts are embedded in the molarity problem:

$$\text{Molarity} = \frac{\text{number of moles of solute}}{\text{volume in one litre of solution}}$$

1. Number of moles of solute $= MV$ where $M =$ Molarity, $V =$ volume in litre
2. Alternatively, the number of moles of solute $= \frac{MV}{1000}$ where $V =$ volume in millilitres (mL)
3. Also, the number of moles of a chemical substance $= \frac{g}{MM}$ where $g =$ mass in grams, $MM =$ molar mass of the chemical

In effect, the problem structure of the molarity problem above comprises: (1) molarity (M), (2) volume (V in mL), and (3) number of moles of solutes (i.e. $\frac{g}{MM}$) where g and MM are mass and molar mass, respectively. Each of these four chemical concepts (M, V, g and MM) represents an aspect of the problem structure. Based on the analysis of the chemical concepts, we can form an equation such as $\frac{g}{MM} = \frac{MV}{1000}$ (where the left-hand and the right-hand sides equal to the number of moles of solutes) to solve for M, the molarity of the acetic acid in vinegar (Ngu, Yeung & Phan, 2015).

The solution for the vinegar problem is:

$g = 15$ grams, $V = 360$ mL, MM $=$ molar mass of the vinegar, $CH_3COOH = (2 \times 12 + 1 \times 4 + 2 \times 16) = 60$

$$\frac{g}{MM} = \frac{MV}{1000}$$

$$\frac{15}{60} = \frac{M \times 360}{1000}$$

$$M = \frac{15}{60} \times \frac{1000}{360}$$

$$M = 0.69 \text{ moles/L or } 0.69 \text{ M}$$

Once again, the power of algebra is such that we can use the equation to solve any unknown variable provided that the other chemical concepts (or variables) are given. The molarity problems pose a challenge to secondary students because the traditional method (Thickett, 2000) for teaching and learning molarity problems does not emphasise the integration of relevant chemistry concepts in a single equation to solve for the unknown variable.

PAUSE AND THINK

Underline the irrelevant information in the problem below:

20 g of sodium sulphate, Na_2SO_4, were dissolved in sufficient water to obtain 500 mL of a solution with a density of 1.11 g/mL. Calculate the molarity of the solution. (Na $= 23$, S $= 32$, O $= 16$).

What have you learned by underlining the irrelevant information?

SHORT-ANSWER QUESTIONS

1. An NaCl solution is 2.0 M. How many millilitres (mL) of this solution will contain 10 g of NaCl? (Na $=$ 23, Cl $=$ 35.5).
2. Phenobarbitone, $C_{12}H_{12}N_2O_3$, is commonly used in medicine as a long-acting sedative. A patient may take 5.0 mL of phenobarbitone as one dose. If the concentration of the phenobarbitone is 0.017 M, what mass of the drug is present in one dose? (C $=$ 12, O $=$ 16, N $=$ 14, H $=$ 1) (Ngu et al., 2009).

Teaching algebra in a chemistry context

IN PRACTICE

1. To apply algebraic problem solving in a chemistry context, it is important to identify the chemical concepts involved in the problem situation.
2. Scrutinise the problem text of the chemistry problems and extract values and variables, express these in an equation, and then solve for the unknown variable.

Conclusion

It is important to differentiate the concepts of the unknown, pronumeral and variable. In particular, it is important to distinguish between a variable and an object in algebraic problem solving. The teaching and learning of algebraic expressions can be achieved through analogical comparison between the manipulation of arithmetic problems and algebraic expressions. Unpacking the complexity of linear equations using relational and operational lines has provided a much-needed framework for teaching and learning of linear equations in a hierarchical order of complexity. The diagram in the equation approach provides aid in translating the problem situation; subsequently, the formulation of an equation aids in solving word problems in real-life contexts. More importantly, prior knowledge of equation-solving skills is critical to ensure success in using the equation approach in solving real-life problems. The power of algebra problem solving is manifested in the teaching and learning of molarity chemistry word problems.

BRINGING IT TOGETHER

1. Why is it important to help students distinguish the unknown, pronumeral and variable?
2. The concept of a variable is critical to algebraic thinking. Explain how you might scaffold a variable to students.
3. The algebraic expressions represent an important aspect of the prior knowledge of learning how to solve linear equations. How might you strengthen students' prior knowledge of algebraic expressions?

4. Will this linear equation $3(n + 7) = 5(n + 3)$ pose a challenge for secondary students? Discuss some considerations when teaching this linear equation.

5. What are the factors that contribute to students' understanding of word problems? Discuss.

6. Explain how you can use algebra to teach problem solving in a physics context.

ACKNOWLEDGEMENTS

The author would like to acknowledge the use and reference to their previous work throughout this chapter, particularly Ngu and Phan (2016b) and Ngu and Phan (2016a).

Further resources

REFERENCES

Australian Institute for Teaching and School Leadership (AITSL) (2011). *Australian Professional Standards for Teachers*. Melbourne: AITSL.

Ayres, P.L. (2001). Systematic mathematical errors and cognitive load. *Contemporary Educational Psychology*, 26(2), 227–48. https://doi: 10.1006/ceps.2000.1051

Baratta, W., Price, B., Stacey, K., Steinle, V. & Gvozdenko, E. (2010). Percentages: The effect of problem structure, number complexity and calculation form. In L. Sparrow, B. Kissane & C. Hurst (eds), *Proceedings of the 33rd Annual Conference of the Mathematics Education Research Group of Australasia (MERGA)* (61–8). Fremantle: MERGA.

Bassok, M. & Holyoak, K.J. (1989). Interdomain transfer between isomorphic topics in algebra and physics. *Journal of Experimental Psychology: Learning, Memory, and Cognition*, 15(1), 153–66. http://dx.doi.org/10.1037/0278-7393.15.1.153

Cai, J. (2005). U.S. and Chinese teachers' constructing, knowing, and evaluating representations to teach mathematics. *Mathematical Thinking and Learning*, 7(2), 135–69. https://doi: 10.1207/s15327833mtl0702_3

Cai, J., Lew, H.C., Morris, A., Moyer, J.C., Ng, S.F. & Schmittau, J. (2005). The development of students' algebraic thinking in earlier grades: A cross-cultural comparative perspective. *ZDM – The International Journal on Mathematics Education*, 37, 5–15.

Chen, O., Kalyuga, S. & Sweller, J. (2017). The expertise reversal effect is a variant of the more general element interactivity effect. *Educational Psychology Review*, 29(2), 393–405. http://dx.doi.org/10.1007/s10648-016-9359-1

Clement, J. (1982). Algebra word problem solutions: Thought processes underlying a common misconception. *Journal for Research in Mathematics Education*, 13(1), 16–30. https://doi: 10.2307/748434

Cowan, N. (2001). The magical number 4 in short-term memory: A reconsideration of mental storage capacity. *Behavioral and Brain Sciences*, 24(1), 87–114; discussion 114–85.

Cramer, K. & Wyberg, T. (2009). Efficacy of different concrete models for teaching the part-whole construct for fractions. *Mathematical Thinking and Learning*, 11(4), 226–57. https://doi: 10.1080/10986060903246479

Gentner, D. (1983). Structure-mapping: A theoretical framework for analogy. *Cognitive Science*, 7(2), 155–70.

Hegarty, M., Mayer, R.E. & Monk, C.A. (1995). Comprehension of arithmetic word problems: A comparison of successful and unsuccessful problem solvers. *Journal of Educational Psychology*, 87(1), 18–32. http://dx.doi.org/10.1037/0022-0663.87.1.18

Kalyuga, S. (2007). Expertise reversal effect and its implications for learner-tailored instruction. *Educational Psychology Review*, 19(4), 509–39. https://doi: 10.1007/s10648-007-9054-3

Kieran, C. (1992). The learning and teaching of school algebra. In D. Grouws (ed.), *Handbook of research on mathematics teaching and learning* (390–419). New York: Macmillan.

Matsuda, N., Yarzebinski, E., Keiser, V., Raizada, R., Cohen, W.W., Stylianides, G.J. & Koedinger, K.R. (2013). Cognitive anatomy of tutor learning: Lessons learned with SimStudent. *Journal of Educational Psychology*, 105(4), 1152–63. http://dx.doi.org/10.1037/a0031955

Mayer, R.E. (1985). Mathematical ability. In R.J. Sternberg (ed.), *Human abilities: An information-processing approach* (127–50). New York: Freeman.

McSeveny, A., Conway, R. & Wilkes, S. (2004). *New signpost mathematics 8*. Melbourne: Pearson Education Australia.

National Research Council (2001). *Adding it up: Helping children learn mathematics*. Washington, DC: National Academy Press.

Ngu, B.H., Chung, S.F. & Yeung, A.S. (2015). Cognitive load in algebra: Element interactivity in solving equations. *Educational Psychology*, 35(3), 271–93. https://doi: 10.1080/01443410.2013.878019

Ngu, B.H., Mit, E., Shahbodin, F. & Tuovinen, J. (2009). Chemistry problem solving instruction: A comparison of three computer-based formats for learning from hierarchical network problem representations. *Instructional Science*, 37(1), 21–42. https://doi: 10.1007/s11251-008-9072-7

Ngu, B.H. & Phan, H.P. (2016a). Comparing balance and inverse methods on learning conceptual and procedural knowledge in equation solving: A cognitive load perspective. *Pedagogies: An International Journal*, 11(1), 63–83. https://doi: 10.1080/1554480X.2015.1047836

Ngu, B.H. & Phan, H.P. (2016b). Unpacking the complexity of linear equations from a cognitive load theory perspective. *Educational Psychology Review*, 28, 95–118. https://doi: 10.1007/s10648-015-9298-2

Ngu, B.H. & Phan, H.P. (2017). Will learning to solve one-step equations pose a challenge to 8th grade students? *International Journal of Mathematical Education in Science and Technology*, 48(6), 876–94. https://doi: 10.1080/0020739X.2017.1293856

Ngu, B.H., Phan, H.P., Hong, K.S. & Usop, H. (2016). Reducing intrinsic cognitive load in percentage change problems: The equation approach. *Learning and Individual Differences*, 51, 81–90. https://doi.org/10.1016/j.lindif.2016.08.029

Ngu, B.H. & Yeung, A.S. (2013). Algebra word problem solving approaches in a chemistry context: Equation worked examples vs. text editing. *The Journal of Mathematical Behavior*, 32(2), 197–208. http://dx.doi.org/10.1016/j.jmathb.2013.02.003

Ngu, B.H., Yeung, A.S. & Phan, H.P. (2015). Constructing a coherent problem model to facilitate algebra problem solving in a chemistry context. *International Journal of Mathematical Education in Science & Technology*, 46(3), 388–403. https://doi: 10.1080/0020739x.2014.979899

Ngu, B.H., Yeung, A. & Tobias, S. (2014). Cognitive load in percentage change problems: Unitary, pictorial, and equation approaches to instruction. *Instructional Science*, 42(5), 685–713. https://doi: 10.1007/s11251-014-9309-6

Parker, M. & Leinhardt, G. (1995). Percent: A privileged proportion. *Review of Educational Research*, 65(4), 421–81. https://doi: 10.3102/00346543065004421

Pawley, D., Ayres, P., Cooper, M. & Sweller, J. (2005). Translating words into equations: A cognitive load theory approach. *Educational Psychology*, 25(1), 75–97. https://doi: 10.1080/0144341042000294903

Richland, L.E., Zur, O. & Holyoak, K.J. (2007). Cognitive supports for analogies in the mathematics classroom. *Science*, 316(5828), 1128–9. https://doi: 10.2307/20036317

Rittle-Johnson, B. & Star, J.R. (2007). Does comparing solution methods facilitate conceptual and procedural knowledge? An experimental study on learning to solve equations. *Journal of Educational Psychology*, 99(3), 561–74. https://doi: 10.1037//1082-989x.7.2.147

Stacey, K. & MacGregor, M. (1999). Learning the algebraic method of solving problems. *The Journal of Mathematical Behavior*, 18(2), 149–67. http://dx.doi.org/10.1016/S0732-3123(99)00026-7

Sweller, J., Ayres, P. & Kalyuga, S. (2011). *Cognitive load theory*. New York: Springer. https://doi:10.1007/978-1-4419-8126-4.

Sweller, J., van Merrienboer, J. & Paas, F. (2019). Cognitive architecture and instructional design: 20 years later. *Educational Psychology Review*, 31(2), 261–92. https://doi: 10.1007/s10648-019-09465-5

Thickett, G. (2000). *Macmillan chemistry pathways 1*. South Yarra: Macmillan.

van Merrienboer, J.J.G., Kirschner, P.A. & Kester, L. (2003). Taking the load off a learner's mind: Instructional design for complex learning. *Educational Psychologist*, 38(1), 5–13. https://doi: 10.1207/s15326985ep3801_2

Vincent, J., Price, B., Caruso, N., McNamara, A. & Tynan, D. (2011). *MathsWorld 7 Australian Curriculum edition*. South Yarra: Macmillan.

Warren, E. & Cooper, T. (2005). Young children's ability to use the balance strategy to solve for unknowns. *Mathematics Education Research Journal*, 17(1), 58–72. https://doi: 10.1007/bf03217409

Wilkie, K. & Clarke, D. (2016). Developing students' functional thinking in algebra through different visualisations of a growing pattern's structure. *Mathematics Education Research Journal*, 28, 223–43. https://doi.org/10.1007/s13394-015-0146-y

Measurement and geometry

Gregory Hine

LEARNING OBJECTIVES

The learning objectives of this chapter are directly linked to the Australian Professional Standards for Teachers (APST) (Australian Institute for Teaching and School Leadership [AITSL], 2011). After studying this chapter, you should be able to:

- outline the importance of measurement and geometry concepts within the secondary mathematics classroom (APST 2.1, 2.3)
- develop an understanding of the concept of geometric proof as articulated in the Australian Curriculum (APST 2.3, 3.1, 3.2, 3.3, 3.4)
- develop an understanding of the concept of geometric transformation as articulated in the Australian Curriculum (APST 2.3, 3.1, 3.2, 3.3, 3.4)
- describe how technology can be used to support learning concepts in measurement and geometry (APST 2.6, 4.1).

Introduction

This chapter commences with an overview of the importance of teaching measurement and geometry at the secondary school level. Underpinned by the Australian Curriculum and research-driven insight, attention is then turned to outlining various key concepts in measurement and geometry education. These concepts include geometric proof and reasoning, and transformation. Supporting the explanations of these concepts are effective approaches used nationally and internationally in teaching geometric proofs and transformations to secondary students. Finally, some considerations for incorporating digital technology in measurement and geometry lessons are offered.

Within the Australian Curriculum, the topics Measurement and Geometry are presented together as one strand to emphasise their relationship to each other, and to enhance their practical relevance. Within this strand, it is required that:

> Students develop an increasingly sophisticated understanding of size, shape, relative position and movement of two-dimensional figures in the plane and three-dimensional objects in space. They investigate properties and apply their understanding of them to define, compare and construct figures and objects. They learn to develop geometric arguments ... make meaningful measurements of quantities, choosing appropriate metric units of measurement ... build an understanding of the connections between units and calculate derived measures such as area, speed and density. (Australian Curriculum, Assessment and Reporting Authority [ACARA], 2020)

The Measurement and Geometry sub-strands, together with the year levels for which each sub-strand is taught, are listed in Table 11.1.

Table 11.1 Australian Curriculum: Measurement and Geometry (sub-strands)

Measurement and Geometry: sub-strands	Year levels each strand is taught
Using units of measurement	F–10
Shape	F–7
Geometric reasoning	3–10
Location and transformation	F–7
Pythagoras and trigonometry	9–10

At an international level, measurement and geometry questions also enjoy a place of prominence within high-stakes mathematics testing. For instance, the Programme for International Student Assessment (PISA) makes explicit mention of measurement, geometry, reasoning and spatial awareness in all of the six proficiency levels (see Thomson, de Bortoli & Buckley, 2013). For the 2012 testing round, Thomson, de Bortoli and Buckley (2013, p. 62) noted that Australia was 'among one of the countries that achieved a mean score on the space and shape sub-scale that was lower than on the overall mathematical literacy scale, indicating students found this content area relatively more difficult'. In the PISA 2018 testing round, Australia scored 29th out of 58 countries with a mean score of 491 (on the lower end of the

'Middle Performance' Proficiency Level), a position which is not significantly different from previous years but is indicative of consistent decline (Thomson et al., 2019). In a similar vein to PISA, the Trends in International Mathematics and Science Study (TIMSS) outlines explicitly how geometry and measurement topics are tied to student achievement in the advanced, high and intermediate international benchmarks (see Thomson et al., 2017, p. 44). The TIMSS content domains at Year 8 level are comprised of Number (30%), Algebra (30%), Geometry (20%), and Data and Chance (20%). The Geometry content domain is comprised of topic areas: Geometric Shapes, Geometric Measurement, and Location and Movement. Across Australian schools, Thomson, Hilmann and Wernert (2012, p. 33) reported that 'Year 8 students' performance was clearly better in data and chance and number than in algebra and geometry'. For the TIMSS 2015 testing round, an identical comment for the mathematics content domains was noted by Thomson et al. (2017, p. 74). Ostensibly, it would appear that Australian schools could benefit from increased attention to improved teaching and learning practices within this essential mathematical strand.

PAUSE AND THINK

Why are Measurement and Geometry such prominent mathematical topics within the Australian Curriculum and internationally (e.g. for PISA/TIMMS)?

The importance of learning geometry

Geometry – commonly regarded as the study of space and spatial relationships – is an important and essential branch of the mathematics curriculum at all year levels (Singhal, Henz & McGee, 2014). According to Singhal, Henz and McGee (2014), the study of geometry develops both logical and deductive thinking, which in turn helps learners expand mentally and mathematically. Conversely, geometry is not concerned merely with learning definitions and properties, nor is it learning geometric proofs via rote memorisation (Siemon et al., 2011). Goldenberg, Cuoco and Mark (2012) make two broad claims about the role of geometry within general education. The first claim is that geometry can help students connect with mathematics. Moreover, 'students connect well with properly selected geometric studies. The many hooks include no less than art, physical science, imagination, biology, curiosity, mechanical design, and play' (p. 3). The second claim is that geometry can be an ideal vehicle for building the 'habits of mind' perspective. For instance, Goldenberg, Cuoco and Mark (2012) asserted that:

> within mathematics, geometry is particularly well placed for helping people develop these ways of thinking. It is an ideal intellectual territory within which to perform experiments, develop visually based reasoning styles, learn to search for invariants, and use these and other reasoning styles to spawn constructive arguments. (pp. 4–5)

Additionally, these authors contend that in a very broad sense, geometry is also ideally placed for helping students connect richly with the rest of mathematics.

Geometry: this important study of space and spatial relationships develops logical reasoning and deductive thinking, which in turn helps learners expand both mentally and mathematically.

The importance of learning measurement

Measurement: the process of identifying the relationship of numbers that can be expressed in terms of length, area, volume, capacity, time, money, mass or weight.

Measurement is a key component of the mathematics curriculum, and it is considered the most practical and hands-on application of mathematics in everything from occupational tasks to day-to-day-life (Siemon et al., 2011; Van de Walle, Karp & Bay-Williams, 2014). The application and variety of measurement concepts surrounding people on a daily basis is considerable – 'from gigabytes that measure amounts of information to font size on computers, from miles per gallon to recipes for a meal' (Van de Walle, Karp & Bay-Williams, 2014, p. 397). Siemon et al. (2011, p. 624) contend that in order to extend mathematical concepts, students should be 'provided with the opportunity to apply their understandings by using number processes and geometric reasoning'. Doing so enables students to integrate mathematics into other learning areas, to appreciate relationships existing between measurement attributes (e.g. length, area) and various units of measurement relevant to each attribute, and to fluently manage measurement tasks both in the classroom and in the outside world. Despite its importance, measurement is not an easy topic for students to understand. Data from international studies consistently indicate that US students are weaker in the area of measurement than any other topic in the mathematics curriculum (Thompson & Preston, 2004). This finding would coincide with recent high-stakes achievement data for Australian students, noted earlier in the chapter.

SHORT-ANSWER QUESTIONS

1. Based upon your reading in this chapter, how would you explain to someone what the topic of geometry comprised?
2. Based upon your reading in this chapter, how would you explain to someone what the topic of measurement comprised?
3. Describe how learning geometric concepts helps students inside and outside of the mathematics classroom.
4. Describe how learning measurement concepts helps students inside and outside of the mathematics classroom.

The concept of geometric proof

The literature exhorts the concept of proof as central to the mathematics discipline and, consequently, authors contend that it should feature prominently in mathematics education (Ball et al., 2002; Siemon et al., 2011). Because of this centrality, proof is considered an essential tool for promoting mathematical understanding in students (Ball et al., 2002; Reid, 2011) and, at the same time, for providing educators with insight about how students learn mathematics (Wilkerson-Jerde & Wilensky, 2011). Singh (2005, p. 21) offered insight into the idea of a classical mathematical proof as:

> to begin with a series of axioms, statements which can be assumed to be true or which are self-evidently true. Then by arguing logically, step by step, it is possible to arrive at a

conclusion. If the axioms are correct and the logic is flawless, then the conclusion will be undeniable. This conclusion is the theorem. Mathematical theorems rely on this logical process and once proven are true until the end of time. Mathematical proofs are absolute.

In the 1990s de Villiers suggested a framework to justify the role of proof in the mathematics classroom. The framework posits that this role can be a means of (a) verification, (b) explanation, (c) systematisation, (d) discovery and (e) communication (de Villiers, 1990). According to Yopp (2011), a majority of papers written since the advent of de Villiers' framework could be characterised by one or more of the suggested roles – although the role of proof in mathematics classrooms has yet to reach a full and exhaustive description. Following on from ideas concerning proof, geometric proof is one particular type of mathematical justification whereby students' deductive reasoning and creative thinking can be developed through systematic argumentation (Kunimune, Fujita & Jones, 2009). According to several commentators, geometric proof once held a more prominent place in the secondary context (Ball et al., 2002; Siemon et al., 2011). Despite the centrality of deductive reasoning to successful progress in mathematics, it was found that teaching geometric proofs to lower secondary students was an extremely difficult task (Battista, 2007; Mariotti, 2007). Consequently, many teachers and students resorted to learning proofs via rote memorisation for later recitation in assessments – omitting key steps in logical reasoning and understanding along the way. It remains the case at an international level that students have great difficulty in constructing and understanding geometric proofs (Healy & Hoyles, 1998; Lin, 2000; Senk, 1985).

Euclid: father of geometry

CONNECTION

The first systematic discussion of geometry was presented in the text, *The Elements*, organised and written by the Greek mathematician Euclid of Alexandria (ca 300 BCE). While many of the axioms presented in this text had been stated by earlier Greek mathematicians, Euclid is credited as being the first person to demonstrate how propositions could work together as a comprehensive deductive and logical system. Within *The Elements*, Euclid outlined, derived and summarised the geometric properties of objects that exist in a two-dimensional plane. Consequently, Euclidean geometry is also known as *planar geometry*.

Questions

1. Conduct an internet search on Euclid. What are Euclid's five postulates from *The Elements*?
2. Again, and after conducting an internet search, outline and describe three real-life applications of Euclidian geometry.

Approaches to teaching geometric proof
Geometric proof and reasoning in the Australian Curriculum

The concept of geometric proof and reasoning is taught implicitly and explicitly in middle year levels (Years 7–10A) of the Australian Curriculum. In Year 7, students begin by classifying triangles according to their side and angle properties, and then apply this knowledge to quadrilaterals. They also investigate various conditions for lines to be parallel and solve problems using logic and reasoning. Year 8 students investigate the concept of congruency in planar shapes using transformations, in triangles using conditions (e.g. Side-Angle-Side), and then solve related numerical problems using reasoning. The concept of similarity features prominently in the Year 9 curriculum, with students using principles of similarity to investigate trigonometric ratios, enlargement ratios and scale factors. In Year 10, students formulate proofs with angles, triangles and circles, and apply logical reasoning to proofs and numerical exercises. Geometric proof and reasoning is also taught explicitly and implicitly – and to varying degrees – in the senior years of mathematics courses, namely: Essential Mathematics, General Mathematics, Mathematics Methods and Specialist Mathematics. Perhaps this topic features most prominently in General Mathematics (Unit 1, Topic 3: Shape and Measurement) and in Specialist Mathematics (Unit 1, Topic 3: Geometry). Table 11.2 displays how geometric proof and reasoning is presented in the Australian Curriculum courses for students in Years 7–10.

Table 11.2 Geometric proof and reasoning in the Australian Curriculum Years 7–10

Year 7	Classify triangles according to their side and angle properties and describe quadrilaterals (ACMMG165)	Demonstrate that the angle sum of a triangle is 180° and use this to find the angle sum of a quadrilateral (ACMMG166)	Identify corresponding, alternate and co-interior angles when two straight lines are crossed by a transversal (ACMMG163)	Investigate conditions for two lines to be parallel and solve simple numerical problems using reasoning (ACMMG164)
Year 8	Define congruence of plane shapes using transformations (ACMMG200)	Develop the conditions for congruence of triangles (ACMMG201)	Establish properties of quadrilaterals using congruent triangles and angle properties, and solve related numerical problems using reasoning (ACMMG202)	
Year 9	Use similarity to investigate the constancy of the sine, cosine and tangent ratios for a given angle in right-angled triangles (ACMMG223)	Use the enlargement transformation to explain similarity and develop the conditions for triangles to be similar (ACMMG220)	Solve problems using ratio and scale factors in similar figures (ACMMG221)	
Year 10/10A	Formulate proofs involving congruent triangles and angle properties (ACMMG243)	Apply logical reasoning, including the use of congruence and similarity, to proofs and numerical exercises involving plane shapes (ACMMG244)	Prove and apply angle and chord properties of circles (ACMMG272)	

Source: ACARA, 2020.

SCENARIO   The structure of deductive proof: a student perspective

In 'Students' understanding of the structure of deductive proof', a journal article by Miyazaki, Fujita and Jones (2017), three mathematics teacher educators propose a theoretical framework based on secondary students learning the structure of deductive proofs. The findings presented by these scholars can inform how teaching approaches may be improved so that junior secondary students develop a more robust understanding of deductive proofs and proving in geometry. Read the article and answer the questions that follow.

Questions

1. Name and describe the three levels of sophistication regarding students' understanding of deductive proofs.
2. According to these scholars, how does their proposed framework identify causes of difficulties of proof in (i) the introductory stage of learning to construct deductive proofs, and in (ii) the bridging process of students developing the understanding of proof?

Three perspectives of geometric proof and reasoning

It has already been argued in this chapter that teaching geometric proof and reasoning is very important for learning mathematics. Furthermore, Huetinck and Munshin (2008, p. 157) assert that 'the concept of proof as justification of assertions through analytic reasoning should be developed informally over time, beginning long before high school'. As such, these authors contend that verbal and written activities focused on probing mathematical questions (e.g. those requiring proof and analytic reasoning) are central within primary and middle school mathematics programs. Providing these opportunities leads students to develop a deeper understanding of and appreciation for the nature of mathematical proof at secondary school level. Huetinck and Munshin (2008) suggest that mathematics educators consider three approaches – and not least the benefits and limitations associated with each approach – concerning the notion of formal proof, namely: synthetic, analytic and transformational. Each of these approaches is illustrated below through the application of the side-side-side (SSS) triangle congruence to a given situation. The situation is to prove that within a parallelogram *ABCD*, triangles *ABC* and *CDA* (formed by drawing a line from *A* to *C* along the 'long' diagonal) are congruent (see Figure 11.1). In addition, Table 11.3 outlines various ways to present the three approaches.

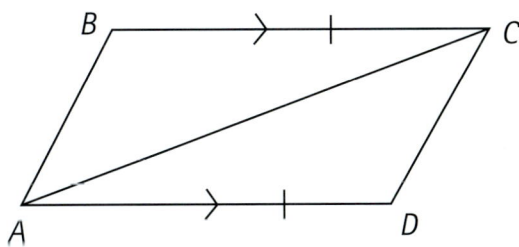

Figure 11.1 Synthetic approach

Synthetic geometry

According to Huetinck and Munshin (2008), the synthetic approach to geometry dates back to Euclid and the Greeks. Essentially, the synthetic approach requires the user to prepare a well-structured and formal, logical system, whereby geometric relationships are proven using rational sequences of definitions, postulates and theorems. Using this approach with middle school students enables them to 'experience the challenge and satisfaction of dealing with probing questions that ask them to justify conclusions, explain how they know a given pattern always works, or find a counterexample to show that a pattern does not always work and is therefore disproved' (Huetinck & Munshin, 2008, p. 159).

Table 11.3 Three ways to present geometric concepts

Synthetic	Analytic	Transformational
$m(\text{length } AB) = \|B - A\|$ Parallel lines are two lines on the plane whose distance from each other is constant. Parallelograms have two sets of parallel sides.	$D = \sqrt{(x_1 - x_2)^2 + (y_1 - y_2)^2}$ Parallel lines on the coordinate plane have the same gradient.	
Congruent planar figures have corresponding sides and angles equal.	Congruent figures in the coordinate plane have corresponding sides of the same measure and absolute value of corresponding gradients equal.	Congruence is a one-to-one correspondence between two geometric figures that preserves distance and angle measure (Okolica & Macrina, 1992).
Perpendicular lines are two lines on the same plane that form right angles.	Perpendicular lines on the same plane have gradients whose product equals -1.	Two lines are perpendicular if one is the reflection line of the other line in itself.
The midpoint of a line segment is equidistant from the endpoints of the segment.	The midpoint of a line segment is found by $M = \left(\dfrac{x_1 + x_2}{2}, \dfrac{y_1 + y_2}{2}\right)$	One way of determining the midpoint of a line segment is to see if it lies on the line of reflection of one endpoint of the segment in the other.

Source: Adapted from Huetinck & Munshin, 2008, p. 159.

Given Figure 11.1, the task is to prove that within parallelogram *ABCD*, triangle *ABC* is congruent to triangle *CDA*. The synthetic approach to proving this triangle congruence could be:

Given $\overline{BC} // \overline{AD}$ and $\overline{BC} = \overline{AD}$.

Required to Prove: $\triangle ABC \cong \triangle CDA$.

Statements	Reasons
1. $BC \cong DA$	1. Given information
2. $AB \cong CD$	2. Properties of a parallelogram (opposite sides are parallel and congruent)
3. $AC \cong AC$	3. Identity (same line)
4. $\triangle ABC \cong \triangle CDA$ QED	4. Side-side-side congruence (SSS)

Analytic geometry

A second perspective of formal proof is through the use of analytic geometry. This approach was first used in the ninth century, but the concept as it is presently construed is rooted in the work of later mathematicians, notably Descartes (after whom the Cartesian plane is named) and Fermat (Huetinck & Munshin, 2008). Although a contemporary mathematics classroom would typically have graphing calculators for students to recreate and dynamically move planar figures – as well as to explore various relationships of shapes and their transformations – the mastery of skills using a manual technique is still very important for conceptual understanding.

To prove that triangle *ABC* is congruent to triangle *CDA* using analytic geometry, we can place the parallelogram on the Cartesian plane and use the distance formula to show corresponding sides have the same length. The coordinate points are $A(-2, -2)$, $B(-1, 1)$, $C(3, 1)$, $D(2, -2)$ (Figure 11.2):

$$AD : \sqrt{(-2 - 2)^2 + [-2 - (-2)]^2} = \sqrt{16} = 4$$
$$BC : \sqrt{(-1 - 3)^2 + (1 - 1)^2} = \sqrt{16} = 4$$
$$AB : \sqrt{[-2 - (-1)]^2 + (-2 - 1)^2} = \sqrt{10}$$
$$DC : \sqrt{(2 - 3)^2 + (-2 - 1)^2} = \sqrt{10}$$

Similar to the synthetic approach, we can easily establish that *AC* is a common side. Thus, with the three known side lengths we can use the side-side-side (SSS) congruence to conclude that triangle *ABC* is congruent to triangle *CDA*.

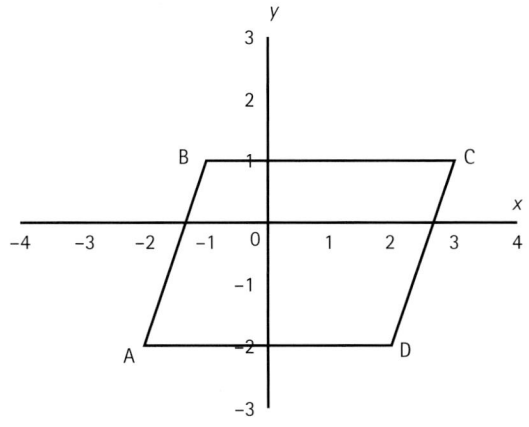

Figure 11.2 Analytic approach

Transformational geometry

Transformational geometry techniques can be demonstrated through graphing technology (e.g. GeoGebra) or via a plane mirror. With either of these tools, students can see that rotations are compositions of reflections by doing the appropriate constructions. Hollebrands (2004) investigated how Year 10 students in the United States constructed the images of a polygon under a reflection, rotation and translation. For the reflection activity given to the research participants, Hollebrands noted that students may have had difficulty with the task as the line of reflection was neither vertical nor horizontal, a finding consistent with the work of Schultz and Austin (1983). Another key finding was that all students seemed to think of reflection as 'flipping', which was evidenced through their explanations of attempting to draw the image congruent to the pre-image. However, the students 'did not attend to relationships between corresponding pre-image and image points and the line of reflection' (Hollebrands, 2004, p. 208). While this research focuses on transformations – which will be discussed later in this chapter – this particular finding has implications for teachers who design geometric proof lessons using a transformational approach.

To prove that triangle *ABC* is congruent to triangle *CDA* using the side-side-side congruence, we begin by drawing diagonal *AC* and letting *P* be the midpoint of *AC* (Figure 11.3). Then we rotate triangle *CDA* 180° (in either direction) about point *P* until it rests completely on triangle *ABC*.

1. Rotate △*CDA* 180° about point *P*.
2. Now since *P* is the midpoint of *AC*, $PA \cong PC$ and both *A* and *A* rotate on to each other.

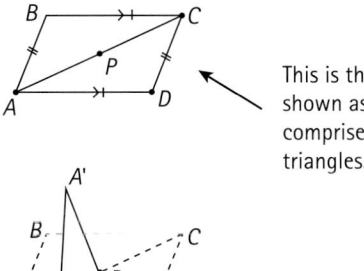

This is the commencing shape, shown as a parallelogram comprised of two congruent triangles.

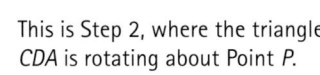

This is Step 2, where the triangle *CDA* is rotating about Point *P*.

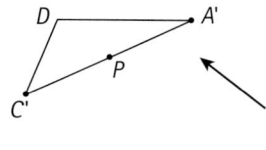

This is the final step, where triangle *CDA* has been rotated 180° about Point *P*. As such, triangle *CDS* now rests on top of triangle *ABC*.

Figure 11.3 Transformational approach

3. Since (by definition of a parallelogram) $AB//CD$ and $AD//CB$, $\angle BAC \cong \angle DCA$ and $\angle CAD \cong \angle ACB$. Therefore, the two pairs of angles ($\angle CAB$ & $\angle ACD$, and $\angle CAD$ & $\angle ACB$) rotate on to each other.

4. Because the angles $\angle CAD$ and $\angle ACB$ coincide, the lines $\overline{AD}$ and $\overline{CB}$ coincide. Similarly, the lines $\overline{AB}$ and $\overline{CD}$ coincide because the angles $\angle BAC$ and $\angle DCA$ coincide.

5. The side-side-side (SSS) congruence can now be applied, as $\triangle CDA$ completely overlaps $\triangle ABC$. Therefore, the two triangles are congruent.

PAUSE AND THINK

Geometric proof is often considered by students as a difficult, inaccessible and onerous topic to master. Think back to your time spent in the classroom coming to terms with any method of geometric proof. Which method did you find the most accessible, and why? For what reasons do you feel students struggle with this important topic, and how do you plan to help them understand it well?

Research-driven insight

Researchers in Japan were interested in discerning lower secondary students' cognitive needs in learning geometric concepts. In Japanese schools, the concept of formal proof is taught intensively to lower secondary students (Japanese grades 7–9). Following their research, Kunimune, Fujita and Jones (2009) suggested an analytical framework that would drive best classroom practice and satisfy student learning needs in geometry. The researchers investigated across a 10-year period how Japanese Year 8 and 9 students successfully completed geometric proof questions – specifically in terms of construction of proof and generality of proof. Considering these two aspects of geometric proof, the researchers presented their findings as a three-level analytical framework developed from the van Hiele model (1959). The framework is found in Table 11.4.

Construction of proof: recognition of how to create deductive arguments in geometry using sufficient knowledge about definitions, assumptions, proofs, theorems and logical circularity.

Generality of proof: recognition of the universality and general applicability of geometric theorems (proven statements), the roles of figures, and the difference between formal proof and experimental verification.

Table 11.4 Student behaviours as a function of geometric proof

Level	Student behaviours
1	Students consider experimental verifications as sufficient to demonstrate that geometrical statements are true.
1a	Students do not achieve generality of proof and construction of proof.
1b	Students achieve generality of proof and construction of proof.
2	Students understand that proof is required to demonstrate geometrical statements are true.
2a	Students achieve generality of proof without understanding logical circularity.
2b	Students understand logical circularity.
3	Students understand simple logical chains between theorems.

The researchers' key findings indicated that the students achieve reasonably well in terms of construction of proof, but not necessarily as well in terms of generality of proof. In other words, students demonstrated they could construct a formal proof (Level 1), yet they may not appreciate the significance of such formal proof in geometry (Level 1a). According to the collected data, 80 per cent of Year 9 students remained at Level 1 in terms of understanding proof. The researchers expressed concern at this finding, as previous research had indicated that by this time in their education, students have studied formal proof in Year 8 using textbooks where 90 per cent of relevant intended lessons can be devoted to justifying and proving geometric facts (Fujita & Jones, 2003). Moreover, students 'may believe that a formal proof is a valid argument, while, at the same time, they also believe experimental verification is equally acceptable to "ensure" universality and generality of geometrical theorems' (Kunimune, Fujita & Jones, 2009, p. 761).

An analytic framework for teaching lower secondary school geometry   IN PRACTICE

Based on over 10 years of classroom-based research, Kunimune et al. (2007) proposed the following analytical framework for teaching lower secondary school geometry (Years 7–9). This framework is designed to help students appreciate the need for formal proofs (in addition to the students being able to construct such proofs):

- Year 7: Commence geometric proof from problem-solving situations (e.g. consider how to draw diagonals of a cuboid)
  - This approach develops students' geometric thinking and provides experiences of mathematical processes, useful in studying deductive proofs in Years 8 and 9
- Year 8: Geometric constructions to be taught, and at the same time these constructions are to be proven
  - This decreases the existing gap between geometric constructions and their proofs (according to research)
- Year 8: Examine differences between experimental verifications and deductive proof
  - Students will gain an appreciation of the differences between experimental verifications and deductive proof
- Year 8: Lessons commence with teaching deductive geometry with a set of already learned properties which are shared and discussed with the classroom, and used as a form of axiom
  - This approach ensures that students are provided with known starting points for their proofs.

In a similar vein, Japanese researchers found that while geometric constructions (e.g. with ruler and compass) can be taught in Year 7, the constructions themselves are often not proved until Year 8 (Shinba, Sonoda & Kunimune, 2004). Moreover, it is in Year 8 that Japanese students learn how to prove simple geometric statements. Through a series of teaching experiments Kunimune, Fujita and Jones (2009) investigated the use of more complex geometric constructions (and proofs of these constructions) with Year 8 students.

For instance, one lesson commenced with the direction '*Let us consider how we can trisect a given straight line AB*'. From their classroom observations, the researchers noted that students were able to first investigate theorems and properties of geometrical figures through construction activities, which led them to a consideration of why the construction worked. Following the appropriate teacher-initiated instruction, students commenced proving the geometric constructions. In addition, the active learning environment enabled students to experience some important processes which led from conjecture to proof (Kunimune, Fujita & Jones, 2009).

SHORT-ANSWER QUESTIONS

1. Recall and summarise the key topics for Geometric Proof and Reasoning in the Australian Curriculum (Years 7–10A).

2. List and distinguish between the three perspectives of presenting geometric proof and reasoning concepts to students.

3. Towards the end of this section of the chapter, some research-driven insights from Japanese mathematics classrooms were shared. Describe two take-aways which could be used for when you teach geometric concepts.

The concept of transformation

Transformation: the changes in positions or size of shapes, which collectively comprise the study of translations, reflections, rotations, symmetry and similarity.

In mathematics the term transformation refers to the changes in positions or size of shapes, which collectively comprise the study of translations, reflections, rotations (slides, flips and turns), symmetry and similarity (Van de Walle, Karp & Bay-Williams, 2014). Another way of viewing transformational geometry is to consider what remains invariant (or the same) as a process happens to an object (Siemon et al., 2011). Broadly speaking, there are two categories under which all geometric transformations belong – rigid transformations and non-rigid transformations – and these will be discussed subsequently. Hollebrands (2004) asserted important reasons to teach geometric transformations within school mathematics curricula. First, transformations provide an opportunity for students to conceptualise important mathematical concepts (e.g. functions, symmetry), and to view mathematics as an interconnected discipline. Second, transformations also engage students in higher-level reasoning activities using a variety of representations. Moreover, Guven (2012, p. 366) argued that the study of transformation can 'lead students to exploration of the abstract mathematical concepts of congruence, symmetry, similarity, and parallelism; enrich students' geometrical experience, thought and imagination; and thereby enhance their spatial abilities'.

Rigid transformations: these transformations (also known as congruence transformations) preserve size and shape, and their movement can be recorded with coordinates.

Rigid transformations

Transformations that do not change the size or shape of an object moved are called rigid transformations (also referred to as rigid motions, isometries or congruence transformations). The movement of rigid transformations can be recorded with coordinates, which enables learning activities to be placed on the Cartesian plane. Typically, three rigid-motion transformations are taught at high school level: translations (or slides), reflections (or flips) and rotations (or turns) (Van de Walle, Karp & Bay-Williams, 2014). Because rigid transformations preserve the size and shape of objects, the study of symmetry is also included under the study of transformations. A popular example of rigid transformations is the tessellation, which is a repeated shape that covers a plane without any gaps or overlaps. As Siemon et al. (2011, p. 633) noted, 'for there to be no gaps or overlaps, the sum of angles of shapes meeting at each vertex must be exactly 360 degrees'. As such, there are only three regular polygons (two-dimensional shapes that have congruent sides and angles) that tessellate: triangles, quadrilaterals and hexagons.

Rigid transformations can also be applied to functions (e.g. polynomials, relations) which can allow for a richer appreciation of how modifications to a rule can lead directly to the movement of a function's graph. For instance, Figure 11.4 shows how the function $g(x) = -\frac{1}{x+2} - 3$ is a composition transformation of the original function $f(x) = \frac{1}{x}$.

Composition transformations are multiple transformations that can be applied to a shape, one after the other. The transformations for $f(x)$ in Figure 11.4 are:

- a reflection in the x-axis
- horizontal translation 2 units left
- vertical translation 3 units down.

In senior years, the topic of matrices is taught in ATAR courses and non-ATAR courses. One of the key topics studied in ATAR courses concerns transformations, where objects (on the Cartesian plane, and hence which have their shape governed by coordinates) are reflected or rotated as a result of matrix multiplication. For instance, the unit square has coordinates (0, 0), (1, 0), (1, 1) and (0, 1). The two figures, Figure 11.5 and Figure 11.6, display how the unit square has undergone a reflection across the y-axis and a rotation of 90 degrees clockwise, respectively. Accompanying each figure is the transformation matrix and mathematical operations which move the object from the pre-image stage to the image.

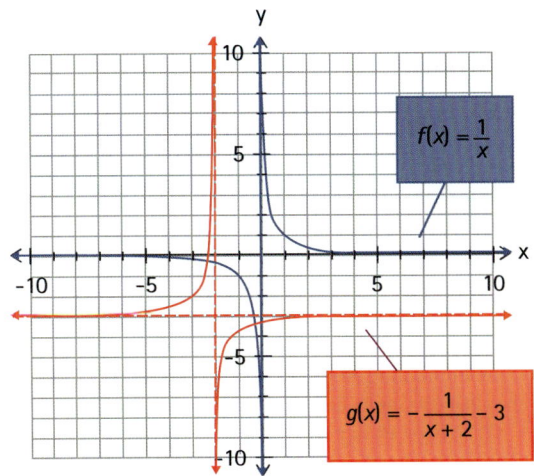

Figure 11.4 Composition transformation in a quadratic function

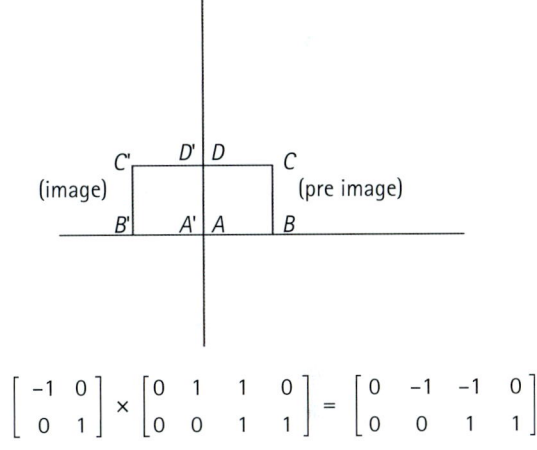

Figure 11.5 Reflection across the y-axis

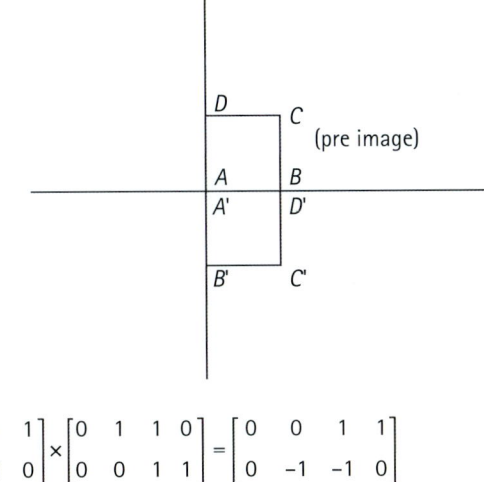

Figure 11.6 Rotation of 90° clockwise

Non–rigid transformations

Non–rigid transformation: there are two types of transformations where an object's size or shape changes. Similar transformations keep the object the same shape but the size changes. Shear transformations keep the area the same, but the shape changes.

Transformations that change the size or shape of an object are called non–rigid transformations. Chiefly, there are two types of non-rigid transformations: similar transformations and shear transformations.

Similar transformations

Two shapes are said to be similar if all of their corresponding angles are congruent and the corresponding sides are proportional (Van de Walle, Karp & Bay–Williams, 2014). Using a scale factor, shapes can be enlarged or reduced – although, technically, images that are smaller than the pre-images are still considered enlargements! Applying a scale factor greater than 1 will make the image larger than pre-image; applying a scale factor less than 1 will make the image smaller. Of course, applying a scale factor of 1 will render the image congruent to the pre-image. One common method of enlarging a figure geometrically is explained by Siemon et al. (2011, p. 635):

> Many children's activity books contain examples or using a grid to enlarge a simple cartoon. A grid is drawn over the shape. A second grid is then drawn with larger or smaller squares, depending on the chosen scale factor. Points where the drawing crosses grid lines are marked on the first and copied onto a similar location on the second. Lines are then joined to construct the enlarged image.

A second method taught in high school uses a point of enlargement, which can lie inside or outside the original object. In Figure 11.7, ΔABC is enlarged by a scale factor of 3 to produce $\Delta A'B'C'$. Notice how the point of enlargement lies outside both triangles; the distance from the point of enlargement to each of the vertices on ΔABC is one-third the distance to $\Delta A'B'C'$.

Dilations are considered to be non-rigid transformations that produce similar two-dimensional figures (Van de Walle, Karp & Bay-Williams, 2014). In Figure 11.8, we can see how the graph of $y = x^2$ has been transformed to $y = 4x^2$. This transformation is described as a dilation parallel to the y-axis, scale factor 4. Popular ways of explaining this transformation to students are to say the pre-image has been 'pulled or stretched' along the y-axis by a factor of 4, or that all of the output values of $f(x)$ have been multiplied by 4.

Figure 11.9 displays how the unit square has undergone two transformations simultaneously: a dilation parallel to x-axis scale factor 3, and a dilation parallel to y-axis scale factor 2. Accompanying this figure is the transformation matrix and mathematical operations which move the object from the pre-image stage to the image.

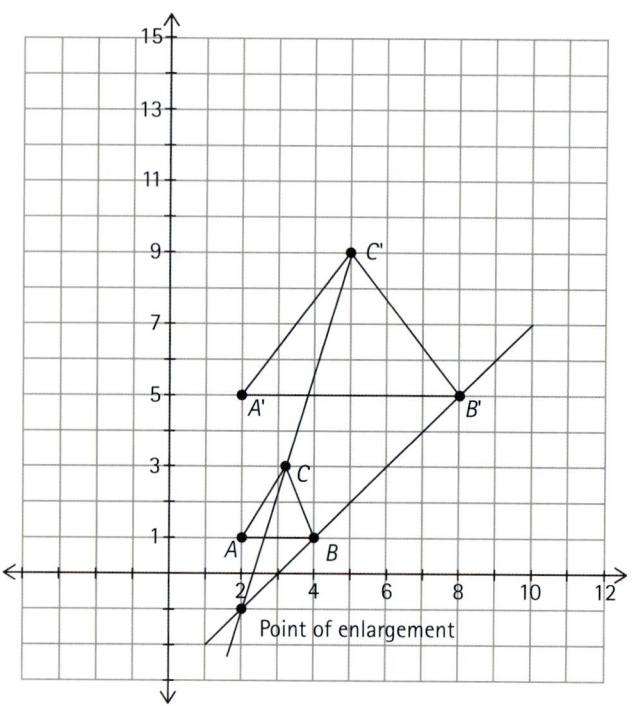

Figure 11.7 Geometric dilation: point of enlargement

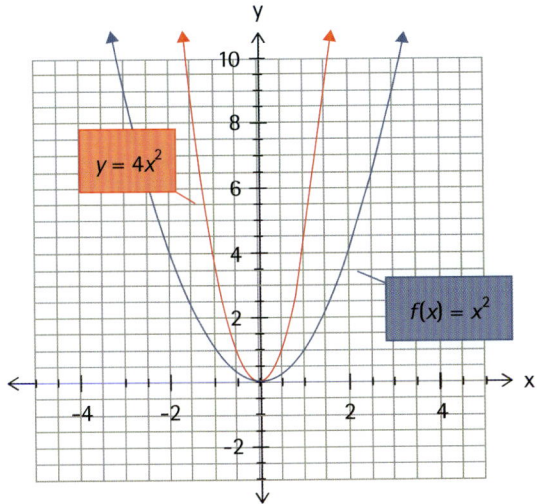

Figure 11.8 Geometric dilation: function transformation

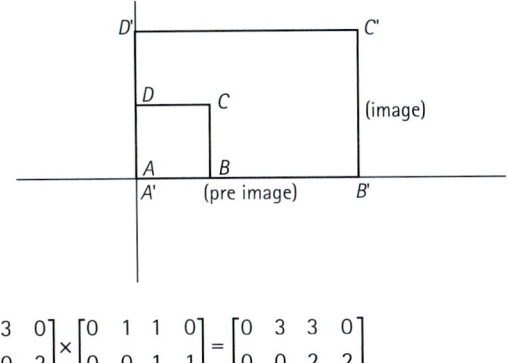

Figure 11.9 Dilation transformations

Shear transformations

Shear transformations (or shears) are transformations that do not preserve the shape of the pre-image. With shears, the area of objects remains constant but the perimeter changes. Shears have been described as the pre-image being pulled in one or more directions, or as 'slices' of the original shape sliding over each other (Siemon et al., 2011). One common application of shears is the italicisation of letters from standard letter format.

Figure 11.10 displays how the unit square has undergone a shear transformation, parallel to y-axis scale factor 4. Accompanying this figure is the transformation matrix and mathematical operations which move the object from the pre-image stage to the image.

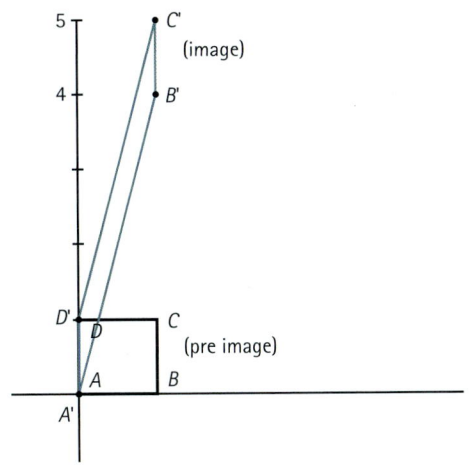

$$\begin{bmatrix} 1 & 0 \\ 4 & 1 \end{bmatrix} \times \begin{bmatrix} 0 & 1 & 1 & 0 \\ 0 & 0 & 1 & 1 \end{bmatrix} = \begin{bmatrix} 0 & 1 & 1 & 0 \\ 0 & 4 & 5 & 1 \end{bmatrix}$$

Figure 11.10 Shear transformation

PAUSE AND THINK

Where in the 'real world' have you seen a rigid, non-rigid, similar or shear transformation?

Approaches to teaching transformation
Transformation in the Australian Curriculum

The topic of transformation is taught across both Number and Algebra and Measurement and Geometry strands. Although this focuses predominantly on the latter strand, we will consider in this section how transformation is presented in both strands. In Year 7, students transform planar objects according to translations, rotations and reflections – as well as identifying line and rotational symmetries. Year 8 students plot linear relationships on the Cartesian plane, and begin to examine how components of linear equations affect corresponding graphs. In Year 9, concepts of scale factor, similarity and enlargements are applied to planar figures; students also begin to graph non-linear relationships with and without technology. Year 10 students explore the connections between algebraic and graphical representations of functions, polynomials and relations. The transformation topics taught to students in Years 7–10 are displayed in Table 11.5. In senior secondary school, students learn about transformations in all four courses: Essential Mathematics (Topic 2: Scale, Plans and Models), General Mathematics (Unit 1, Topic 3: Shape and Measurement; Unit 2, Topic 3: Linear Equations and their Graphs), Mathematics Methods (Unit 1, Topic 1: Function and Graphs and Topic 2: Trigonometric Functions; Unit 2, Topic 1: Exponential Functions) and Specialist Mathematics (Unit 2, Topic 1: Trigonometry and Topic 2: Matrices).

Table 11.5 Transformation in the Australian Curriculum (Years 7–10)

Year 7	Describe translations, reflections in an axis, and rotations of multiples of 90° on the Cartesian plane using coordinates. Identify line and rotational symmetries (ACMMG181)		
Year 8	Plot linear relationships on the Cartesian plane with and without the use of digital technologies (ACMNA193)		
Year 9	Use the enlargement transformation to explain similarity and develop the conditions for triangles to be similar (ACMMG220)	Solve problems using ratio and scale factors in similar figures (ACMMG221)	Graph simple nonlinear relations with and without the use of digital technologies and solve simple related equations (ACMNA296)
Year 10/10A	Explore the connection between algebraic and graphical representations of relations such as simple quadratics, circles and exponentials using digital technology as appropriate (ACMNA239)	Apply understanding of polynomials to sketch a range of curves and describe the features of these curves from their equation (ACMNA268)	

Source: ACARA, 2020.

The van Hiele levels of geometric thought

van Hiele levels of geometric thought: a five-level hierarchy describing the thinking processes used to understand geometric ideas and contexts.

In the 1950s Dina and Pierre van Hiele developed a hierarchical framework, now known as the van Hiele levels of geometric thought, to suggest how students learn geometry. This sequential, five-level framework describes the thinking processes used by learners in geometric and spatial contexts, with the types of ideas referred to as 'objects of thought' (van Hiele, 1959). According to Crowley (1987), the products of thoughts (or the learning that has taken place) at any given level become the objects of thought at the next level. Assisted by appropriate instructional experiences, learners progress through the following five levels, each of which depends on successful achievement of previous levels (Fuys, Geddes & Tischler, 1988). These levels are now considered as: *visualisation*, *analysis*, *abstraction*, *deduction* and *rigour*.

Level 1: Visualisation

Students at this level recognise and name figures based solely on the global visual characteristics of the figure, often comparing the figure to a known prototype. For instance, a square may be defined by a Level 1 student as a square because 'it looks like a square' (Van de Walle, Karp & Bay-Williams, 2014). The general goal for learners is to explore that 'shapes are alike and different and to use these ideas to create classes of shapes (both physically and mentally)' (Mason, 2014, p. 4).

Level 2: Analysis

In Level 2, learners analyse figures in terms of their components and properties, discover properties and rules of a class of shapes empirically, but do not explicitly interrelate figures or properties (Guven, 2012). To illustrate, learners are able to identify what makes a rectangle a rectangle (e.g. four sides, opposite sides parallel, opposite sides same length, four right angles, congruent diagonals) (Van de Walle, Karp & Bay-Williams, 2014).

Level 3: Abstraction

Following the objects of thought from Levels 1 and 2, students begin to perceive relationships between particular objects and the properties of these objects. Certain informal deductive statements (for instance, in the form of if–then reasoning) are elicited by learners when asked to create meaningful definitions and to justify their reasoning (Mason, 2014). Such a statement may be 'if all four angles are right angles, the shape must be a rectangle. If it is a square, all angles are right angles. If it is a square, then it must be a rectangle' (Van de Walle, Karp & Bay-Williams, 2014, p. 429). At this stage, the role and significance of formal deductive reasoning is not understood by learners.

Level 4: Deduction

Level 4 'learners are able to work with abstract statements about geometric properties and make conclusions based more on logic than intuition' (Van de Walle, Karp & Bay-Williams, 2014). To achieve this, learners are able to construct proofs, understand the role of axioms, definitions, theorems, corollaries and postulates, and appreciate the meaning of necessary and sufficient conditions for establishing geometric truth (Mason, 2014). At this level, students would be expected to construct proofs such as those typically found in a Year 11 Mathematics Methods class.

Level 5: Rigour

At the highest level of the hierarchy, the objects of thought are axiomatic systems themselves, not merely the deductions within a system (Van de Walle, Karp & Bay-Williams, 2014). Learners – typically at the university level – can understand the use of indirect proof, contrapositive proof and non-Euclidean systems (Mason, 2014). For instance, a learner having already understood the formal aspects of deduction may establish theorems in different axiomatic systems, before analysing and comparing those systems (Guven, 2012). An example of this may be an analysis of spherical geometry, which is based on lines drawn on a sphere rather than in a plane or ordinary space (Van de Walle, Karp & Bay-Williams, 2014).

CONNECTION — Van Hiele: phases of learning

Within the five levels of geometric thought, and supported with the appropriate instruction, learners progress through each level of thought via five phases of learning (van Hiele, 1959). These phases of learning are *information*, *guided orientation*, *explicitation*, *free orientation* and *integration*.

Phase 1: Information

In the initial phase, teachers engage students in conversations and activities about their prior topical knowledge. As a result of these engagements, students become oriented to the new topic and discover what direction further study will take (Mason, 2014).

Phase 2: Guided orientation

Students are provided with the opportunity to explore the new topic of study through carefully structured and sequenced tasks, and learning materials. According to Crowley (1987, p. 5), the tasks should be designed to elicit specific responses that 'gradually reveal to the students the structures characteristic of this level'. For instance, a teacher may ask students to use a tablet app to create a parallelogram with the diagonals included, and then a larger and smaller version of the original parallelogram.

Phase 3: Explicitation

During explicitation, students describe what they have learned about the topic in their own words (Mason, 2014). Compared with the preceding phases of learning, the teacher's role is minimal, other than to introduce relevant mathematical terms and to ensure that students use these terms accurately.

Phase 4: Free orientation

In this phase, students encounter more complex tasks where they can apply the relationships that they are learning to solve problems and investigate more open-ended tasks (Mason, 2014). These tasks may be open-ended in nature, they may contain multiple steps, or they could be flexibly designed for solution using various methods. Herein, Crowley (1987, p. 6) illustrates such a task: 'Fold a piece of paper in half, then in half again. Try to imagine what kind of figure you would get if you cut off the corner made by the folds on the corners at a 30 degree angle'.

Phase 5: Integration

In the final phase, students review and summarise what they have learned. As a result of this review and summary, the goal is to integrate new knowledge with existing knowledge (Crowley, 1987; Mason, 2014).

Questions

1. Would you consider the statements 'all squares are rectangles' and 'all parallelograms have congruent diagonals' to be true or false? Why or why not?

2. Create some of your own geometric statements so that some are true and some are false. Then ask yourself, 'which of the van Hiele levels of geometric thought are supported to creating and evaluating these statements?'

Hollebrands: teaching Year 10 geometric transformations

The work of Hollebrands (2004) was mentioned earlier in this chapter. This researcher conducted task-based interviews with six Year 10 students who were selected from a class of 17 students. The students – none of whom had studied geometric transformations in primary or middle school – were purposely selected to reflect the range of mathematical abilities present within the class. Interviews took place before, during and after a seven-week instructional unit on geometric transformations. According to Hollebrands (2004, p. 208), the purpose of the initial interview 'was to gain insights into students' prior understandings of the four transformations that were going to be taught: translations, reflections, rotations, and dilations'.

Students were presented with various transformative tasks on paper which comprised a picture of a polygon and a parameter (e.g. line of reflection, centre of dilation or rotation, translating vector). For instance, one type of task required students to draw the image of a polygon following a reflection, rotation or translation. The researcher used the terms 'flip', 'slide' or 'turn' if students displayed uncertainty about the words presented. Although it was expected that students would find translations to be the easiest transformation to understand, the researchers' interviews revealed that these were the most difficult. Following an analysis of collected interview data, the key findings indicated:

> In general, students envisioned transformations as actions or motions performed on a figure; were unfamiliar with the role of parameters (lines of reflection, translating vectors, centres of rotation); and did not focus on relationships among pre-images, images, and parameters (for example, corresponding pre-image and image points are equidistant from the centre of rotation. (Hollebrands, 2004, p. 113)

According to the researcher, these conceptions were evidenced in students' strategies, which appeared to rely on visual cues. Moreover, students who seemed to rely only on visual information rather than on properties of transformations were more likely to draw images incorrectly. Again, while this research reported on students' understandings of geometric transformations (i.e. rotations, reflections, translations), the key findings can be considered by teachers preparing to teach any units of work involving transformations of any kind (e.g. functions and graphs, matrices).

SHORT-ANSWER QUESTIONS

1. Using examples, explain the difference between a rigid and a non-rigid transformation.
2. Distinguish between a similar and a shear transformation with an example of each to illustrate your explanation.
3. Recall and briefly describe the five van Hiele levels of geometric thought.
4. Create a list of topics in secondary mathematics where transformations of any kind are used. When you have finished, check all middle and senior years courses in the Australian Curriculum to ensure that all topics have been included. Alongside each topic, write a statement indicating

how students may experience conceptual difficulties in mastering the skills and concepts. This statement could be written in the form of 'common student misconceptions/errors' associated with the topic (e.g. confusing dilations parallel to the *x*-axis with those parallel to the *y*-axis). Which topic did you decide would be the most difficult for students to learn?

Using technology in measurement and geometry

The literature contains substantial evidence concerning the benefits of technology when learning mathematical concepts (Brown, 2010; Hopper, 2009; Kissane, 2007). Researchers and teachers alike recall that several decades ago, hand-drawn diagrams were the only constructions possible for geometry students. Since the advent and increasing development of the digital age there are many dynamic geometry environments (DGE) within which to engage high school students. From a United States perspective, Goldenberg, Cuoco and Mark (2012, p. 5) argued that geometry represents the only visually oriented mathematics offered to students, and curricula tend to present an 'otherwise visually impoverished, nearly totally linguistically mediated mathematics, a mathematics that does not use, train, or even appeal to the metaphorical right brain'. As such, many students who desire a visually rich mathematics curriculum – where visualisation and visual thinking serve not only as potential hooks, but as the first opportunities to participate – have already dropped out before they can experience mathematics in an accessible way (Goldenberg, Cuoco & Mark, 2012). These comments represent a prevalent attitude for teachers to provide DGE for students of geometry. Doing so enables students to develop visualisation skills – which are reliant on considerable practice (Siemon et al., 2011) – and to explore geometrical problems (e.g. looking at what changes and what remains invariant after transformations). However, various factors impede the successful implementation and navigation of technology within the classroom (Goos & Bennison, 2008). In the two research summaries some of these factors are presented, concomitant with data-driven recommendations for best practice.

PAUSE AND THINK

Every four years, the International Conference for Mathematics Education (ICME) is convened. At the time of writing, ICME-13 (Hamburg, Germany) was the most recent of these conferences held. Access the summary report by Sinclair et al. (2017) 'Geometry education, including the use of new technologies: A survey of recent research' presented at that conference on the theme of research in geometry education and answer the following questions.

1. What are the seven themes of geometry education as summarised in this report? Which of these themes would you consider to be 'traditional' or 'contemporary' in nature?

2. The authors of this paper offer a challenge for mathematics educators and researchers. How do you think geometric ways of thinking and geometric understanding – coupled with technological advancements – can broaden the range of learners who might become interested in and excel at mathematics?

Research–driven insight

GeoGebra

Aventi, Serow and Tobias (2014) investigated the conceptual development of 25 Year 9 Australian students as they participated in a unit of work that used DGS. Specifically, the researchers looked at (a) the characteristics of student responses when exploring linear relationships using GeoGebra, and (b) the nature of student interaction when using GeoGebra as an exploration tool. GeoGebra is free, open-source mathematics education software released in 2004. This software enables teachers and students to interact dynamically with mathematical topics such as geometry, algebra, statistics and calculus. According to researchers in mathematics education, one of GeoGebra's key strengths is that it can facilitate the mathematical learning of students at every year level from primary school to university (Hohenwarter et al., 2008; Kllogjeri & Shyti, 2010). Aventi, Serow and Tobias (2014) used a pre-experimental research design which involved a pre-test, teaching intervention and end-of-topic test with a delayed post-test (six weeks delay). Relating to the first research question posed, the findings from this study indicate that:

> GeoGebra assisted students in increasing the complexity of their responses when exploring linear relationships. At the completion of the teaching sequence, the use of correct terminology was evident with students linking the concepts of y-intercept and gradient to form the equation of a line. (Aventi, Serow & Tobias, 2014, p. 84)

Regarding the second research question, students reported enjoyment at the dynamic nature of GeoGebra, as they could explore mathematical tasks quickly. Although students appeared initially reluctant to engage with the various DGS tools offered, the researchers noted that student familiarity and proficiency quickly increased as the teaching sequence progressed. Aventi, Serow and Tobias (2014) concluded that there are multiple benefits of using GeoGebra in the Linear Relationships sub-strand in the middle years. Besides students demonstrating an overall increase in recall knowledge from pre-test to post-test, benefits include students being assisted by GeoGebra to increase the complexity of their responses and using mathematical terminology correctly. However, the researchers underscore that when using this DGS it is essential to build familiarity with the tools so students can focus on the exploration at hand, rather than the tools they are using to explore it.

Assessment items

Lowrie et al. (2014) explored how 807 Singaporean Grade 6 students (aged 11–12 years) processed spatially demanding graphic tasks using either pencil and paper or iPad. The researchers divided students into groups according to their spatial visualisation ability (low, medium and high). Then samples of each ability group were asked to complete two tasks – namely, the Symmetry task and the Street Map task – using either pencil and paper or an iPad. The Symmetry task asked students to consider an image that required folding across a line of reflection. For the Street Map task, students had to superimpose and rotate a visual compass from its usual north position on the given graphic. For both tasks, the researchers noted that students with high spatial visualisation ability performed at a much higher level than those who possessed medium or low spatial visualisation ability.

These researchers discovered significant differences in student performance and strategy across the test modes and tasks. With reference to the Symmetry task, students using iPads outperformed those students not using those technological devices. For this task, outperformance referred to the extent to which students were able to mentally reflect an object. For instance, and because the students could not physically fold the object across the line of symmetry on the iPad (as they could have done on paper), the researchers speculated that the digital mode prompted them to mentally reflect the given object. By contrast, Lowrie et al. (2014, p. 434) noted:

> those students who completed the Street Map task in a pencil-and-paper form scored higher than those students who solved it on iPad. The iPad students used a variety of strategies to solve the task, with the highest proportion using imagery to evoke a mental representation of a compass indicating the North direction.

The pencil-and-paper mode tended to encourage students to draw a compass on the diagram. This strategy saw students produce a higher proportion of correct responses relative to iPad students. While the iPad students had working-out paper to use, the researchers noted not only their comparative unwillingness to use a paper-and-pencil strategy, but a lower success rate when they elected to use this strategy.

The results from this study hold several important educational implications. First, the differences in test mode (i.e. pencil and paper, iPad) appear to influence students' mathematical performance on graphic tasks with spatial demands. Given the National Assessment Program – Literacy and Numeracy was implemented in 2016 using a digital online environment, it may be difficult to compare the digital-based performance of students to earlier cohorts of students who used non-digital tests. Second, there may be scope for the Australian Curriculum to include visuospatial reasoning skills, especially given the differences in students' strategy in completing both tasks. Third, the finding that iPad users were less likely to monitor their thinking via diagram sketching or annotation suggests that educators need to consider carefully their choice of format.

SHORT-ANSWER QUESTIONS

1. What are DGEs? How can the use of DGEs within the mathematics classroom be beneficial for students?

2. Using as few words as possible, explain how the findings of Aventi, Serow and Tobias (2014) regarding GeoGebra could be applied to your own teaching.

3. Using as few words as possible, explain how the findings of Lowrie et al. (2014) regarding either the Street Map task or the Symmetry task could be applied to your own teaching.

4. The Year 8 Content Description ACMMG202 (ACARA, 2020) states 'Establish properties of quadrilaterals using congruent triangles and angle properties, and solve related numerical problems using reasoning'. Propose how you would teach this content description using a technology-based approach.

Conclusion

Geometry is an essential branch of the mathematics curriculum at all year levels, as it develops both logical and deductive thinking, which in turn helps learners to expand mentally and mathematically. Measurement is a key component of the mathematics curriculum, and it is considered the most practical and hands-on application of mathematics in everything from occupational tasks to day-to-day-life. The concept of proof is considered central to the mathematics discipline and consequently it should feature prominently in mathematics education. Because of this centrality, proof is an essential tool for promoting mathematical understanding in students and, at the same time, for providing educators with insight about how students learn mathematics. When teaching geometric proof, mathematics educators consider three approaches, namely: synthetic, analytic and transformational.

In mathematics the term 'transformation' refers to the changes in positions or size of shapes, which collectively comprise the study of translations, reflections, rotations (slides, flips and turns), symmetry and similarity. Broadly speaking, there are two categories under which all geometric transformations belong – rigid transformations and non-rigid transformations. Transformations that do not change the size or shape of an object moved are called rigid transformations. Examples of rigid transformations include translations (or slides), reflections (or flips), rotations (or turns) and symmetry. Non-rigid transformations change the size or shape of an object, and two categories of these are similar transformations and shear transformations. Existing literature contains substantial evidence concerning the benefits of technology when learning mathematical concepts. Researchers and teachers alike recall that several decades ago hand-drawn diagrams were the only constructions possible for geometry students. Since the advent and increasing development of the digital age, there are many DGE within which to engage high school students. However, various factors impede the successful implementation and navigation of technology within the classroom, and these factors must be carefully considered when supporting any learning experiences with technology.

BRINGING IT TOGETHER

1. Briefly describe and justify the importance of teaching measurement and geometry concepts to students.
2. Outline some reasons to support the claim that teaching geometric proofs to lower secondary students is an extremely difficult task.
3. Explain how the van Hiele hierarchical framework for levels of geometric learning could be useful in your lesson planning.
4. What are some considerations for planning geometric lessons which are supported by digital technology?
5. Your task is to plan two lessons for senior secondary students to demonstrate the utility of technology in geometric proofs and mathematical reasoning. In the first lesson, students are given a structured lesson to investigate similarity in triangles. For the second lesson, your senior secondary students must creatively design a learning

experience (e.g. podcast, video clip, PowerPoint presentation) where they can showcase their technological proficiency and mathematical expertise. The learning experience should be for a small group of their peers, and a marking rubric should be available to provide feedback and a mark for each student.

6. The acronym ICT is generally understood to be expanded as 'Information and Communications Technology', although an interesting alternative for educators is 'It Can't Teach'. In light of this latter expansion of the ICT acronym, outline your philosophy of using technology to support student learning across Years 7–12. Also describe what you believe to be your strengths and areas for improvement regarding the integration of technologies into a secondary mathematics classroom.

REFERENCES

Further resources

Australian Curriculum, Assessment and Reporting Authority (ACARA) (2020). *Structure: Content strands*. Sydney: ACARA. Retrieved from https://www.australiancurriculum.edu.au/f-10-curriculum/mathematics/structure/

Australian Institute for Teaching and School Leadership (AITSL) (2011). *Australian Professional Standards for Teachers*. Melbourne: AITSL.

Aventi, B., Serow, P. & Tobias, S. (2014). Linking GeoGebra to explorations of linear relationships. In J. Anderson, M. Cavanagh & A. Prescott (eds), *Curriculum in focus: Research guided practice. Proceedings of the 37th Annual Conference of the Mathematics Education Research Group of Australasia (MERGA)* (79–86). Sydney: MERGA.

Ball, D.L., Hoyles, C., Jahnke, H.N. & Movshovitz-Hadar, N. (2002). The teaching of proof. *Proceedings of the International Congress of Mathematics*, 3, 907–22.

Battista, M.T. (2007). The development of geometric and spatial thinking. In F. Lester (ed.), *Second handbook of research on mathematics teaching and learning*. Charlotte, NC: National Council of Teachers of Mathematics/Information Age Publishing.

Brown, R. (2010). Does the introduction of the graphics calculator into system-wide examinations lead to change in the types of mathematical skills tested? *Educational Studies in Mathematics*, 73(2), 181–203.

Crowley, M.L. (1987). The van Hiele model of development of geometric thought. In M.M. Lindquist (ed.), *Learning and teaching geometry: K–12*. 1987 Yearbook of the National Council of Teachers of Mathematics (1–16). Reston, VA: National Council of Teachers of Mathematics.

de Villiers, M. (1990). The role and function of proof in mathematics. *Pythagoras*, 24, 17–24.

Fujita, T. & Jones, K. (2003). *Interpretations of national curricula: The case of geometry in Japan and the United Kingdom. Proceedings of the British Educational Research Association Annual Conference*. Edinburgh, Scotland.

Fuys, D., Geddes, D. & Tischler, R. (1988). The van Hiele model of thinking in geometry among adolescents. *Journal for Research in Mathematics Education Monograph*, 3. Reston, VA: National Council of Teachers of Mathematics.

Goldenberg, E.P., Cuoco, A.A. & Mark, J. (2012). A role for geometry in general education. In R. Lehrer & D. Chazan (eds), *Designing learning environments for developing understanding of geometry and space*. Mahwah, NJ: Lawrence Erlbaum Associates.

Goos, M. & Bennison, A. (2008). Surveying the technology landscape: Teachers' use of technology in secondary mathematics classrooms. *Mathematics Education Research Journal*, 20(3), 102–30.

Guven, B. (2012). Using dynamic geometry software to improve eighth grade students' understanding of transformation geometry. *Australasian Journal of Educational Technology*, 28(2), 364–82.

Healy, L. & Hoyles, C. (1998). *Justifying and proving in school mathematics. Technical report on the nationwide survey*. London: Institute of Education, University of London.

Hohenwarter, M., Hohenwarter, J., Kreis, Y. & Lavicza, Z. (2008). *Teaching and learning calculus with free dynamic mathematics software GeoGebra*. 11th International Congress on Mathematical Education. Mexico: Monterrey, Nuevo Leon.

Hollebrands, K.F. (2004). High school students' intuitive understandings of geometric transformations. *The Mathematics Teacher*, 97(3), 207–14.

Hopper, S. (2009). The effect of technology use on student interest and understanding in geometry. *Studies in teaching 2009 research digest* (37–42).

Huetinck, L. & Munshin, S.N. (2008). *Teaching mathematics in the 21st century: Methods and activities for grades 6–12* (3rd edn). Upper Saddle River, NJ: Pearson Education.

Kissane, B. (2007). *Exploring the place of hand-held technology in secondary mathematics education*. 12th Asian Technology Conference on Mathematics, 16–20 December, Taipei, Taiwan.

Kllogjeri, P. & Shyti, B. (2010). GeoGebra: A global platform for teaching and learning math together and using the synergy of mathematicians. *International Journal of Teaching and Case Studies*, 2(3), 225–36.

Kunimune, S., Egashira, N., Hayakawa, T., Hatta, H., Kondo, H. Matsumoto, S., Kumakura, H. & Fujita, T. (2007). *The teaching of geometry from primary to upper secondary school*. Shizuoka: Shizuoka University [in Japanese].

Kunimune, S., Fujita, T. & Jones, K. (2009). *Strengthening students' understanding of 'proof' in geometry in lower secondary school*. Proceedings of CERME 6 (756–65). Lyon, France.

Lin, F.L. (2000). An approach for developing well-tested, validated research of mathematics learning and teaching. In T. Nakahara & M. Koyama (eds), *Proceedings of the 24th Conference of the International Group for the Psychology of Mathematics Education*, 1, 84–88.

Lowrie, T., Ramful, A., Logan, T. & Ho, S.Y. (2014). Do students solve graphic tasks with spatial demands differently in digital form? In J. Anderson, M. Cavanagh & A. Prescott (eds), *Curriculum in focus: Research guided practice. Proceedings of the 37th Annual Conference of the Mathematics Education Research Group of Australasia (MERGA)* (429–36). Sydney: MERGA.

Mariotti, M.A. (2007). Proof and proving in mathematics education. In A. Guitierrez & P. Boero (eds), *Handbook of research on the psychology of mathematics education*. Rotterdam: Sense Publishers.

Mason, M. (2014). The van Hiele levels of geometric understanding. In *Professional handbook for teachers* (4–8). Evanston, IL: McDougal Littell Inc.

Miyazaki, M., Fujita, T. & Jones, K. (2017). Students' understanding of the structure of deductive proof. *Educational Studies in Mathematics*, 94, 223–39.

Okolica, S. & Macrina, G. (1992). Integrating transformational geometry into traditional high school geometry. *Mathematics Teacher*, 85(9), 716–19.

Reid, D.A. (2011). Understanding proof and transforming teaching. In L.R. Wiest & T. Lamberg (eds), *Proceedings of the 33rd Annual Meeting of the North American Chapter of the International Group for the Psychology of Mathematics Education (PME)*. Reno, NV: PME.

Schultz, K. & Austin, J. (1983). Directional effects in transformation tasks. *Journal for Research in Mathematics Education*, 14, 95–101.

Senk, S.L. (1985). How well do students write geometry proofs? *Mathematics Teacher*, 78(6), 449–56.

Shinba, S., Sonoda, H. & Kunimune, S. (2004). *Teaching similarity of geometrical figures: An emphasised learning process from constructing to proving*. Memoirs of Center for Educational Research and Teacher Development, Shizuoka University, 10, 11–22 [in Japanese].

Siemon, D., Beswick, K., Brady, K., Clark, J., Faragher, R. & Warren, E. (2011). *Teaching mathematics: Foundations to middle years*. South Melbourne: Oxford University Press.

Sinclair, N., Bartolini Bussi, M.G., de Villiers, M., Jones, K., Kortenkamp, U., Leung, A. & Owens, K. (2017). Geometry education, including the use of new technologies: A survey of recent research. In G. Kaiser (ed.), *Proceedings of the 13th International Congress on Mathematical Education*. ICME-13 Monographs. Cham: Springer. https://doi.org/10.1007/978-3-319-62597-3_18

Singh, S. (2005). *Fermat's last theorem*. London: Fourth Estate.

Singhal, R., Henz, M. & McGee, K. (2014). Automated generation of geometry questions for high school mathematics. In S. Zvacek, M.T. Restivo, J. Uhomoibhi & M. Helfert (eds), *Proceedings 6th International Conference on Computer Supported Education* (14–25). Barcelona, Spain.

Thompson, T.D. & Preston, R.V. (2004). Measurement in the middle grades: Insights from NAEP and TIMSS. *Mathematics Teaching in the Middle School*, 9(9), 514–19.

Thomson, S., de Bortoli, L. & Buckley, S. (2013). *PISA 2012: How Australia measures up*. Melbourne: Australian Council for Educational Research.

Thomson, S., de Bortoli, L., Underwood, C. & Schmid, M. (2019). *PISA 2018: Reporting Australia's results (Volume I student performance)*. Melbourne: Australian Council for Educational Research.

Thomson, S., Hilmman, K. & Wernert, N. (2012). *Monitoring Australian Year 8 student achievement internationally: TIMSS 2011*. Melbourne: Australian Council for Educational Research.

Thomson, S., Wernert, N., O'Grady, E. & Rodrigues, S. (2017). *TIMSS 2015: Reporting Australia's results*. Melbourne: Australian Council for Educational Research.

Van de Walle, J.A., Karp, K.S. & Bay-Williams, J.M. (2014). *Elementary and middle school mathematics: Teaching developmentally* (8th international edition). Essex: Pearson Education.

van Hiele, P.M. (1959). A child's thought and geometry. In D. Geddes, D. Fuys & R. Tischler (eds), *English translation of selected writings of Dina van Miele-Geldof and Pierre M. van Hiele*. Washington, DC: Research in Science Education Program.

Wilkerson-Jerde, M.H. & Wilensky, U.J. (2011). How do mathematicians learn math?: Resources and acts for constructing and understanding mathematics. *Journal of Mathematical Behavior*, 78, 21–43.

Yopp, D. (2011). How some research mathematicians and statisticians use proof in undergraduate mathematics. *Journal of Mathematical Behavior*, 31, 115–30.

CHAPTER 12

Statistics and probability

Robyn Reaburn

LEARNING OBJECTIVES

The learning objectives of this chapter are directly linked to the Australian Professional Standards for Teachers (APST) (Australian Institute for Teaching and School Leadership [AITSL], 2011). After studying this chapter, you should be able to carry out a valid statistical investigation in the classroom. This process requires knowledge of:

- why data are needed and how data can be produced (APST 2.1)
- the processes of, and the difficulties and misconceptions students may have in, the production of summary statistics, graphs and tables (APST 1.2, 2.1)
- misconceptions students may have about probability and how these might be addressed (APST 1.2, 2.1)
- how data collection, data summary and probability combine so that statistical inferences can be made (APST 2.1, 2.2).

Introduction

Quantitative information is everywhere, and many decisions that we make, or are made for us, are based on statistical data. For example, whether a vaccine is approved depends on an analysis of its efficacy and safety. Quantitative data can also be used to persuade us to alter our behaviour, such as to vote for a particular political party or to buy a certain product (Ben-Zvi & Garfield, 2004).

The discipline of statistics is important because it allows for decision making based on evidence (Watson & Neal, 2012). Unfortunately, the teaching of statistics often emphasises computational procedures and does not promote understanding of how to collect and interpret data – to carry out a statistical investigation. The ability 'to generate and evaluate knowledge' forms part of the general capabilities in the Australian Curriculum, in particular, critical and creative thinking (Australian Curriculum, Assessment and Reporting Authority [ACARA], 2020a). Many adults in our society cannot think statistically about the information they are given, which affects their lives (Jones, Langrall & Mooney, 2007). There is also a growing need for graduates such as statisticians and data analysts, who are both statistically and computer literate, to work in industry.

The Australian Curriculum: Mathematics F–10 (ACARA, 2020b) has three strands: Number and Algebra, Measurement and Geometry, and the one of relevance to this chapter, Statistics and Probability. In turn, the Statistics and Probability strand has two sub-strands: Data Representation and Interpretation, and Chance. In this chapter these sub-strands will be addressed in turn. Later in the chapter how these sub-strands are combined into the topic of statistical inference (in Years 11 and 12) will be described.

It cannot be overemphasised that the practice of statistics in a classroom should be more than the mere carrying out of procedures such as drawing graphs and calculating means, but carrying out meaningful investigations. Watson and Neal (2012) describe five steps in a statistical investigation:

1. Asking the question
2. Data collection
3. Data reduction
4. Data representation
5. Inference.

In this chapter we take a walk through the processes of a statistical investigation, highlighting areas where students have difficulties and misconceptions. We are also going to look at some of the issues in the teaching and understanding of probability, and how probability is used in the process of statistical inference.

Statistics (discipline of): the study of the collection, organisation, analysis and interpretation of data.

Asking the question

Data are collected because we need to know something. Will the vaccine work? What types of cereal should my supermarket stock? In this chapter we look at some of the characteristics of street trees grown in Western Victoria (the 'Trees Data Set', officially named the Colac Otway Shire Trees data set).[1] It would be suggested to students that planners are concerned about shade, so they want to know which variety of tree grows fastest. The original data set contains over 3000 trees of over 30 varieties. Therefore, the data have been reduced to three types: the Double Pink Flowering Gum, the Red Flowering Gum and the Mancherian Pear[2] (a total of 419 trees). The teacher would need to reduce the data to a level that is manageable for the students at their point of learning. This will prevent students from being overwhelmed while still experiencing the messiness of real data.

Data collection: sampling

PAUSE AND THINK

Before students consider data collection it can be useful for you and your students to consider how much information you have given to the social media applications you use. What do the companies who own these applications know about you? What do these companies do with the data you have given them (e.g. your birth date, the sites you visit)?

If valid decisions are to be made, then the data need to be collected in a suitable manner. Therefore, it is important that students are exposed to the principles of data collection from a young age.

Sampling appears in the Australian Curriculum: Mathematics in Years 7, 8 and 9.

> Year 7: Identify and investigate issues involving numerical data collected from primary and secondary sources. (ACMSP169)
>
> Year 8: Investigate techniques for collecting data, including census, sampling and observation. (ACMSP284)
>
> Year 9: Investigate reports of surveys in digital media and elsewhere for information on how data were obtained to estimate population means and medians. (ACMSP227)
> (ACARA, 2020b)

Young students are usually very interested in themselves, so early data collection usually involves students collecting data about themselves and their classmates. As they progress

---

[1] The 'Trees Data Set' data used throughout this chapter is a selection of 419 trees from the data set 'Colac Otway Shire Trees' (2015) from data.gov.au. The full data set is available from https://data.gov.au/dataset/ds-dga-3ce1805b-cb81-4683-8f46-e7bd2d2a3b7c/details. Licensed under the Creative Commons Attributes 3.0 Australia Licence (https://creativecommons.org/licenses/by/3.0/au/).

[2] Note: In this chapter we are using the names of the trees from the data set, but some of the names are incorrect. (Prunus blierana is a double-flowering plum [not Double Pink Flowering Gum] and Manchurian Pear [not Mancherian Pear])

through school, they should be encouraged to broaden their basis to more extensive data sets (Curcio, 1989). The development of the internet means that there are now sources of extensive real data sets available, some of which are listed at the end of this chapter.

Students should also come to realise that much of the data that are collected consist of samples, incomplete selections from populations, and this leads to one of the main difficulties with making decisions from these data. For example, say the mean height of one class of Year 5 students was 140 cm. Will this be true for all Year 5 classes in the country? Individuals vary and even if several samples came from the same population their characteristics will differ. Conclusions based on samples, therefore, always contain an element of uncertainty.

Students also need to know the importance of proper sample collection. If their favourite TV show asks viewers to phone in to answer 'yes' or 'no' to a question, is the answer going to be representative of Australia as a whole? They could also be asked to survey their school for the students' opinions on a topic. How should they conduct this survey if they only have the time and resources to ask, say, 40 students? Problems like this lead to an understanding of stratified sampling. Surveys also require careful wording. This is illustrated here.

Sample: 'part of a population. It is a subset of the population, often randomly selected for the purpose of estimating the value of a characteristic of the population as a whole' (ACARA, n.d.).

Population: 'the complete set of individuals, objects, places etc. about which we want information' (ACARA, n.d.).

> ### PAUSE AND THINK
>
> Sometimes organisations ask questions so that they are likely to receive the answers that suit their purposes. Choose some controversial examples from the media at the time of your teaching. Examples could include climate change or logging. Ask your students to design biased and unbiased questions for a survey. Compare the phrasing of these questions.
>
> What action do you think companies should take in relation to carbon emissions?
>
>   Or
>
> Do you think companies should be allowed to continue to pollute our atmosphere?

Data reduction

Once the data have been collected it is important that they are represented in a way so that the investigator can make sense of them. A quick look at an extract of the Trees Data Set in Table 12.1 is enough to show us that we cannot make much sense of the data in its raw form. The process of changing the representation of data to tell a story is known as transnumeration and is a vital part of statistical analysis (Pfannkuch & Wild, 2004). In this section we look at data reduction by using descriptive statistics.

Transnumeration: the process of transforming 'raw data into multiple graphical representations, statistical summaries and so forth, in a search to obtain meaning from the data' (Pfannkuch & Wild, 2004, p. 17).

Descriptive statistics in the curriculum

The commonly used descriptive statistics are found in the secondary curriculum in Year 7 to Year 10 (ACARA, 2020b).

Table 12.1 Extract of the raw data from the Trees Data Set

Red Flowering Gums (RFG)			Double Pink Flowering Gums (DPFG)			Mancherian Pear (MP)		
Tree ID	Height (m)	Year planted	Tree ID	Height (m)	Year planted	Tree ID	Height (m)	Year planted
6025	6	1993	10662	2	1985	6689	3	2003
6051	1	2007	10682	2	1985	8258	2	1995
6054	1	2006	10694	2	1985	8301	2	2000
6068	1	2007	12222	1	1985	8309	3	1995
6134	6	1983	12294	2	1985	8312	2	1995
6194	6	1988	12305	2	1985	8315	2	1995
6234	5	1983	12308	3	1985	8318	2	1995
6283	1	2007	12311	3	1985	8321	2	1995
.	.	.	.	.	.	.	.	.
6362	2	2003	8269	10	1970	8257	3	1995

Year 7: Calculate mean, median, mode and range for sets of data. Interpret these statistics in the context of data. (ACMSP171)

Year 8: Investigate the effect of individual data values, including outliers, on the mean and median. (ACMSP207)

Year 9: Compare data displays using mean, median and range to describe and interpret numerical data sets in terms of location (centre) and spread. (ACMSP283)

Year 10A: Calculate and interpret mean and standard deviation of data and use these to compare data sets. (ACMSP279) (ACARA, 2020b)

These descriptors place an emphasis on interpretation. Yet in many classrooms the concentration is on the teaching of the procedures. It is important that such calculations are carried out in the context of investigations, so that the use of these statistics is meaningful.

Measures of centre

Measures of centre are used to find a number that is representative of the entire data set. The most used measures of centre, or averages, are the mean, median and mode. It is essential that students should not only know how these statistics are calculated, but also know the different properties of each of these statistics and come to appreciate why one might be used in preference to another.

The mean

The formula to calculate the **mean** is well known. Unfortunately, many students will have no further understanding and will not be able to explain why anyone would perform the calculation in the first place (Groth & Bergner, 2006; Mokros & Russell, 1995). Because

Mean: 'The arithmetic mean of a list of numbers is the sum of the data values divided by the number of numbers in the list. In everyday language, the arithmetic mean is commonly called the average; for example, for the following list of five numbers, {2, 3, 3, 6, 8}, the mean equals (2 + 3 + 3 + 6 + 8)/5 = 22/5 = 4.4' (ACARA, n.d.).

these students do not realise that a mean is somehow representative of data, they do not use it when it would be useful, such as to describe data or to compare data sets (Konold & Pollatsek, 2002; Reaburn, 2012).

Some of the properties of the mean are not well known. For example, the mean is the value that results if all the objects of interest are shared equally; this is illustrated in Figure 12.1.

Figure 12.1 The distribution of apples between five baskets (left) and the number in each basket when they are evenly shared (right)

It is the author's experience that students will assume that the mean is the normal, or most common, value. They can also get confused about fractional/decimal means (Mokros & Russell, 1995). How do you make sense of a statement that says the mean number of children in a family is 1.5? An activity to overcome this confusion is illustrated in Figure 12.2.

Here is a street with 8 houses. Can you find a way of putting children into the houses so that the street has a mean of 1.5 children per house? Draw the children in their gardens.

Figure 12.2 An example of an activity that builds understanding of fractional means

Median: 'the value in a set of ordered data that divides the data into two parts. It is frequently called the "middle value"' (ACARA, n.d.).

The median

The median is relatively simple to calculate. The question arises as to why the median may be preferred to the mean. The activity outlined in Figure 12.3 regarding the sodium content of some foods demonstrates why this might be. The mean of the data is 1012 mg/100 g and the median is 283 mg/100 g. The data were obtained from the Australian Food Composition Data Base (Food Standards Australia & New Zealand, 2019) and the foods were deliberately selected so that a large difference between the mean and median would be evident.

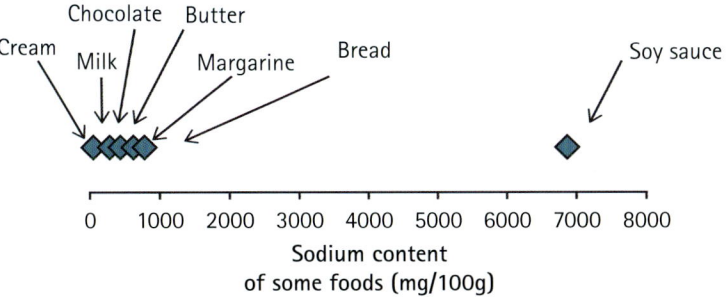

1. Why is there a difference between the mean and median?
2. Which statistic, the mean or median, best represents the data as a whole? Explain your answer.
3. What would happen to the values of the mean and median if the soy sauce was removed?

Figure 12.3 The activity to demonstrate why the median can be the preferred statistic in some circumstance
Source: Idea from Top Drawer Teachers (n.d.).

The median in use

It is useful for students to see real-life examples of where a median is used. Here is an example from a newspaper article (Henrique-Gomes, 2020):

> The Australian Council of Social Service and University of New South Wales's Poverty in Australia report found that more than one in eight people (13.6%) lived below the poverty line ... The report adopts what is known as the relative poverty line, set at 50% of the median household disposable income, and finds 3.24 million people in Australia were living below that marker of $457 per week in 2017–18.

Why is the median used in this report?

The mode

The mode is often used when describing a data set but is not often used when comparing data sets. However, the Australian Curriculum requires that students be familiar with its use.

Characteristics of the mean, median and mode

Usually, examples in statistics involve numbers in a context. Here is one exercise where the numbers are not contextualised, but it can be invaluable in helping students to develop knowledge and

IN PRACTICE

Mode: 'a measure calculated by identifying the value that appears with greatest frequency in a set of data. If two numbers occur in a set with equal frequency, the set is said to contain bimodal data. If there are more than two numbers in a set that occur with equal frequency, the set is said to contain multimodal data' (ACARA, n.d.).

SCENARIO

▶ ▶

understanding of the characteristics of the commonly used measures of centre. This example is also interesting because every student may come up with different answers. In addition, there is no prescribed formula that can be used.

Choose a set of five numbers where (A different set of numbers is required for each question):

1. The mean, median and mode are all equal.
2. The mean is greater than the median.
3. The median is greater than the mean.
4. The mean is greater than the mode.

Measures of spread

Variability and variation

Without variability, the discipline of statistics would not exist. If all objects of interest (people, animal, objects on an assembly line) were the same, then measuring one would measure all. The words variability and variation are often used interchangeably, but in this chapter the words will be used as recommended by Shaughnessy (2007). Variability describes the tendency of a quantity to have different values. Variation is the measure of this variability. Variation is important in our continuing quest to tell the story of our data. Students should be able to see that two data sets could have the same mean but vary greatly in spread. Commonly used measures of spread include the range and the standard deviation.

Range: 'in statistics, the range is the difference between the largest and smallest observations in a data set' (ACARA, n.d.).

The standard deviation

The formula to calculate the sample standard deviation is:

$$s = \sqrt{\frac{\sum_{i=1}^{N}(x_i - \bar{x})^2}{N-1}}$$

Standard deviation: 'a measure of the variability or spread of a data set. It gives an indication of the degree to which the individual data values are spread around their mean' (ACARA, n.d.).

If the data belong to a sample, the divisor is one less than the number of values in the sample. If the data belong to an entire population, then the divisor is N, the number of values in the population. Unfortunately, student textbooks do not always make this difference clear. To add to the confusion, their scientific calculators often use the incorrect notation:

$$\sigma_n \text{ and } \sigma_{n-1}$$

In reality, the letter 's' should be used for sample standard deviations, with the Greek letter 'σ' reserved to describe entire populations.

Numbers in the media   CONNECTION

One problem found in statistical education is that students can calculate the measures and yet be oblivious to the poor use of statistics in the media (Utts, 2003). One way to combat this problem is to encourage students to analyse articles from the media. One framework for doing this is the Critical Numeracy Framework (Watson, n.d.). The students identify the mathematical terminology and define it within the context of the article. They then consider the information they have and do not have and determine if they need more information to be convinced by the subject matter. Here is an example (Malo, 2018):

> Melbourne house prices have taken their biggest hit since 2012, falling by almost 2 per cent in the past three months, with experts predicting prices will continue to slide.
>
> The city's median house price dropped by $16 000 in the June quarter to $882 000, according to data from Domain group, which showed the leafy inner east was the city's worst performing region. Melbourne recorded the steepest quarterly drop across all capital cities, followed closely by Sydney, where the median price dropped by 1.4 per cent.

Here is part of a student's assignment critiquing the article. As you read it note that the student defined the mathematical terms and explained their use in the article. In addition, why the median was the statistic of choice in this context was explained. The student (who gave permission for the work to be used) also took note of the emotive language, considered the source of the data, and considered the accuracy of the claims. It would have been useful to query who the 'experts' were but overall, the marker was impressed with this critique.

> The major conclusion in [this] article is that house prices have fallen by a large amount over a quarter in Melbourne. Emotive language is used such as 'dive', 'falling', 'slide', 'worst performing' and 'steepest drop' and this creates a perception that the house price is decreasing extremely rapidly. The reason given for this is that the median house price for houses sold over this time was $882,000, which is a decrease of $16,000 compared to the quarter before that.
>
> The mathematical terms used ... are per cent, median and quarter. Median ... refers to 'the point in a distribution where half are above and half below' (Blastland & Dilnot, 2008, p. 82). In this article median is used to refer to the middle price of houses in the city of Melbourne, there is an equal number of houses that are less expensive and more expensive than the median of $882,000 ... A quarter means one of four parts of something. The article refers to the June quarter which is a specialised term used in the financial sector. In this context, the word 'quarter' refers to a quarter of the year and the word June means the three months to the month of June.
>
> The median is used as a way to determine the average price of houses in Melbourne and it is appropriate. This is due to the massive diversity of houses in the city. If the mean was used to work out an average then very expensive mansions could skew the results.
>
> The article initially states Melbourne's median house price fell by almost 2 per cent. This is an approximation and is [used] by the writers to emphasise their point that prices have fallen ... Taking the price drop of $16,000 and dividing it by the original price of $898,000 and multiplying it by 100 we can get the actual percentage drop of around 1.78%. We can conclude that Melbourne's median house price drop is a larger per cent drop than Sydney's.
>
> The data came from the Domain group who is a large real estate industry business. Domain group was founded by the Fairfax business that, at the time of publication, also owned *The Age* newspaper (Fairfax media, 2019). This weakens the reliability of the data, as it is less likely for the newspaper to question the source of the data when it comes from within the same organisation. There are no reasons given for the price drop.

Descriptive statistics for the trees

Table 12.2 shows some of the summary statistics for the Trees Data Set – the mean, median, standard deviation, and the range for the heights of the three types of trees in the Trees Data Set. What do these descriptive statistics tell us about the trees? What more would you (and your students) like to know? Why is there such a difference in the standard deviations? Think about this before going to the next section where graphs and tables will be examined. The graphs and tables will give us more information than the summary statistics alone.

Table 12.2 Descriptive statistics for the Trees Data Set (m)

Tree type	Mean (m)	Median (m)	Standard deviation (m)	Range (m)
DPFG	2.82	3	1.53	13
RFG	3.77	4	2.33	11
MP	2.78	3	0.97	4

SHORT-ANSWER QUESTIONS

1. At the end of each week you want to have walked an average (mean) of 8000 steps per day. Your step count for the week so far is in this table:

Day	Sunday	Monday	Tuesday	Wednesday	Thursday	Friday
# Steps	11 300	6700	8220	5400	7420	6200

   a. How many steps do you need to walk on Saturday to achieve your goal of a mean number of 8000 steps per day?

   b. What features will a student need to know about a mean to be able to carry out this calculation?

2. Think of an imaginary data set for each of the following scenarios. A separate answer is required for each:

   a. The mean is the most appropriate measure of centre to summarise the data.

   b. The mode is the most appropriate measure of centre to summarise the data.

   c. The median is the most appropriate measure of centre to summarise the data.

   In your answers, explain the characteristics of your imaginary data sets that make each statistic the most appropriate.

3. A researcher has collected reading data from 600 students in Year 6 across your state. The following graph was produced.

   The researcher claims that this graph illustrates that the boys in the study performed higher than the girls.

   a. Do you think this is correct? Explain your reasoning.

   b. What criteria should students consider in coming to their decisions?

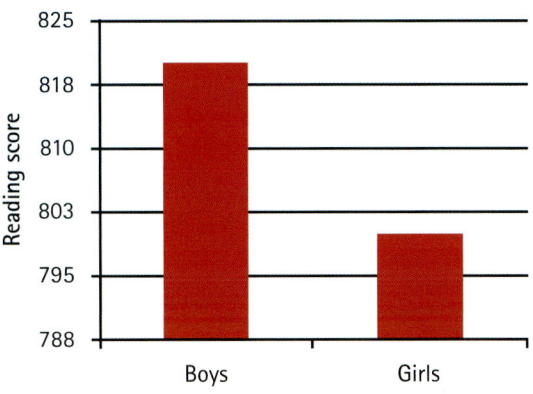

Data representation

In this section we look at further methods for data transnumeration – the making of tables and the drawing of graphs.

Summary tables can be used as a form of data display, or they can be used as an intermediate step in data representation by graphs (Friel, Curcio & Bright, 2001). For smaller data sets the students can use tallying to assist with this summary. The data for the trees are summarised in Table 12.3.

Table 12.3 Frequency table for the Trees Data Set

Height (m)	Frequency		
	DPFG	RFG	MP
0–1	0	0	1
>1–2	28	29	4
>2–3	80	14	24
>3–4	65	10	20
>4–5	43	19	14
>5–6	11	19	2
>6–7	0	17	0
>7–8	1	3	0
>8–9	0	4	0
>9–10	0	0	0
>10–11	1	2	0
>11–12	0	0	0
>12–13	1	1	0
>13–14	0	0	0

Table 12.3 (*cont.*)

Height (m)	Frequency		
	DPFG	RFG	MP
>14–15	1	0	0
Total	231	118	65

It is still difficult to see if there is a difference between the Red Flowering Gums, the Double Pink Flowering Gums and the Mancherian Pears. What should be done next? Ask the students for suggestions.

Graphs

One of the options available is to graph the data. The most important feature of any graph is that it transforms numerical or other data into a form that uses spatial characteristics (e.g. height or length) to represent quantity (Curcio, 1989). Graphs may be used to describe data, summarise data, compare and contrast two or more data sets, and, depending on the type of graph, predict the next case. Graphs also allow the display of mathematical relationships that cannot be recognised in numerical form (Curcio, 1989).

Most graphs have similar structural components such as titles, axes, scales and grids that give information about the measurements being used and the data being measured. Knowing how to label graphs and to draw scales is important, but unfortunately, much teaching of graphing emphasises the mechanics of graphing at the expense of interpretation.

Graphs in the Australian Curriculum

Graphing appears in Years 7, 9 and 10 in the Data Representation and Interpretation area of the Australian Curriculum: Mathematics (ACARA, 2020b).

> Year 7: Construct and compare a range of data displays including stem-and-leaf plots and dot plots. (ACMSP170)
> Year 9: Construct back-to-back stem-and-leaf plots and histograms and describe the data, using terms including 'skewed', 'symmetric' and 'bimodal'. (ACMSP282)
> Year 10: Construct and interpret box plots and use them to compare data sets. (ACMSP249)
> Compare shapes of box plots to corresponding histograms and dot plots. (ACMSP250)
> Use scatter plots to investigate and comment on relationships between two numerical variables. (ACMSP251) (ACARA, 2020b).

In Chapter 2 it was pointed out that it should not be assumed that students can read and interpret graphs without instruction (Kemp & Kissane, 2010). It is important that teachers ask questions to monitor students' understanding.

The Trees Data Set

The histograms for the trees data are displayed in Figure 12.4. These give a visual indication of the variation in standard deviations represented in Table 12.2. The height of the MPs show less variation in height than the other two types.

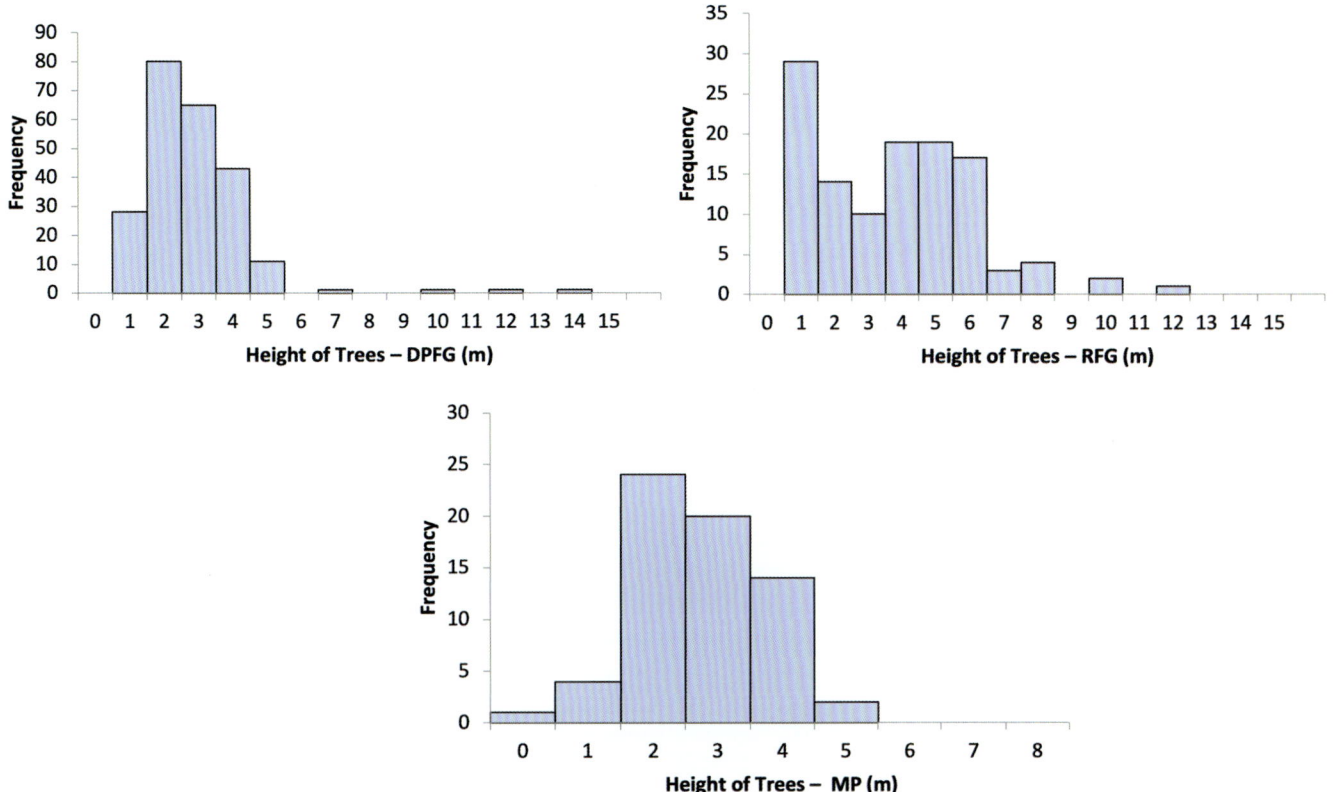

Figure 12.4 Histograms for the Double Pink Flowering Gums (DPFG), the Red Flowering Gums (RFG) and the Mancherian Pears (MP). (Data from The Trees Data Set. Produced in Microsoft Excel.)

PAUSE AND THINK

The computer program in Figure 12.4 has not set the scale to be equal for all three species. How does this affect interpretation?

Year 10 students are required to use box-and-whisker plots (ACMSP250). Figure 12.5 shows the box plot for the DPFG trees in detail. The other box plots are shown in Figure 12.6. Encourage students to relate these plots to the value of the standard deviations.

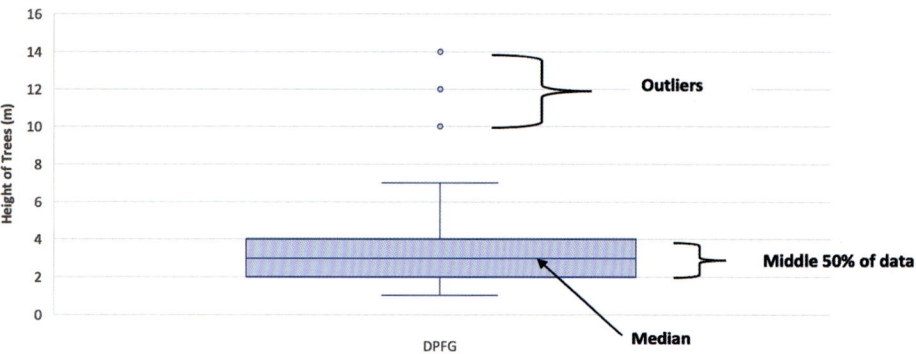

Figure 12.5 An example of a box-and-whisker plot using the DPFG trees data. The lower line on the 'box' represents the 25% percentile (Q1), the point that cuts off the lower 25% of the values. The upper line on the 'box' represents the 75% percentile (Q3), the point that cuts off the lower 75% of the data. Therefore, the box represents the middle 50% of the data. The 'whiskers' are drawn at either the minimum or maximum of the data, or to a point that is 1.5 times longer than the IQR. The individual points that are further away from the box than this are known as outliers. (Data from the Trees Data Set. Plot produced in Microsoft Excel.)

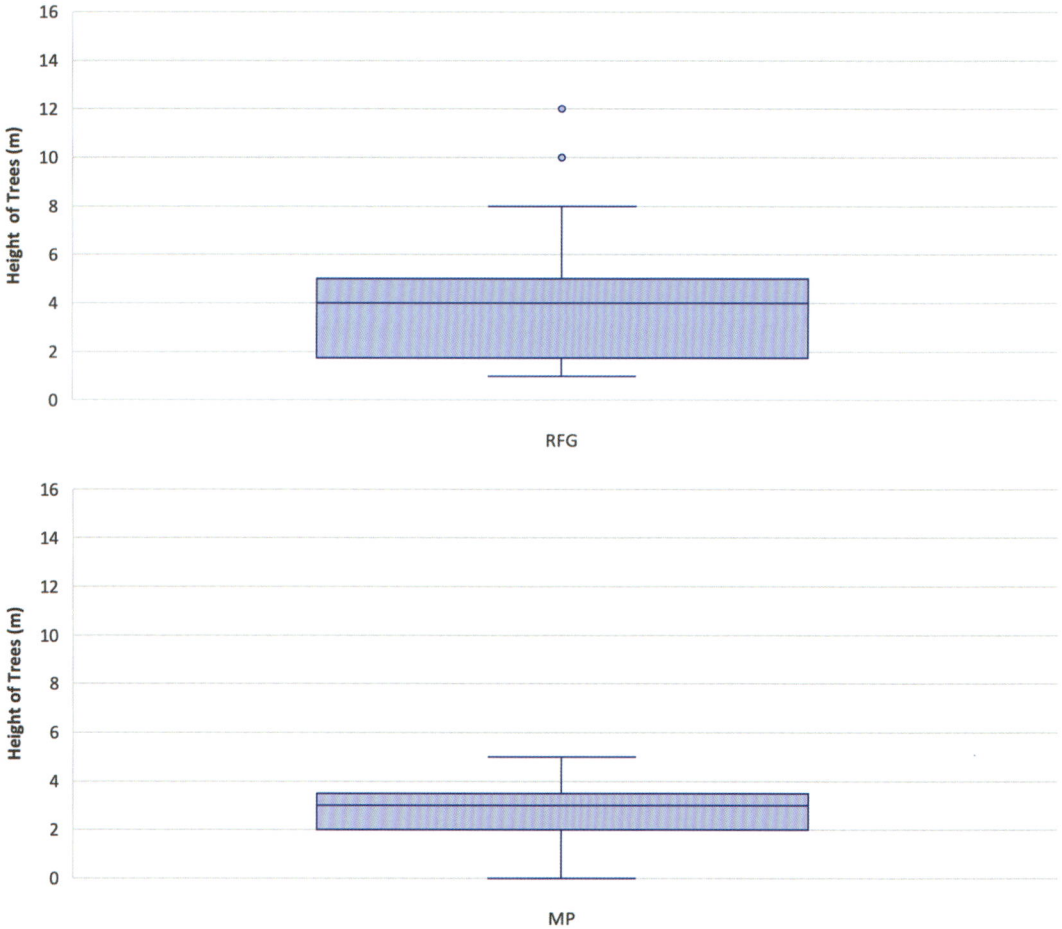

Figure 12.6 Box-and-whisker plots for the Red Flowering Gums (RFG) and the Mancherian Pears (MP). (Data from the Trees Data Set. Produced in Microsoft Excel.)

The curriculum for Year 10 also states that the students should draw scatterplots. These graphs can be the most difficult to interpret as they deal with relationships between two variables. The scatterplots of height versus time are found in Figure 12.7.

Variable (noun): 'In statistics, a variable is something measurable or observable that is expected to either change over time or between individual observations. Examples of variables in statistics include the age of students, their hair colour or a playing field's length or its shape' (ACARA, n.d.).

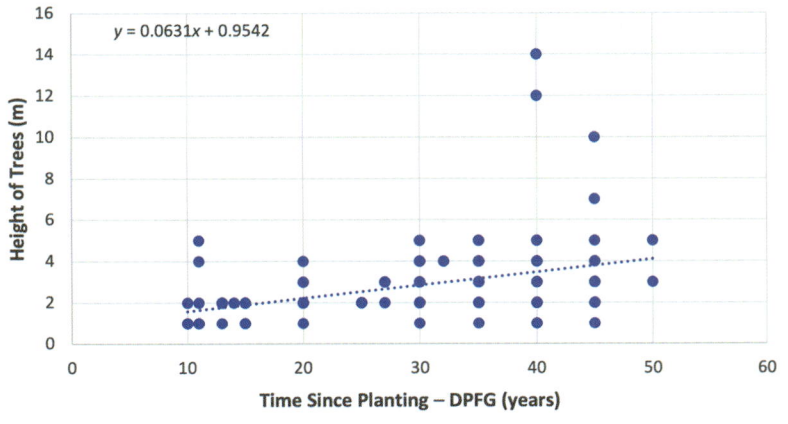

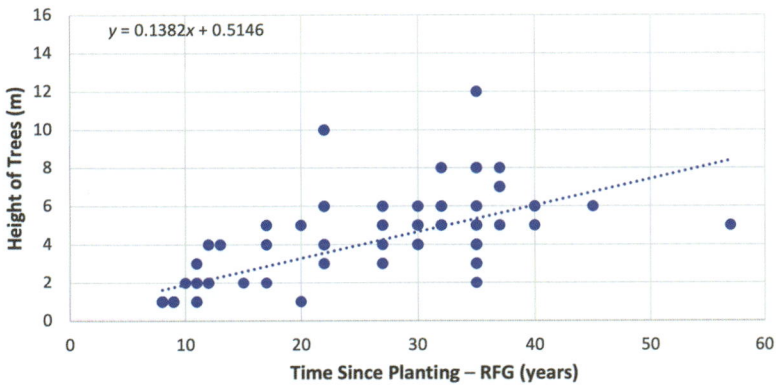

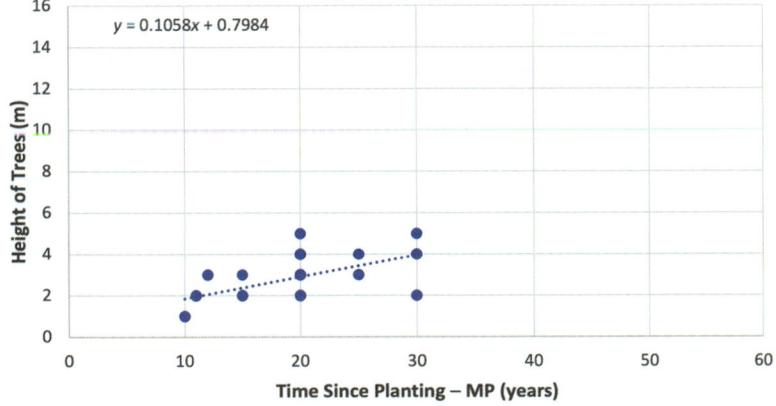

Figure 12.7 Scatterplots of height (m) versus time (years) for the Trees Data Set. Microsoft Excel allows for addition of the line of best fit with its equation.

Now our original question can be answered – which tree grows the fastest? According to our data, the RFG trees grow more quickly than the others at a rate of 0.14 m/year. Yet, there is more to be considered before a final recommendation and this is described below.

What other questions arise from these data?

What other questions have arisen from our analysis? Before professional statisticians can write their final reports, they often need to go back to the client to find out more information. One question that arises is the relatively younger age of the Mancherian Pear trees. Is this because they were introduced later by the planners, or are they short-lived trees?

It would be expected that older students would be able to suggest another analysis – a comparison of the tree growth rates. Students of lower grades, who are not as experienced, would be guided to this decision by the teacher so that time can be factored into the analysis. Yet, if growth rates only are used, the variation in the age of the trees between the species would not be noticed. Teachers need to decide which features of the data are suitable for their students' stage of learning and adjust them accordingly.

Probability: 'The probability of an event is a number between 0 and 1 that indicates the chance of that event happening; for example, the probability that the sun will come up tomorrow is 1, the probability that a fair coin will come up "heads" when tossed is 0.5, while the probability of someone being physically present in Adelaide and Brisbane at exactly the same time is zero' (ACARA, n.d.).

Event: 'An event is a subset of the sample space for a random experiment; for example, the set of outcomes from tossing two coins is {HH, HT, TH, TT}, where H represents a "head" and T a "tail"' (ACARA, n.d.).

Sample space: 'A sample space is the set of all possible outcomes of a chance experiment; for example, the set of outcomes (also called sample points) from tossing two heads is {HH, HT, TH, TT}, where H represents a "head" and T a "tail"' (ACARA, n.d.).

Probability

In this section we take a detour from our statistical investigation and examine some issues around the teaching of probability (chance). Probability is an area that is worth studying in its own right, not only because many of the decisions we make are based on probabilistic data, but because of the dramatic increase in gambling advertising in recent years. Probabilistic thinking is also used in the processes of formal statistical inference. In the last section of this chapter we will see how probability and the other processes of data analysis come together in statistical inference, the last step in a statistical investigation (Watson & Neal, 2012).

Probability is the mathematics describing real-world events that cannot be predicted with certainty (Jones, Langrall & Mooney, 2007). If students are used to doing mathematics with a 'correct' answer they can find dealing with probability very frustrating!

Probability calculations using the classical approach require identifying the events of interest and dividing by the total number of events in the sample space. The empirical approach to probability involves collecting data (e.g. voting intention) or using historical data (e.g. breast cancer records) and assigning the probabilities from the results. Probabilities may also be assigned from expert knowledge. For example, how would a person assign the chance of an accident at a nuclear power site? Thankfully, there are not many previous examples for an empirical approach to be used. Giving a comprehensive description of teaching probability is not possible here, but the following sections give an overview of common problems students find in this area.

Issues in the understanding of probability

In the secondary mathematics curriculum (ACARA, 2020b) there is little mention of chance experiments. This is unfortunate, as experimenting with materials such as coins, spinners and dice is extremely important for a good understanding of probability. Some examples of when this is useful are included in the next sections.

Randomness

One of the seemingly simple, but poorly understood ideas in probability is that of randomness. In everyday conversation the word 'random' may refer to something that is disordered, irregular, patternless or unpredictable (Peterson, 1998), or something that has no cause (Batanero, Green & Serrano, 1998). In reality, random events have a long-term regular pattern of outcomes even though each individual outcome is unpredictable (Moore, 1990). A simple example is that of tossing a coin.

IN PRACTICE

Heads vs. tails

This example is based on Green (1982). A teacher asked Clare and Susan each to toss a coin a large number of times and to record every time whether the coin landed Heads or Tails. For each 'Heads' a 1 is recorded and for each 'Tails' a 0 is recorded.

Here are the two sets of results:

Clare: 010110011001010110110100011100011011010101100010001
0101001110011010110010110010110010010111011001110011
010100101100101011000100110101100111011101011001100011

Susan: 1001110111101001110010011100100011101111111010101
11100000010001010010000010001100010100000000011001
0000000111110000110101001001001111110100110001100011000

Now one girl did it properly, by tossing the coin. The other girl cheated and just made it up.

Which girl cheated?

It is helpful if students should first decide on what criteria they will use to make their choice and then make a decision. After they have made their choice, the students should toss a fair coin 20 times and then reconsider. They are often surprised by the number of Heads or Tails they might get in a row, and how far the observed proportions may vary from the expected 50%.

Independence

Another important concept in probability is that of independence. Formally, two events are independent if the outcome of one event has no influence on the probability of the next event. If we toss a coin, the probability of a head or tail remains unchanged from one toss to the next no matter what has gone on before. Students may wonder why they need to know this, but the onset of extensive gambling advertising has made this of personal as well as theoretical importance. If a roulette ball lands on red four times in a row, it is common to believe that it is more likely that the ball will fall on black the next time (the Gambler's fallacy). The best way for students to avoid such beliefs is to experiment with chance processes.

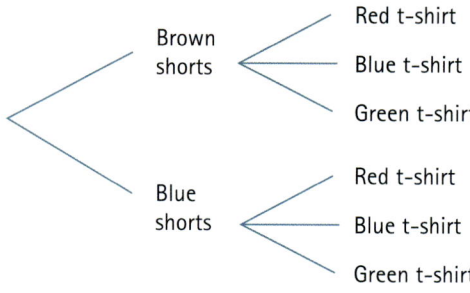

Figure 12.8 An example of a tree diagram, suitable for young children, to determine the different combinations of clothes to wear

Identifying the sample space

If students are to use the classical probability approach to calculate probabilities accurately, they need to define all the outcomes of the random process. This can be difficult. One strategy that students find useful is the odometer strategy. With this strategy one variable is held constant and then matched with all items of the second variable until all the combinations are formed, and then repeated with the other variables (English, 2005). A tree diagram, illustrated in Figure 12.8, is one way to achieve this.

SCENARIO   ## If you toss two coins, how many outcomes are there?

An activity I like to use is to ask the students to toss two coins 20 times. If the result is two Heads, Player A gets a point. If the result is two Tails, Player B gets a point. If the result is One Head and One Tail, Player C gets a point. The question is asked: Is the game fair?

I like to use this activity because the trend is clearer after the whole class data are combined, illustrating the importance of needing large sample sizes to be sure of trends in random processes. It also corrects a common misconception that when two coins are tossed there are only three, equally likely outcomes.

When I first used this activity in a professional development session for teachers, I was astonished to find that several of the teachers also held the same 'three outcome – equally likely' misconception. I have learned never to assume that people, even mathematics teachers, have accurate beliefs about probability. Figure 12.9 shows an illustration of the four possible outcomes of tossing two coins.

R.D. A mathematics teacher educator.

Figure 12.9 The four possible outcomes of tossing two coins

Understanding conditional probability

A conditional probability, probability of A given B, reduces the sample space to events in B. The formula for a conditional probability A, given B, P(A|B) is:

$$P(A|B) = \frac{P(A \cap B)}{P(B)}$$

Teachers of mathematics may concentrate on numerical results, but it is an effective strategy to start with worded problems. For example, look at the problem in Figure 12.10.

When this question was given to school and university students 51 per cent of the school students and 14 per cent of the university students could not tell the difference between (a) and (b) (Watson & Kelly, 2007; Reaburn, 2013). Once the students have understood the difference in sample space between (a) and (b), then it is a good strategy to progress to numerical problems in tables before progressing to more difficult problems. An example of a suitable table is found in Figure 12.11.

Which probability do you think is bigger?
a. The probability that a woman is a school teacher.
OR
b. The probability that a school teacher is a woman.
c. Both (a) and (b) are equally likely.

Figure 12.10 An example of a conditional probability expressed in words
Source: Watson & Kelly, 2007.

The table below shows the number of defective TVs produced every week at two factories by the day shifts and by the night shifts.

	Factory A	Factory B
Day	40	30
Night	40	60

a. How many defective TVs are produced at Factory B every week?
b. How many defective TVs are produced by a night shift every week?
c. If you were told that a defective TV was produced by Factory A, what is the probability it was produced by a night shift?

Figure 12.11 An example of a conditional probability in table form
Source: Watson & Kelly, 2007.

Conditional probability: the probability of an event A given that another event B has occurred. The notation is P(A|B). For this conditional probability, the sample space is reduced to B.

PAUSE AND THINK

This is a question with which secondary students often have difficulty: 'The weather report for yesterday stated that there was a 90 per cent probability of rain, but it did not rain'. Students often believe that if it did not rain then the forecast was incorrect. How can you demonstrate to students that the forecast was indeed 'correct'? Is there a simulation you can devise so that students can see how such processes work?

Some aspects of probability in the senior secondary curriculum
The binomial distribution: an example of a probability distribution

In late secondary school, students come to more formal probability concepts including probability distributions. The distribution may be represented by a graph, table or mathematical formula. An example using the binomial distribution (ACMM149) is illustrated in Figure 12.12 and Table 12.4. The question was: There are seven red and three green balls in a bag. You are going to select three balls, one after the other, with replacement. What is the probability of getting 0, 1, 2 or 3 red balls?

The situation tells us that the probability of drawing a red ball is 0.7 and the probability of drawing a green ball is 0.3. Because each draw is independent (a consequence of sampling

Probability distribution: the list of all possible outcomes in a random experiment with their accompanying probability of occurrence.

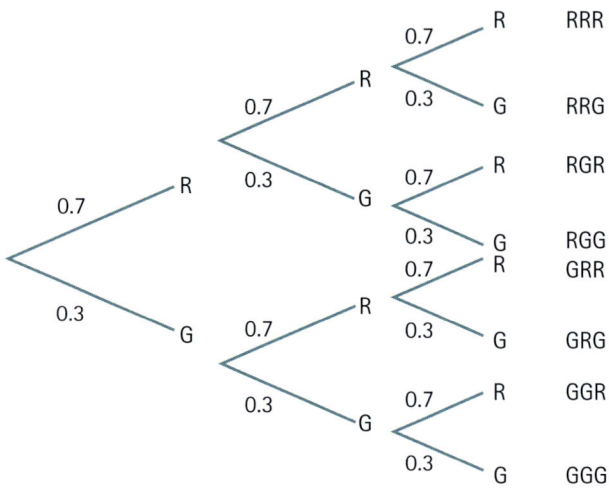

with replacement) to calculate the probability of three red (3R) we multiply along the RRR branch to get the answer of 0.7^3. Similarly, the probability of the second branch (RRG) is $0.7^2 0.3^1$. An examination of the tree tells us that there are three ways of getting two red balls overall. Therefore, $P(2R) = 3 \times 0.7^2 0.3^1$. We can then find the probability of getting two red balls, one red ball, and no red balls. This results in the probability distribution for this situation illustrated in Table 12.4. From examples like these the students can be led to deduce the formula for the binomial probabilities:

$$P(X; n, p) = {}^nC_x p^x (1-p)^{(n-x)}$$

Figure 12.12 An example of a binomial probability using a tree diagram to determine all the outcomes

Table 12.4 The probability distribution of the scenario in Figure 12.12

Outcome	Calculation of probability	Value
3R	$1 \times 0.7^3 0.3^0$	0.343
2R	$3 \times 0.7^2 0.3^1$	0.441
1R	$3 \times 0.7^1 0.3^2$	0.189
0R	$1 \times 0.7^0 0.3^3$	0.027
	Total =	1.00

It is important that students are made aware of the terminology that may be used in the asking of the questions. For example, they could be asked to find the probability of 3 successes (exactly 3), greater than 3 successes (does not include 3), greater or equal to 3 successes (includes 3). The terminology is similar for 'less than' questions. It is also helpful for students to be reminded about the complement of the events they are interested in. For example, if asked to find the probability of 5 or fewer successes out of 6 trials, it is quicker to find the probability of 6 successes and then subtract this probability from one.

Complementary events: 'Events A and B are complementary events if A and B are mutually exclusive (have no overlap) and $Pr(A) + Pr(B) = 1$, where the symbol $Pr(A)$ denotes the probability of event A occurring' (ACARA, n.d.).

A continuous probability distribution: the normal distribution

There is one extremely important distribution that often applies to continuous data. This is the normal distribution, represented by the formula:

$$P(X) = \frac{1}{\sigma\sqrt{2\pi}} e^{-\frac{(x-\mu)^2}{2\sigma^2}}$$

Teachers can choose to show a normal distribution to the students, but it is recommended that the students 'discover' this distribution and its properties for themselves. For this example, a random sample of the heights of 500 males, downloaded from the CensusAtSchool (NZ) website, was combined into a histogram (Figure 12.13).

If a line is drawn connecting the tops of each histogram the familiar 'bell curve' results. The students can now explore the properties of this distribution. What is a reasonable guess for the value of the mean? Compare the guess with the mean of the data calculated from the data. Then calculate the standard deviation and mark in the points that are one, two and three standard deviations above and below the mean (see Figure 12.14).

If the students use different data sets and compare their graphs, they will see the similarities between them. They will see that a 'lot' of the data falls between one standard deviation of the mean, 'even more' of the data falls between two standard deviations of the mean, and 'almost all' of the data falls between three standard deviations of the mean. The exact proportions, that apply to any normally distributed data, are listed in Table 12.5. Once the connection is made between areas under the graph and the proportions (and hence probabilities), the students extend their knowledge to the standard normal distribution and calculate the probabilities of normal distributed data (ACMMM168, ACMMM169).

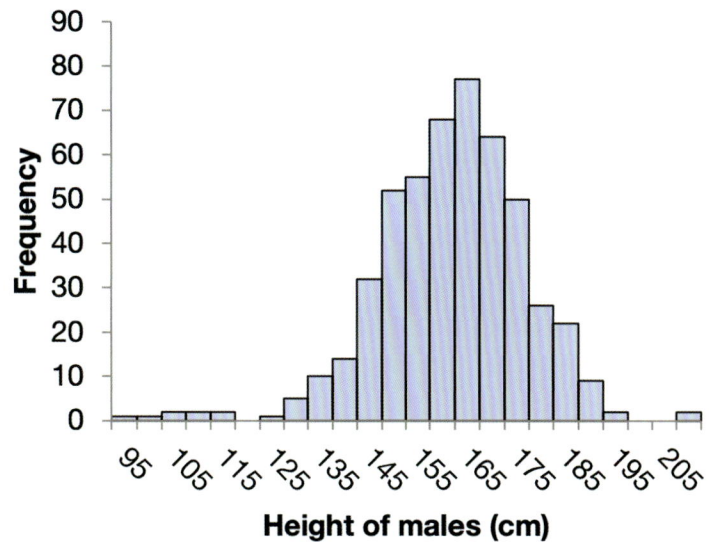

Figure 12.13 The distribution of heights of 500 males chosen by the random sampler on the CensusAtSchool website

Source: © CensusAtSchool (https://new.censusatschool.org.nz/). Produced in Microsoft Excel.

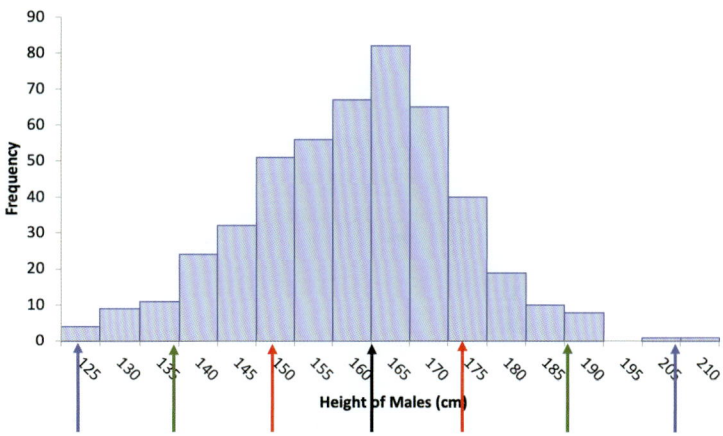

Figure 12.14 The distribution from Figure 12.13 with arrows showing the value of the mean (160 cm – black arrow) and the points showing approximately one (red lines), two (green lines) and three (blue lines) above and below the mean. (Standard deviation = 13.0 cm.)

Source: © CensusAtSchool (https://new.censusatschool.org.nz/). Produced in Microsoft Excel.

Table 12.5 Standard deviation of Figure 12.13

Section of distribution	Percentage of population within this section
Mean ± 1 standard deviation	68.26%
Mean ± 2 standard deviations	95.44%
Mean ± 3 standard deviations	99.72%

SHORT-ANSWER QUESTIONS

1. Here is a practical example of how probability can be used to assist in decision making. Please note that the disease and the figures in this example are not real.

   A common disease of the skin 'rashitis' causes painful red rashes that not only disturb the sufferer's sleep but also are embarrassing to the patient. A visit to a dermatologist resulted in the following options.
   - Patients may decide to have no treatment. If this option is chosen, 70% will have no change to their condition and 30% will notice an improvement.
   - Of those patients who decide to have the treatment on offer, 40% will have major side effects. Of these, 30% will show no change in their condition and 70% will notice an improvement.
   - Of those patients who decide to have the treatment on offer, 50% will have minor side effects. Of these, 40% will show no change in their condition and 60% will notice an improvement.
   - Of those patients who decide to have the treatment on offer, 10% will have no side effects. Of these, 30% will show no change in their condition and 70% will show an improvement.

   Which option should you choose? As you work through these scenarios, decide on how you will represent the information to assist in interpretation. In addition, think about whether everyone with the condition would make the same decision. To what extent does the mathematics determine the decision, and to what extent might personal preference determine the decision?

2. Make sketches of two normal distribution curves:
   a. Where the means are equal, but the standard deviations are not equal.
   b. Where the standard deviations are equal, but the means are not equal.

The final process in statistical investigations: inference

Inference: the process of using a sample to draw conclusions about an entire population.

Statistical inference is the process by which conclusions are made about populations from samples. Unfortunately, the idea of inference is only given limited acknowledgement in the Australian Curriculum where students are asked to interpret the results of their investigations (Watson & Neal, 2012). Despite this, students should be introduced to, and encouraged to use, inferential thinking. For example, if the students measured the height of the trees around their school, what would they be able to say about all the trees in their town?

Inference in the senior mathematics curriculum

Parameter: a descriptive statistic that applies to the entire population.

The senior mathematics curricula require students to explore the distributions of sample means and sample proportions and find confidence intervals for the mean and proportion of the entire population (the parameters) (Mathematical Methods ACMMM174 [ACARA, 2020c]; Specialised Mathematics ACMSM141 [ACARA, 2020d]). Confidence intervals use the information from the sample to give a range of possible values for the parameter of interest. The calculations are easy, but understanding the probabilistic thinking behind these is difficult.

The production of a confidence interval is possible because of the Central Limit Theorem. Understanding this theorem requires a knowledge of the features of the normal distribution and some imagination. The teacher should guide the students through this process:

- Imagine taking a random sample sized 30 or more from a population. Calculate the mean of this sample.
- Imagine that his process has been repeated a very large number of times.
- Record all the sample means and plot a histogram of the results.
- What will this histogram look like?

It is expected that many of the sample mean values will be close to the value of the population mean. It is also expected that the mean of all the sample means would be equal to the value of the population mean. Sometimes a sample mean will not be close to the value of the population mean, and sometimes the value of a sample mean will not be close at all. In fact, together the sample means form a normal distribution. This is true if (1) the sample size is large enough (≥ 30), and (2) even if the first population is not normally distributed.

The next question is how will the standard deviation of the sample means relate to the standard deviation of the original population? Will it be larger or smaller? It is the author's experience that students have difficulty with this, but usually someone comes to the realisation that a collection of 'measures of centre' will not spread as much compared to the original population of individuals. The two standard deviations are connected by this relationship:

Standard deviation of the sample means (known as the 'standard error' – very confusing for students!) = standard deviation of the population/square root of the sample size. That is:

$$s_{\bar{x}} = \frac{s}{\sqrt{n}}$$

The 'standard error' of a proportion is found by the formula, $\sqrt{p\left(\frac{1-p}{n}\right)}$.

To find the confidence interval, the required number of standard errors is added and subtracted from the sample mean or proportion. This is illustrated for a mean in Figure 12.15.

For the mean: $\bar{x} \pm z \times \frac{s}{\sqrt{n}}$. For a proportion: $p \pm z\sqrt{p\left(\frac{1-p}{n}\right)}$.

Random sample: 'A sample is called a random sample (or a simple random sample), if it is selected from a population at random. That is, all the elements of the population had an equal probability of being included in the sample' (ACARA, n.d.).

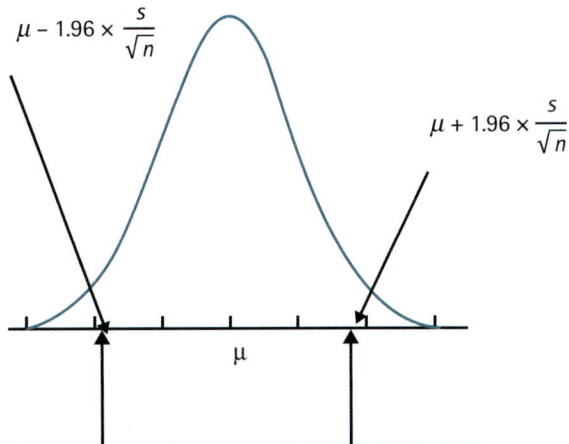

If a sample mean falls between these two numbers, adding and subtracting two standard errors from this mean will give an interval that will include the value of the population mean. If we could undertake this process 100 times, 95 of the resultant intervals would include the value of the population mean.

Figure 12.15 The relationship between the distribution of sample means and the process of finding a confidence interval to estimate the value of the population mean

What determines 'z'? This is determined by the 'confidence' level; that is, how sure you want to be that the value of the parameter will be included in the interval. The common levels of confidence, and z, are found in Table 12.6.

Table 12.6 Commonly used values of z when calculating confidence intervals

Number of standard errors (z)	Level of confidence
1.645	90%
1.96	95%
2.58	99%

How is this *level of confidence* interpreted? We cannot be sure that any one interval does include the value of the parameter, but the level of confidence gives us an idea of how sure we can be. A 95% confidence interval suggests that if the process of sampling and calculating the confidence interval had been carried out 100 times, 95 of the resulting intervals would include the value of the population mean.

It is important that teachers do not underestimate the difficulty students have in understanding the reasoning behind these calculations.

PAUSE AND THINK

The process for finding a confidence interval is difficult to understand. Draw up a lesson plan for teaching this concept. Do you understand the process? What problems might students have in understanding this process?

SHORT-ANSWER QUESTIONS

1. The 95% confidence interval for the mean number of hours of television watched by Australians per day is from 3 to 6 hours.
   a. What is the meaning of this statement?
   b. What does the '95%' refer to?
   c. An individual selected at random watches television for 10 hours on Monday. Is this unexpected? Explain your reasoning.
   d. How would you explain the answer to (c) to your students?
2. This question illustrates a common mistake made by students. Two students were calculating probabilities based on the binomial distribution. They were asked to calculate the probability of getting 15 'successes' out of 150 trials where the probability of a success is 0.3. Student A calculated the probability of 15 successes out of 150 trials using the usual formula. Student

B calculated the probability of getting a success rate of 10%. They got different answers. Why were the answers different, and which student was correct? Explain your reasoning.

3. This question illustrates the importance of encouraging students to use diagrams. Your students have been asked to calculate a probability using standard distribution tables. The mean of the distribution is 50 g and the standard deviation is 12 g. The students are asked to calculate the probability of obtaining an object selected at random that is over 80 g. Which of the following strategies will be best used to answer this problem correctly? Drawing a sketch of the situation will help.

   a. Work out the probability of the object being less than 20 (using the z-score) and subtracting this value from 1.00.

   b. Work out the probability of the object being less than 20 (using the z-score) and subtracting this from 0.5.

   c. Work out the probability of the object being less than 20 (using the z-score) and adding this to 0.5.

   d. Work out the probability of the object being greater than 80 (using the z-score).

Conclusion

Quantitative information is everywhere, and many decisions that we make, or are made for us, are based on statistical data and calculations of probability. If students are going to be able to take part in our society as critical and creative thinkers, they need to understand how statistical investigations are undertaken. Therefore, the teaching of statistics and probability should be much more than practising procedures – it needs to include how to collect data, how to evaluate data, and how to make valid conclusions and inferences.

BRINGING IT TOGETHER

1. Why do we calculate averages (mean, median and mode)? This question appears trivial, but students often cannot answer this question.

2. If we know the mean, the number of scores in that data set, we can calculate the total score. If you know the median and number of scores in a data set, can we calculate the total score?

3. The mean and median of a data set are reported. The median is much higher than the mean. What does this statement tell us about the data? What would the data be like if the median was much lower than the mean?

4. Some tomato plants were grown in two different potting mixes. After three weeks, the mean height of the plants grown in Potting Mix 1 (PM1) was 18.2 cm, and the mean height of the plants grown in Potting Mix 2 (PM2) was 14.2 cm. These data seem to

suggest that Potting Mix 1 should be the preferred mix. The data are summarised in the box plots in Figure 12.16.

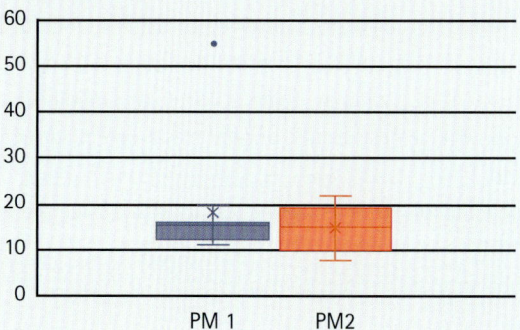

Figure 12.16 The height of tomato plants after 3 weeks (cm) for two potting mixes

a. Describe what you know about these data from looking at the plots.
b. Have the plants in Potting Mix 1 shown more growth than those in Potting Mix 2?
c. What further investigation would you do before accepting these data?

Say your results were like those summarised in the box plots in Figure 12.17. The mean height for the plants in Potting Mix 1 is 18.2 cm, and the mean for the plants in Potting Mix 2 is 14.2 cm (as before).

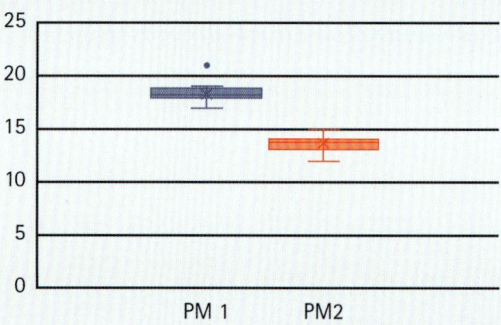

Figure 12.17 The height of tomato plants after 3 weeks (cm) for two potting mixes

d. What can you say about these data from these box plots?
e. Have the plants in Potting Mix 1 shown more growth than those in Potting Mix 2?
f. Explain how the scenarios in Figure 12.16 and Figure 12.17 are different from each other.

5. Go to the mathematics textbooks used in your school and look at the chapters that deal with statistics and probability. How many questions are devoted to procedures, and how many are devoted to analysis and evaluation? How could you add to the procedural questions to shift your students to higher-order thinking?

REFERENCES

Further resources

Australian Curriculum, Assessment and Reporting Authority (ACARA) (n.d.) *Mathematics glossary*. Retrieved from https://www.australiancurriculum.edu.au/f-10-curriculum/mathematics/glossary

Australian Curriculum, Assessment and Reporting Authority (ACARA) (2020a). *General capabilities*. Sydney: ACARA. Retrieved from https://www.australiancurriculum.edu.au/f-10-curriculum/general-capabilities/critical-and-creative-thinking/

Australian Curriculum, Assessment and Reporting Authority (ACARA) (2020b). *The Australian Curriculum, Mathematics*. Retrieved from https://www.australiancurriculum.edu.au/f-10-curriculum/mathematics/

Australian Curriculum, Assessment and Reporting Authority (ACARA) (2020c). *The Australian Curriculum, Mathematical Methods*. Retrieved from https://www.australiancurriculum.edu.au/senior-secondary-curriculum/mathematics/mathematical-methods/

Australian Curriculum, Assessment and Reporting Authority (ACARA) (2020d). *The Australian Curriculum, Specialised Mathematics*. Retrieved from https://www.australiancurriculum.edu.au/senior-secondary-curriculum/mathematics/specialist-mathematics/

Australian Institute for Teaching and School Leadership (AITSL) (2011). *Australian Professional Standards for Teachers*. Melbourne: AITSL.

Batanero, C., Green, D. & Serrano, L. (1998). Randomness, its meanings and educational implications. *International Journal of Mathematical Education in Science and Technology*, 29(1), 113–23.

Ben-Zvi, D. & Garfield, J. (2004). Statistical literacy, reasoning, and thinking; goals, definitions, and challenges. In D. Ben-Zvi & J. Garfield (eds), *The challenge of developing statistical literacy, reasoning and thinking* (3–15). The Netherlands: Kluwer Academic Publishers.

Blastland, A. & Dilnot, M. (2008). *The tiger that isn't: Seeing through a world of numbers*. London: Profile Books.

Curcio, F. (1989). *Developing graph comprehension*. Reston, VA: The National Council of Teachers of Mathematics.

data.gov.au. (2015). *Colac Otway Shire trees*. Retrieved from https://data.gov.au/dataset/ds-dga-3ce1805b-cb81–4683-8f46-e7bd2d2a3b7c/details

English, L. (2005). Combinatorics and the development of children's combinatorial reasoning. In G. Jones (ed.), *Exploring probability in school: Challenges for teaching and learning* (121–44). New York: Springer.

Food Standards Australia & New Zealand (2019). *Australian food composition database*. Retrieved from https://www.foodstandards.gov.au/science/monitoringnutrients/afcd/pages/default.aspx

Friel, S., Curcio, F. & Bright, G. (2001). Making sense of graphs: Critical factors influencing comprehension and instructional implications. *Journal for Research in Mathematics Education*, 32(2), 124–58.

Green, D. (1982). Testing randomness. *Teaching mathematics and its applications*, 1(3), 95–100.

Groth, R. & Bergner, J. (2006). Preservice elementary teachers' conceptual and procedural knowledge of the mean, median and mode. *Mathematical Thinking and Learning*, 8(1), 37–63.

Henriques-Gomes, L. (2020). One in eight people in Australian living in poverty, report finds. *The Guardian*, 21 February. Retrieved from https://www.theguardian.com/australia-news/2020/feb/21/one-in-eight-people-in-australia-living-in-poverty-report-finds

Jones, G., Langrall, C. & Mooney, E. (2007). Research in probability: Responding to classroom realities. In F. Lester (ed.), *Second handbook on mathematics teaching and learning* (909–55). Charlotte, NC: The National Council of Teachers of Mathematics.

Kemp, M. & Kissane, B. (2010). A five step framework for interpreting tables and graphs in their contexts. In C. Reading (ed.), *Data and context in statistics education: Towards an evidence-based society. Proceedings of the Eighth International Conference on Teaching Statistics* (ICOTS 8). Ljubljana, Slovenia; The Netherlands: International Statistics Institute.

Konold, C. & Pollatsek, A. (2002). Data analysis as the search for signals in noisy processes. *Journal for Research in Mathematics Education*, 33(4), 259–89.

Malo, J. (2018). Melbourne median house price falls to $882,082 over June quarter: Domain Group report. *Domain*, 26 July. Retrieved from https://www.domain.com.au/news/melbourne-median-house-price-falls-to-882082-over-june-quarter-domain-group-report-20180726-h134ao-754199/

Mokros, J. & Russell, S. (1995). Children's concepts of average and representativeness. *Journal for Research in Mathematics Education*, 26(1), 20–39.

Moore, D. (1990). Uncertainty. In L. Steen (ed.), *On the shoulders of giants: New approaches to numeracy* (95–137). Washington, DC: National Academy Press.

Peterson, I. (1998). *The jungles of randomness: Mathematics at the edge of uncertainty*. London: Penguin Books.

Pfannkuch, M. & Wild, C. (2004). Towards an understanding of statistical thinking. In D. Ben-Zvi & J. Garfield (eds), *The challenge of developing statistical literacy, reasoning and thinking* (3–15). The Netherlands: Kluwer Academic Publishers.

Reaburn, R. (2012). *Strategies used by students to compare two data sets. Mathematics education – Expanding horizons. Proceedings of the 35th Annual Conference of the Mathematics Education Research Group of Australasia (MERGA)* 633–639. Singapore: MERGA.

Reaburn, R. (2013). Students' understanding of conditional probability of entering university. In V. Steinle, L. Ball & C. Bardini (eds), *Mathematics education: Yesterday, today and tomorrow. Proceedings of the 36th Annual Conference of the Mathematics Education Research Group of Australasia (MERGA)*. Melbourne: MERGA.

Shaughnessy, J. (2007). Research on statistics learning and reasoning. In F. Lester (ed.), *Second handbook on mathematics teaching and learning* (957–1009). Charlotte, NC: The National Council of Teachers of Mathematics.

Top Drawer Teachers (n.d.). *Outliers*. Retrieved from https://topdrawer.aamt.edu.au/Statistics/Misunderstandings/Misunderstandings-of-averages/Outliers

Utts, J. (2003). What educated citizens should know about statistics and probability. *The American Statistician*, 57(2), 74–79.

Watson, J. (n.d.). *Critical numeracy in context*. Retrieved from http://www.nlnw.nsw.edu.au/videos08/critical_numeracy/pdf/jane_watson.pdf

Watson, J. & Kelly, B. (2007). The development of conditional probability reasoning. *International Journal of Mathematics Education in Science and Technology*, 38(2), 213–35.

Watson, J. & Neal, D. (2012). Preparing students for decision-making in the 21st Century – Statistics and probability in the Australian Curriculum. In B. Atweh, M. Goos, R. Jorgensen & D. Siemon (eds), *Engaging the Australian Curriculum: Mathematics – Perspectives from the field* (89–113). Adelaide: Mathematics Education Research Group of Australasia.

CHAPTER 13

Functions and calculus

Robyn Reaburn

LEARNING OBJECTIVES

The learning objectives of this chapter are directly linked to the Australian Professional Standards for Teachers (APST) (Australian Institute for Teaching and School Leadership [AITSL], 2011). After studying this chapter, you should be able to:

- identify the main features of a function and the misconceptions students may have about functions (APST 2.1, 2.2)
- understand the concepts of troublesome knowledge, threshold concepts and concept image as these apply to the teaching of calculus (APST 1.2)
- explain the skills and knowledge needed by students to study calculus and some common misconceptions that can work against students' success in this area (APST 2.1, 2.2)
- describe the alternative ways of understanding derivatives and integrals and describe the common misconceptions students may have about these concepts (APST 2.1, 2.2, 2.6, 3.3).

Introduction

How do we find the area of a region bounded by two curves? How do we find the area of a circle? The ancients solved problems such as these with methods of exhaustion. Mathematicians such as Fermat, Descartes, Kepler and Barrow used geometry (Katz, 1993). In the seventeenth century Isaac Newton (1642–1727) in England and Gottfried Leibniz (1646–1716) in Germany revolutionised mathematics with the invention of calculus. During their lifetimes Newton and Leibniz had a bitter argument about who had come to these ideas first; in modern times, primacy is usually given to Newton, but it is Leibniz to whom we owe the notation used in modern calculus (Motz & Weaver, 1993).

This chapter reviews the key concepts that are needed for students to understand calculus, including rates of change, tangents, limits and the notion of infinity. Then this chapter moves onto derivatives and integrals, including their applications. None of these ideas can be understood without a thorough understanding of the function, and it is with this concept the chapter begins.

The most important concept of them all: the function

Function: a rule that associates with each element x in a set S a unique element $f(x)$ in a set T. We write $x \mapsto f(x)$ to indicate the mapping of x to $f(x)$. The set S is called the domain of f and the set T is called the codomain. The subset of T consisting of all the elements $f(x) : x \in S$ is called the range of f. If we write $y = f(x)$, we say that x is the independent variable and y is the dependent variable.

A function is a relation that associates members of one set ($a \in A$) to a unique member of another set ($b \in B$) (Wolfram MathWorld, 2020). Whereas several members of **A** can be associated to a member of **B**, each member of **A** is associated with only one member of **B**. In late primary and early secondary school, students encounter functions in terms of rules connecting one set of numbers in another, numbers in tables, and as graphs (Australian Curriculum, Assessment and Reporting Authority [ACARA], 2010 to present). As they progress through to senior secondary school, they will encounter functions represented as a mapping from one space (the domain) to another (the range).

It is important that students come to see all of these forms as functions. Ferrini-Mundy and Graham (1991) found that students do not always regard a graphical representation of a function to be a function if there is no accompanying formula. In addition, students may also believe that functions can only be described by one formula (Dreyfus & Eisenburg, cited in Ferrini-Mundy & Graham, 1991). It is also important that students come to see a function as a whole entity. Those students who only consider a function point by point (the static view), will have problems when they are introduced to derivatives in their later work. Students in senior high school should also appreciate that the two sets **A** and **B** do not have to consist of numbers for a function to exist.

Learning functions at school

It is common for students to be introduced to functions via the function machine (an example can be found in Figure 13.1). Here they are asked to work out what the 'machine' does to the numbers. Representations such as this emphasise the input to one output nature of a function. Students are often encouraged to set puzzles for their classmates to work out the rule between given inputs and outputs. These exercises also assist students to develop pattern recognition, an essential feature of mathematics.

As students progress to Year 7, they will start to convert these rules into algebraic language (ACMNA176).

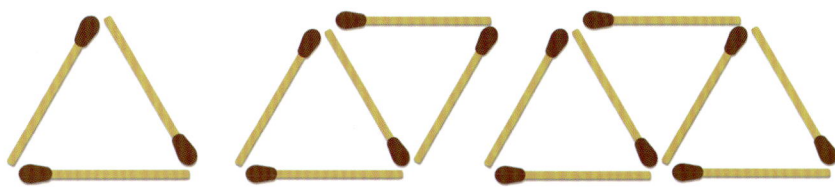

Figure 13.1 An example of a function machine. This machine is equivalent to $y = 2x + 3$

Linear functions

Linear relationships appear in the curriculum in Year 7 (ACMNA178, ACMNA179, ACMNA180). One way of introducing such relationships is to use an exercise similar to that in Figure 13.2. Such exercises encourage students to look for patterns and understand that algebra can represent real situations. In this example it also helps if the students use real matchsticks.

Step 3 is extremely important. It is not unusual for students to believe that they have to write in 'mathematics' before they can fully explain the relationship. Note the emphasis in Step 4 that the letters represent the numbers of the objects, and not just the objects themselves. In addition, note that in Step 6 the students need to have found the relationship if the question is to be answered.

1. Look at the pattern above this text made with matchsticks.
2. Make the next three terms of the pattern and fill out a two-column table where the headings are 'number of triangles' and 'number of matchsticks'.
3. Write a sentence that describes the relationship between the number of triangles and the number of matches.
4. Write an equation that shows this relationship. Use 't' to represent the <u>number of</u> triangles, and 'm' to represent the <u>number of</u> matchsticks.
5. Use this equation to predict how many matches it takes to make 10 triangles. Then check to see if you are correct.
6. If you used 151 matches, how many triangles would you have?

Figure 13.2 An example of a linear function in a real setting

With appropriate learning experiences, students learn to appreciate that graphs, tables and algebraic equations can all describe the same relationship (ACMNA177, Year 7; ACMNA193, Year 8). Once a few examples have been explored, then they can be introduced to the general form of the linear function: $y = mx + c$ and explore what m and c represent. For example, if in Figure 13.2 they were making squares instead of triangles, they would be adding three matches for each extra square. How is this reflected in the equation of the function? Ultimately, it is desired that students realise that the gradient, m, reflects the change in the quantity of y for each unit change in x – that is, that the gradient represents a rate of change (Bezuidenhout, 2006; ACMNA214, Year 9).

Non-linear functions

Quadratic functions and circles appear in the Australian Curriculum in Grade 9 (ACMNA296). It is important that students are aware why quadratic functions are functions whereas circles are not. Real-life examples include the handshake problem and the skeleton tower (see links at the end of the chapter). These problems are not straightforward, so teachers may prefer an easier example such as plotting the area of a square or rectangle against the length of one side. The concept of reflection can be used to deal with the negative *x*-values. It is also useful to look for parabolic shapes found in the community, for example, in bridges and time-lapse photographs of basketball shots. In Year 10 students will work more extensively on solving quadratic equations (ACMNA241) and the sine, cosine and tangent functions in the context of the unit circle (ACMNA274).

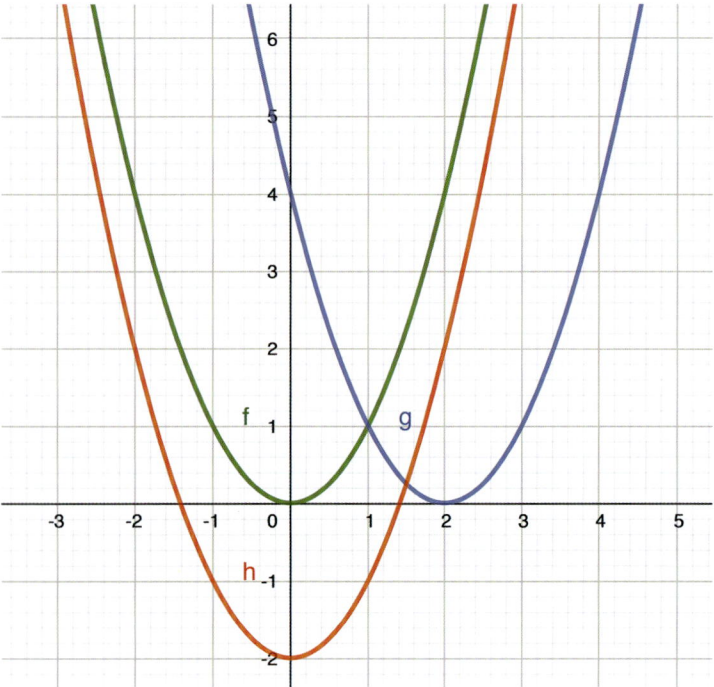

Figure 13.3 Exploring transformations with the use of GeoGebra.
$f(x) = x^2$; $g(x) = (x - 2)^2$; $h(x) = x^2 - 2$
Source: produced using GeoGebra.

Transformations

Students first become accustomed to the idea of translations, reflections and rotations in Year 6 but not in the context of functions in the Cartesian plane (ACMMG142). Transformations in this context appear in the curriculum in Year 10 (ACMNA239). It is very important that students learn that any one transformation has the same effect on all functions. For example, the function $y = f(x - a)$ is the horizontal translation of $y = f(x)$ through *a* units regardless of the function type.

The teaching of this topic is greatly enhanced with the use of technology. Plotting examples of the different transformations is tedious and students can quickly lose sight of the purpose of the exercise. With the use of technology, the transformations are created instantly. An example is shown in Figure 13.3, where GeoGebra (www.GeoGebra.org/classic?lang=en) was used to explore translations of a quadratic equation.

Covariational reasoning

The knowledge that functions consist of a process that accepts inputs to produce outputs is essential to students' learning of functions (Carlson et al., 2002). What is more difficult, and can be forgotten, is that students should also develop covariational reasoning. Carlson et al. (2002) defined covariational reasoning as '[t]he cognitive activities involved in two varying quantities while attending to the way in which they change in relation to each other' (p. 355). If students do not develop this reasoning, they will not be able to understand fully what such relationships are saying and will have great difficulty in understanding the concepts of limits and differentiation in their later work.

Summary

The concept of a function underlies much of mathematics and understanding that is essential if students are to be successful in calculus. It is recommended that students start to learn the concept with real-life examples. As students progress through secondary school, they should gradually come to see functions as algebraic entities that can be worked on without this real-life connection. It is also important that students see the different representations of a function as all representing functions, that gradients reflect rates of change, and that they should develop covariational reasoning.

PAUSE AND THINK

Think about activities you have carried out over the last week that have required mathematics. Which of these could be described by a function? Which ones could be used to assist students' learning about functions?

SHORT-ANSWER QUESTIONS

1. The relationship in Figure 13.2 can be described as $m = 2t + 1$, where m is the number of matches, and t is the number of triangles. How does this function differ from the function $y = 2x + 1$? How would their graphs differ?
2. Taxi fares are based on a fixed initial charge at the start of the journey (flag fall) and then a fixed rate per kilometre when moving. Say your taxi fare came to $45. What might have been the flag fall, and what might have been the rate per kilometre? How far did you travel? As you consider this, think about how many solutions you might find, and how this question could be used in the classroom. What challenges might the students face?

Calculus

By the time students are introduced to calculus, they should be able to describe the most common function types, be competent in the manipulation of algebraic formulas, and have a good understanding of exponents, including those that contain fractions and negative numbers. They also need to be familiar with the shapes of common polynomial graphs and to have good skills in their construction and their meaning (Ubuz, 2007). Orton (1983a) found that limited graphical understanding is one factor in low performance in differentiation. In addition, students also need to be familiar with composite functions.

Troublesome knowledge, threshold concepts and concept image

Calculus, as is the case for other topics in mathematics, is notorious for the ability of students to be able to 'do it', but not to be able to explain what they are doing and why. In fact, to understand concepts, such as the limit, requires considerable imagination and flexible thinking. It is useful, therefore, to consider the ideas of a concept image, cognitive obstacles and threshold concepts.

Tall and Vinner (1981) describe a concept image as 'the total cognitive structure that is associated with a concept' (p. 152). It includes all the mental pictures, properties and processes a person may hold about the concept and is built up as a result of the accumulation of all the experiences the person has had over time. Each person, therefore, may have a concept image that is different from that of his or her peers (Parameswaran, 2006). It is important, therefore, that teachers ask students to explain their reasoning as well as carry out procedural tasks as a check on their understanding.

Bachelard (1938, cited in Cornu, 1991) described what he called *cognitive obstacles*. One form of these obstacles is the epistemological obstacle; this occurs because the mathematical concepts the student is dealing with are difficult. A similar idea is described by Meyer and Land (2003), who refer to *threshold concepts*. These are concepts that act as a portal, in that they open up 'new and previously inaccessible way[s] of thinking about something' (p. 1) and represent a 'transformed way of understanding, or interpreting, or viewing something without which the learner cannot progress' (p. 1). Many threshold concepts are also *troublesome*, in that they may be counter-intuitive, alien or incoherent to the learner (Meyer & Land, 2003). In this context, troublesome knowledge may include rates of change, the nature of tangent lines and the concept of a limit. These concepts will now be examined in turn.

Rates of change

Rate of change: the change in values of a function for a change in one unit of x.

Quantifying rates of change is one of the main features of calculus and as such should be considered a threshold concept. Students need to have a clear idea as to the nature of rates of change in mathematical contexts. Exploring gradients is introduced in the curriculum in Year 9 (ACMNA214) but it is important that students also come to see these as the rate of change of the function.

Orton (1983a) recommended that students should thoroughly explore the ideas of rates of change, average rate, gradient of a line, tangents and secants to the curve before the more algebraic representations are introduced. As they do this, students should also explore where functions are increasing and decreasing, where the average rate of change is constant and where it is changing.

Students also need to understand that every point on a curve may have a different value for the rate of change. Otherwise they may try to treat non-linear functions as linear functions (Dreyfus & Eisenberg, 1983, as cited in Ferrini-Mundy & Graham, 1991). The concept of an instantaneous rate of change appears in the Australian Curriculum in Mathematical Methods (ACMMM084).

Nature of tangent lines

One way of viewing the derivative is that it gives a formula for finding the rate of change at any single point on a curve of a function (Mathematical Methods ACMM084). This rate of change is also the gradient of the tangent line at this point (Mathematical Methods ACMMM085). As finding derivatives is a major task in calculus, students need to have an exact understanding of the nature of these tangent lines. Barnes (1993) gives a useful definition:

> A tangent is a straight line through a point on the curve, going in the same direction as the curve. At the point in question, the gradient of the tangent is the same as the

gradient of the curve, that is, the derivative of the function. Near the point, the tangent
is closer to the curve than any other straight line through the point. (p. 13)

There are other ways of describing tangents that are beneficial to students, depending on
the context. The tangent to the curve at point A can be thought as:

- A line passing through points infinitely close to A on the curve, or the line which the
  curve becomes when one magnifies it in the neighbourhood of A.
- The limit of the secants as the intersection at one point on the curve tends towards
  the other intersection.
- The line passing through A whose slope is given by the derivative at A of the function
  associated with the curve (where the derivative is assumed to exist) (Artigue, 1991,
  p. 174).

As the students become more flexible and experienced, they will tend to use the concept that
best suits for each purpose.

Concept of a limit

Understanding the limit is one of the tasks required for advanced mathematics. However,
owing to its difficulty, the nature of the limit is a form of knowledge that is not only a
threshold concept, but is 'troublesome' (Meyer & Land, 2003). In fact, Davis and Vinner
(1986) stated that as students develop their ideas of limits, misconceptions are probably
unavoidable. To complicate the teaching of this concept, if teachers try to simplify the
concept too much 'serious conceptual problems' may arise (Tall, 1992, p. 502).
Understanding is not always helped by the use of formal definitions (Cornu, 1991). This
problem is illustrated by a formal definition of the limit.

Let $f(x)$ be defined on an open interval about x_0, except possibly at x_0 itself. We say
that $f(x)$ approaches the limit L as x approaches x_0, and write:

$$\lim_{x \to x_0} f(x) = L$$

if for every $\varepsilon > 0$, there exists a corresponding number $\delta > 0$ such that for all x

$$0 < |x - x_0| < \delta \Rightarrow |f(x - L)| < \varepsilon$$

(Thomas & Finney, 1996)

In words, this can be described as: 'A limit is a number that the y-value of a function can be
arbitrarily close to by restricting x-values' (William, 1991, p. 221). It is not expected that a
student new to calculus would relate to such a definition (Robert & Schwarzenberger, 1991).
It is important, therefore, that students should be introduced to the limit in other ways.
One of the most common misconceptions is for students to think that a limit is never
reached and this is possibly owing to the way that the word 'limit' is used in everyday
language; speed limits, for example (Cornu, 1991).

How should students be introduced to the concept of the limit?

Cottrill et al. (1996) recommended that covariational reasoning should be used to introduce
students to the concept of the limit. Here is an example:

What happens to the value of the function $f(x) = \frac{x^2-4}{x-2}$ as x gets close to 2? This function is a straight line of the form $f(x) = x + 2$ with an important difference at $x = 2$ where the denominator equals zero and the function is undefined. It is fruitful to use covariational reasoning to explore the values of the function as x gets closer and closer to 2. This is illustrated in Table 13.1 and was generated by a spreadsheet in Excel.

Table 13.1 Values of $f(x)$ as x approaches 2

x	$f(x)$
1.9	3.9
1.99	3.99
1.999	3.999
1.9999	3.9999
1.99999	3.99999
1.999999	3.999999
2	value undefined
2.000001	4.000001
2.00001	4.00001
2.0001	4.0001
2.001	4.001
2.01	4.01

As the x-values approach 2, the value of the function approaches 4 even though at exactly $x = 2$ the function is undefined. This is an example where the limit is not actually reached. Tall (1992) reported that the context in which students are first introduced to limits may dominate their thinking about limits in future work. Therefore, it is important that a variety of examples with different properties is shown to the students.

Examples such as this that use covariational reasoning illustrate what is known as the *dynamic* idea of the limit. This idea of a limit emphasises motion (for example, as x moves towards a, $f(x)$ moves towards L). Whereas this idea may lead to conflicts with the formal definition (Szydlik, 2000), this form of understanding the limit has advantages (Keene, Hall & Duca, 2014). By using language such as 'getting closer to' and 'approaching' students can be assisted in building their intuitions about the nature of limits and how they are found (Keene, Hall & Duca, 2014, p. 564).

PAUSE AND THINK

How have you used the word 'limit' in your daily life? How could these examples help or hinder a student to understand the mathematical concept of a limit?

In her 2001 paper Mamona-Downs lists some 'approach types' (p. 264) that students have used when attempting to define a limit, and these are listed in Table 13.2.

Table 13.2 Approach types describing the limit

Approach type	Typical informal statement
Dynamic-theoretical	A limit describes how a function moves as x moves to a certain point.
Boundary	A limit is a number or point beyond which a function cannot go.
Formal	A limit is a number, to which the y-values of a function can be made arbitrarily close by restricting x-values.
Unreachable	A limit is a number or point the function gets close to but never reaches.
Approximation	A limit is an approximation that can be made as accurate as you wish.
Dynamic-practical	A limit is determined by plugging in numbers closer and closer to a given number until the limit is reached.

Source: Mamona-Downs, 2001, p. 264.

PAUSE AND THINK

Examine each definition in Table 13.2 and describe how a student may come to this particular view. What are the advantages and disadvantages for each approach? For those you think are incorrect, what counter examples could you find?

Derivatives

The derivative is a formula that can be used to find the rate of change at any single point on a curve of a function. The value of the derivative at a point is equivalent to the value of the gradient of the tangent line at the same point (Barnes, 1993; Kaplan, Ozturk & Ocal, 2015). The value of the gradient at this point is also known as the instantaneous rate of change. These two interpretations, that the derivative is the gradient of the tangent line at a point and the instantaneous rate of change at a point, are requirements of the Senior Mathematics Curriculum, Mathematical Methods (ACMMM084, ACMMM085). Kaplan, Ozturk and Ocal (2015, p. 65) defined the derivative as 'the limit of the ratio of the increase in the value of the function to the change in the independent variable as the change of the independent variable approaches zero'. This idea is represented by the equation:

$$\lim_{h \to 0} \frac{f(x+h) - f(x)}{h}$$

Derivative: is a formula that allows the calculation of the gradient of the tangent at any point on a function.

In Australia, this formula is often introduced to students by allowing them to observe what happens as two points on a secant move closer and closer together on a function. This process is illustrated in Figure 13.4. Here the secant line (in blue) becomes closer and closer to the tangent (in red) of the line at *A* as the horizontal distance between *A* and *B* is narrowed until finally it becomes co-linear with the tangent (Tall, 1992).

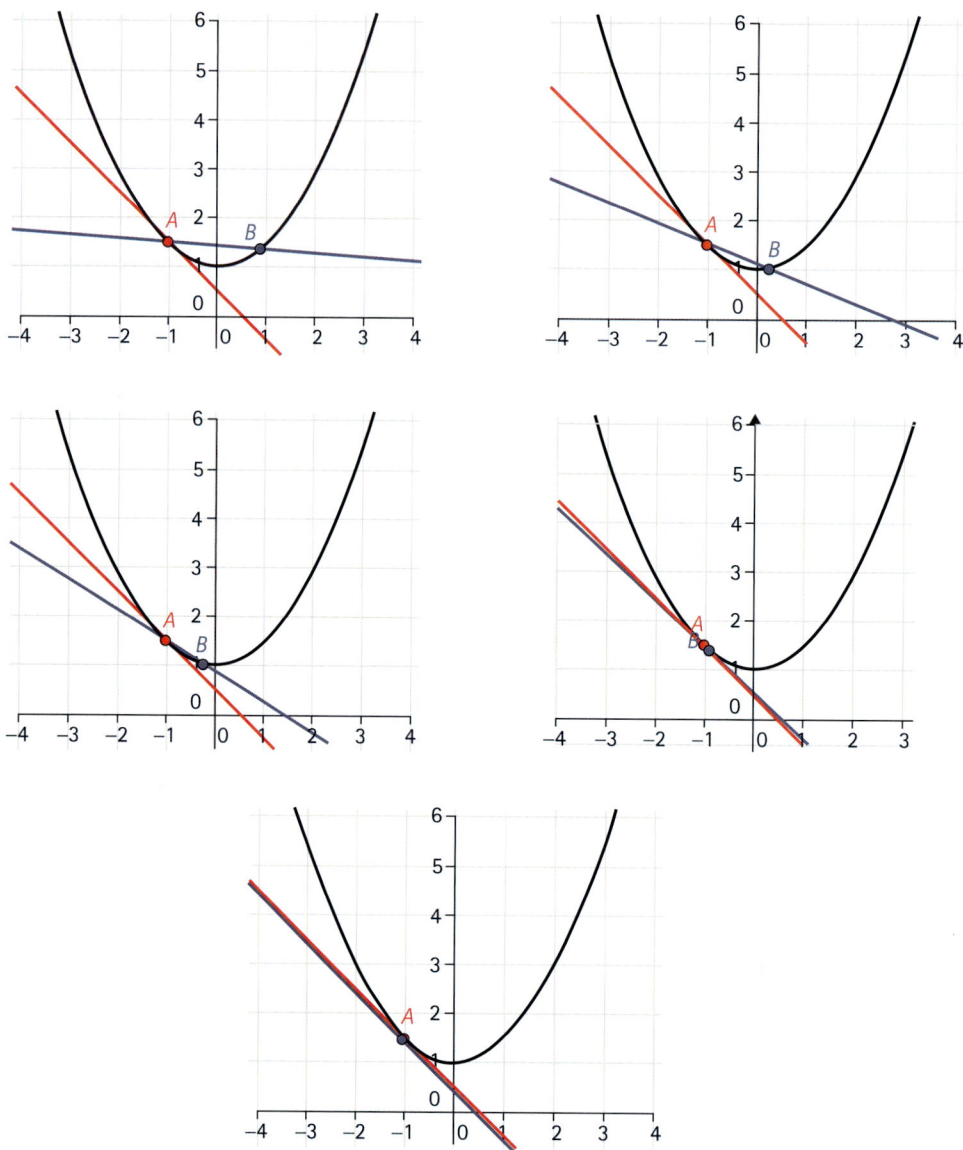

Figure 13.4 The change in the secant line (blue) as the point *B* is moved closer and closer to *A*

Students may find this method of finding the derivative from 'first principles' difficult, but in reality this process is no more than finding the rate of change (or gradient) of a function per unit change in *x*.

This is illustrated with the following equations. In the shift from equation (1) to (3) all that has happened is that function notation has been introduced and the difference in *x*-values is now represented by *h*:

$$m = \frac{y_2 - y_1}{x_2 - x_1} \tag{1}$$

$$m = \frac{f(x_2) - f(x)_1}{x_2 - x_1} \tag{2}$$

$$m = \frac{f(x + h) - f(x)}{h} \tag{3}$$

After this point has been reached, the derivative is just the limit as $h \to 0$ of equation (3) to produce equation (4):

$$f'(x) = \lim_{h \to 0} \frac{f(x + h) - f(x)}{h} \tag{4}$$

This algebraic process can also be accompanied by, or preceded by, the use of a spreadsheet. This is illustrated in Table 13.3 and Table 13.4.

Table 13.3 Values of the gradient of $y = x^2$ at $x = 3$ and $x = 3 + h$ as the values of h decrease

x_1	$x_2 - x_1$	x_2	y_1	y_2	$y_2 - y_1$	Gradient
3	0.1	3.1	9	9.61	0.61	6.1
3	0.01	3.01	9	9.0601	0.0601	6.01
3	0.001	3.001	9	9.006001	0.006001	6.001
3	0.0001	3.0001	9	9.0006	0.0006	6.0001
3	0.00001	3.00001	9	9.00006	0.00006	6.00001
3	0.000001	3.000001	9	9.000006	0.000006	6.000001
3	0.0000001	3.0000001	9	9.000001	0.000001	6.0000001

Table 13.4 Formulas for Table 13.3

	A	B	C	D	E	F	G
1	x_1	$x_2 - x_1$	x_2	y_1	y_2	$y_2 - y_1$	gradient
2	3	0.1	= A2 + B2	9	= C2^2	= E2 − D2	$= \frac{F2}{B2}$
3	3	$= \frac{B2}{10}$	= A3 + B3	9	= C3^2	= E3 − D3	$= \frac{F3}{B3}$
4	3	$= \frac{B3}{10}$	= A4 + B4	9	= C4^2	= E4 − E3	$= \frac{F4}{B4}$

Students' misconceptions about derivatives

Some of the problems that students may have with derivatives can result from mistakes in simple algebra. For example, students may make errors in the expansion of parentheses such as $(x + h)^2$ or $(x + h)^3$. Students may also not realise that the derivative itself is a function (Ubuz, 2007) (Mathematical Methods ACMMM089).

Ubuz (2007) has also found that students might believe that the value of the derivative at a point is equivalent to the derivative of the overall function or that the equation of the tangent line is equivalent to the derivative. Students will also find the concept of a derivative difficult if they have poor covariational reasoning.

In the section on functions, it was stated that students should be familiar with both algebraic and graphical representations and this is equally important when dealing with the derivatives of these functions. For example, students may be able to find the gradients of the tangents of functions that are expressed algebraically, but not when these functions are presented graphically (Orton, 1983a).

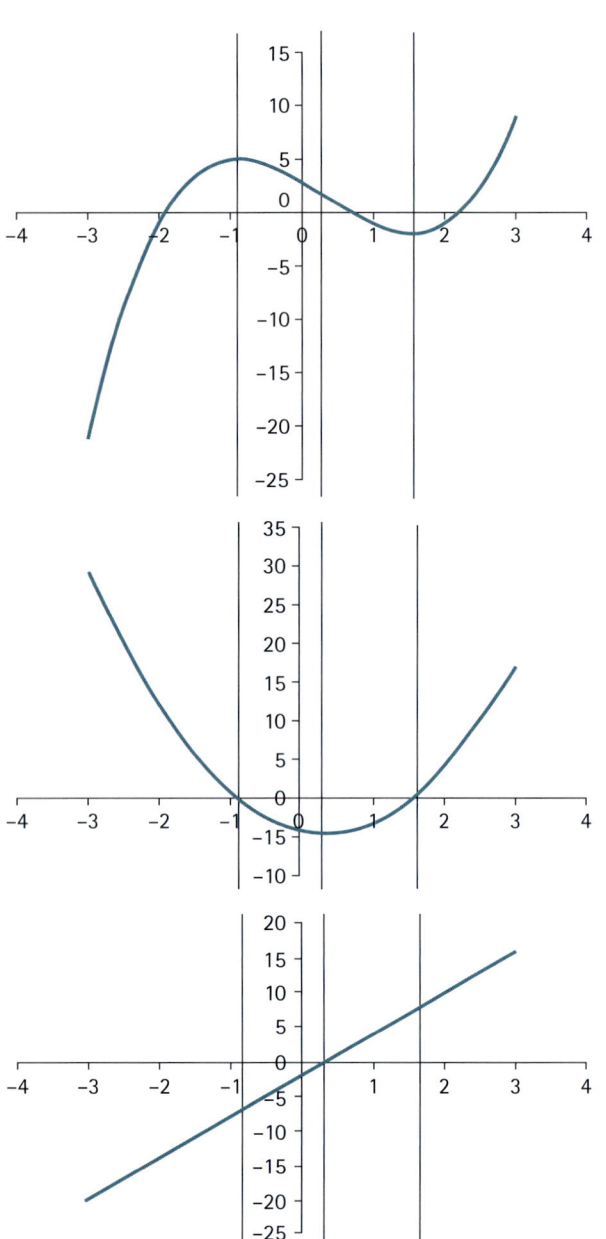

Figure 13.5 A graph of a function with the graphs of its first and second derivatives

Second derivatives

The first derivative is the derivative of an original function that allows the calculation of the rate of change on any point on the function. Second derivatives are found by finding the derivative of the first derivative, which tells us the rate of change of the derivative.

Therefore, if the second derivative at a particular x-value is positive, that means $f'(x)$ is increasing as it passes through that x-value. Similarly, $f'(x)$ is decreasing where the second derivative is negative. This information allows a method for confirming if a particular stationary point of the original function is a local maximum or minimum (Mathematical Methods, ACMMM111). Say that a possible minimum or maximum is found at $x = c$. If at a nearby x-value to the left of c, $f'(x)$ is positive, zero at $x = c$, and negative at a point to the right of c, then the original function will have a maximum at $x = c$. $f'(x)$ will be decreasing at this point and $f''(x)$ will be negative. Similarly, for a minimum point on the original function, $f'(x)$ will be negative, zero and positive and $f''(x)$ will be positive.

It is well known that in general, students may try to accommodate a lack of understanding by learning the rules by rote, and then making mistakes as they do not remember the rules correctly (Chance, del Mas & Garfield, 2004). Therefore, it is important that students gain an understanding of the reasoning that leads to these rules. Figure 13.5 gives an example of how this might be done. The students are given a polynomial function to the power of three in graphical form and asked to draw the graph of the first and second derivatives. Once it is appreciated that the first derivative is a quadratic function, they should be able to reason that the graph of the second derivative will be a straight line. What happens when $f''(x)$ equals zero? From such an exercise, the ideas about inflection points and concavity can be introduced.

Answering those tricky examples

SCENARIO

It is very important that students should not just be asked to carry out procedures but should be asked to explain their reasoning. In this example students were challenged with a non-procedural question. Their answers show areas of uncertainty that would be the basis of fruitful class discussions.

Reaburn et al. (2018) gave three first-year tertiary calculus students, 'Anna', 'Bruce' and 'Cathy', a series of functions in graphical form and asked them if they thought these functions had a limit at, and were differentiable at, all values of x. One of these graphs is shown in Figure 13.6. The graph was left deliberately ambiguous at $x = 0$.

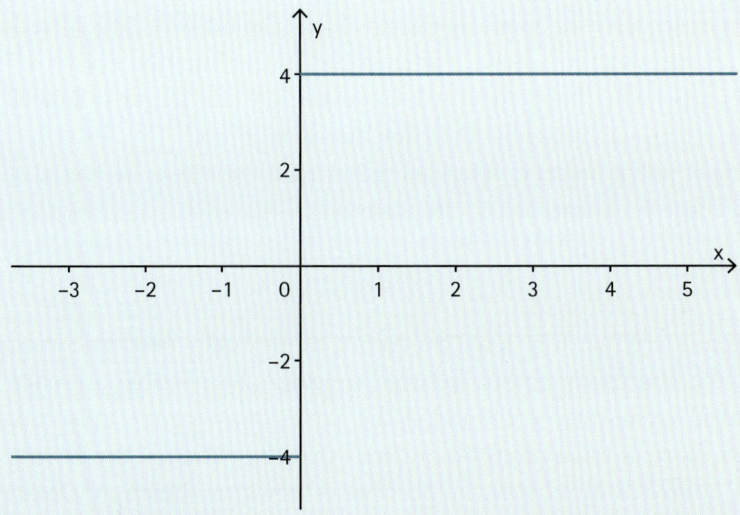

Figure 13.6 Functions in graphical form

Anna stated that the function was discontinuous at $x = 0$, and Cathy stated that the right-hand and left-hand limits do not agree, therefore the function did not have a limit at this point. Bruce was unsure.

All students demonstrated uncertainty about the differentiability of this function. Anna noted the function takes up 'two points at the values $x = 0$, on the y-axis' and that the derivatives of both sides of the function are zero but was ultimately unsure. Bruce stated that 'it doesn't matter if there is a break there . . . from both sides, x still goes to zero'. He then noted that 'the left and right limits won't agree' (reasoning he did not use earlier). After admitting he was confused, he finally concluded that the function was not differentiable at $x = 0$. Cathy considered the possible tangents at $x = 0$. She noted that 'you can sort of rotate the tangent at any, you can pick any tangent around there', concluding that the function was not differentiable at this point.

Uses of derivatives
Graphing of functions

The Australian Curriculum document for Mathematics Methods states that students should sketch curves associated with simple polynomials, find stationary points, and local and global maxima and minima, and examine behaviour as $x \to \infty$ and $x \to -\infty$ (ACMMM095).

When searching for stationary points, students may find the algebra worrisome, but the principle itself is simple. At a stationary point, the gradient of a tangent line is horizontal and has a gradient of zero. Therefore, if the formula for a function is known, finding these stationary points simply requires finding the values of x where the derivative is equal to zero. Students need to be reminded that the stationary points found this way are often only local maxima or minima and may not be the global maximum or minimum of the function. In addition, they also need to be aware that the derivative may also be equal to zero at a point of inflexion. The dominant terms are the x terms with the highest exponent, which therefore will grow more quickly than any other term and are found by using covariational reasoning to consider what happens as the x-values take on very large positive or negative values (that is, as $x \to \infty$ or $x \to -\infty$). For example, in the function $f(x) = x^3 - x^2 + 2x$, we know that no matter what happens at values near $x = 0$, for larger values of x (positive or negative) the function will resemble the cubic function $y = x^3$.

The graphing of polynomials is extended to Rational Functions in Specialist Mathematics (ACMSM100). These functions are in the form of:

$$f(x) = \frac{g(x)}{h(x)}$$

When students are faced with rational functions, they usually need to consider the dominant terms and the stationary points and also consider the possible presence of asymptotes and holes. Vertical asymptotes are found at the values of x where $f(x)$ is undefined because $h(x)$ equals zero. Horizontal asymptotes are lines that the values of the function approach as the values of x approach infinity in the positive or negative direction. Oblique asymptotes occur when the degree of $g(x)$ is greater than the degree of $h(x)$. Because graphs of rational functions do not cross vertical asymptotes, students often think that this applies to horizontal asymptotes as well and teachers need to be aware of this.

There is another complication that students need to deal with – the presence of discontinuities – holes. If both the numerator and denominator are factorised into linear factors, and the same factor appears in both the numerator and denominator, then once this factor is solved for zero there is a discontinuity at that point. For example, consider the function:

$$f(x) = \frac{x^2 - 4}{x^2 - 5x + 6}$$

This can be factorised to:

$$f(x) = \frac{(x + 2)(x - 2)}{(x - 3)(x - 2)}$$

Asymptote: a line is an asymptote to a curve if the distance between the line and the curve approaches zero as they 'tend to infinity'. For example, the line with equation $x = \frac{\pi}{2}$ is a vertical asymptote to the graph of $t = \tan x$, and the line with equation $y = 0$ is a horizontal asymptote to the graph of $y = \frac{1}{x}$.

This tells us that the function $f(x)$ is undefined at $x = 3$ and that there is a vertical asymptote at $x = 3$, but at $x = 2$ the function does not have a vertical asymptote, but a discontinuity. This is illustrated in Figure 13.7.

Successful graphing of rational functions takes time and practice by the students, and teachers should allow for this in their planning. The teacher also needs to provide a variety of examples so that the students can become confident with the different situations that might occur. The careful and planned use of technology also assists in developing understanding.

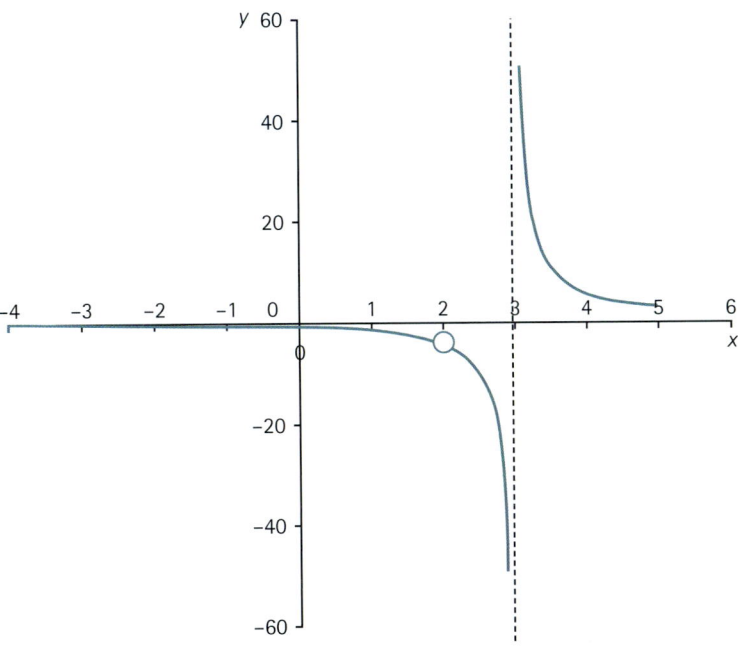

Figure 13.7 The graph of the rational function. Note the 'hole' at $x = 2$

Optimisation

Students are often intrigued to find that their work on derivatives assists them to find the answers to practical questions. What is the maximum volume of a box that can be made with a fixed area of cardboard? If we want to make a soft drink container that can hold 300 mL, what shape will use the least materials? Because derivatives can be used to find maxima and minima of functions, such questions can be answered, and this process of optimisation is one of the useful applications of derivatives. Optimisation in a variety of contexts is one of the requirements of Mathematical Methods (ACMMM096).

For example, the students could be required to find the maximum area of a box (without a lid) that could be made out of a sheet of A4 paper. This is a useful example for an introduction to optimisation as it can be solved in several ways and students can compare the methods. The students can use trial and error, experiment with actual paper or use a spreadsheet. When they use calculus, the students will find the function that describes the volume of the box in one variable, and then use the derivative to find the maximum of the function. No matter the method, they will need to draw a diagram to assist in finding the relationship between the equations and the physical object. With this example a diagram shows that one way to solve the problem is to look at the size of the square that needs to be cut out of each corner of the paper (Figure 13.8).

When this problem is solved using calculus, the derivative is a quadratic equation where only one of the solutions gives a suitable size of the square. It must be admitted, however, that such optimisation problems can be more complicated than this example. It can be difficult for the students to combine the different formulas that describe a particular situation into a function that is differentiable. In addition, students with poor

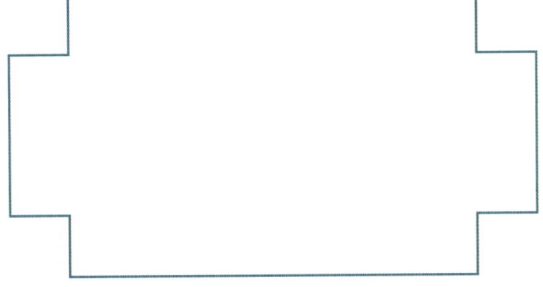

Figure 13.8 The diagram to help solve the box problem

spatial reasoning may find drawing appropriate diagrams difficult and much practice might be needed.

Further derivatives: trigonometric functions

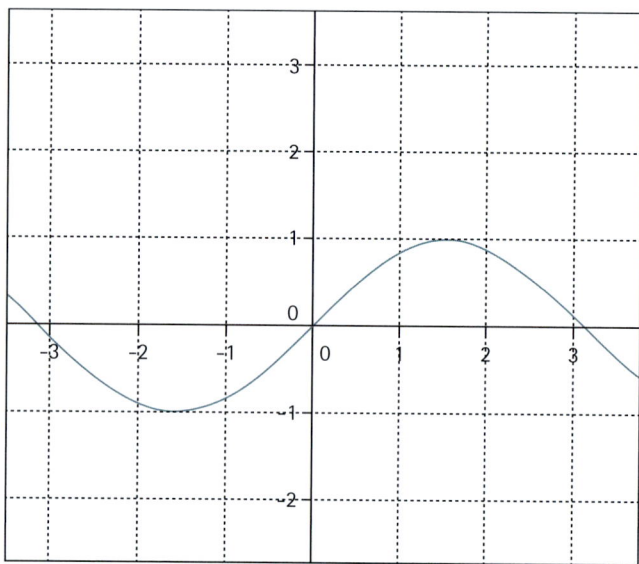

Figure 13.9 Graph of the sine function

In earlier years students use the trigonometric functions as ratios of sides of right-angled triangles. Students in Mathematical Methods are also required to relate these functions to the unit circle, observing what happens to the x- and y-values as we travel around the circle (ACMMM034). They are also required to use the derivatives of these functions (ACMMM034).

Teachers often just tell students the formulas for the derivatives of sine(x), cosine(x) and tangent(x). A more investigative approach using covariational reasoning can lead to better understanding. For example, look at the graph of $y = \sin(x)$ in Figure 13.9.

If the students trace out the figure from left to right, they will see that the initial negative gradient becomes less steep until it becomes zero. After this the gradient is positive, becoming steeper until it again becomes less steep until it is zero, and so on. If they sketch out this pattern, they will find they will get a pattern that is similar to, but out of phase with, the original function. They should not then be surprised that the derivative is actually the cosine function.

If a similar process is used for the $\tan(x)$ function, the students will see that the derivative always has a positive value. Exploration with the quotient rule (next section) will allow the students to see that the derivative of the tangent function is $\sec^2(x)$.

Product, quotient and chain rules

While we know that the derivative of $f(x) + g(x) = f'(x) + g'(x)$, can this rule be extended to when two functions are multiplied together? In other words, is the derivative of $f(x) \times g(x)$ equal to $f'(x) \times g'(x)$? What about the derivative of $\frac{f(x)}{g(x)}$? Is this equal to $\frac{f'(x)}{g'(x)}$? If students experiment with polynomial functions, it should become immediately apparent that the derivative of two functions that are multiplied together is not the product of their derivatives. Neither is the derivative of one function divided by another the quotient of the two derivatives.

Once this is established, then the rules for finding the derivative of the product or quotient of two functions can be introduced. The more capable and curious students can investigate the proofs of these rules. Similarly, students can explore the chain rule. Give the students a function to experiment with, for example:

$$f(x) = (\sin(x))^2$$

What is the derivative of this function? Because we know that the derivative of $\sin(x)$ is $\cos(x)$, and the derivative of x^2 is $2x$, then we know the derivative should have a '2' and '$\cos(x)$' in the answer. However, the derivative is not $2\cos(x)$. This would be the derivative if our original function were to be $f(x) = 2\sin(x)$. If students are prepared to accept that two different functions will not have the same derivative, then they should be ready to accept that something else needs to be done, and in this case this is using the chain rule. The chain rule can be written in two different ways:

1. If $F(x) = (f \circ g)(x)$, then the derivative of $F(x)$ is: $F'(x) = f'(g(x))g'(x)$
2. If $y = f(u)$ and $u = g(x)$, then the derivative of y is: $\frac{dy}{dx} = \frac{dy}{du} \times \frac{du}{dx}$

If students are familiar with both notations, they can use the notation they prefer. The more curious and capable students can investigate the proof of this rule. The chain rule is commonly summarised as the 'inside outside' rule, but care must be taken if the rule is to be taught this way. Students can become so confident in this shortcut that they can be reluctant to go back to using the chain rule when more complicated derivatives arise, and they will make errors.

Connection between function graphs and graphs of their derivative functions

Students can have the idea that doing calculus only involves manipulating symbols and numbers (Hughes Hallet, 1991). Drawing the graph of a derivative function from the original function or drawing an original function from the graph of its derivative function, however, requires higher-order thinking, in particular, a thorough understanding of rates of change.

Here again a graphing tool such as GeoGebra can be helpful. For example, in Figure 13.10 the graph of the derivative (the red points) was formed as the point c was moved along the x-axis for the function $y = x^3$. This figure illustrates that the derivative of this function is always positive, and why the point at $x = 0$ is a point of inflexion, not a stationary point.

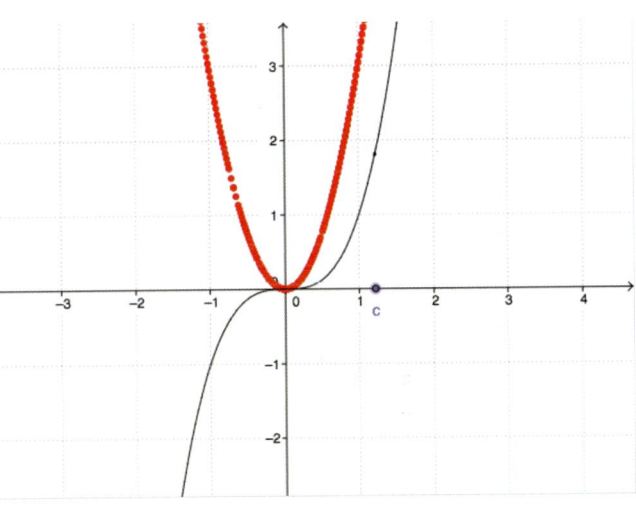

Figure 13.10 The graph of the function $y = x^3$ with its derivative function (red points). The red points are created as the point c is moved along the x-axis

Source: produced using GeoGebra.

SHORT-ANSWER QUESTIONS

1. Students often rush when using the chain rule, regarding it no more than the 'inside outside' rule. How well does this rule work for finding the derivative of this function?

$$y = \sqrt{\sin(x)^2}$$

▶▶

▶▶

2. Commonly, an asymptote is defined as a line to which the graph 'never reaches'. Work out the following example, considering how this example can be used for teaching, and what mistakes students may make. The graph of the function $f(x) = \frac{(x-3)^2}{x^2-5x}$ has an asymptote at $y = 1$. What does the graph of this function do in the region of $x = 9$? Can you find another rational function with similar properties?

3. Students often believe that wherever the second derivative is equal to zero, there must be an inflection point. The second derivative of the function $f(x) = \frac{x^6}{30} - \frac{x^5}{20} - x^4 + 3x + 20$ is equal to zero at $x = 4$, $x = 0$ and $x = -3$. Verify this statement by finding the first and second derivatives. Plot the graph, using a graphing calculator or other means, and investigate these points. As you do this, consider how this example could be used for teaching, and what mistakes students might make.

4. The second derivative of the function $f(x) = \sqrt[3]{x}$ is undefined at $x = 0$. The function itself is defined at this point. Find the first and second derivatives. Then graph $f(x)$. What does this tell you about the role of second derivatives in finding inflection points? What purpose do such examples play in teaching?

Reversing the process: anti-differentiation and integration

If we know the derivative function, what is the original function? For simple polynomials students should easily be able to work out how the original function can be found. With guidance, they should also be able to see that without further information, if there is a constant in the original function its value cannot be determined. Therefore the '+ C' given at the end of the process of anti-differentiation, or integration, is a necessity (Mathematical Methods ACMMM121).

The terminology can be confusing. If f is a function defined on an interval, I, and f has an antiderivative on this interval, the set of the antiderivatives of f is called the indefinite integral and has this notation:

$$\int f(x)dx$$

Integration: the process by which one can get to the original function if the derivative is known.

Finding these antiderivatives is known either as anti-differentiation or integration. The function f is known as the *integrand* of the integral and x is the *variable of integration*. An antiderivative is one of the set of functions that can be a solution to the process of integration.

Students may become confused about the difference between an *indefinite integral* and a *definite integral*. Indefinite integrals have no limits (note that the word 'limit' has a different meaning in this context) and the solution to finding an indefinite integral is a function. In contrast, definite integrals have limits, and while the process of integration is required in finding their solutions, the final answer is a number. For example, to find the solution to $\int_2^4 (x^3 - x^2)\, dx$ we need to first find the integral and then evaluate this integral with the values '4' and '2':

$$\int_2^4 (x^3 - x^2)\,dx = \left[\frac{x^4}{4} - \frac{x^3}{3}\right]$$

which is then evaluated at both $x = 4$ and $x = 2$:

$$= \left(\frac{4^4}{4} - \frac{4^3}{3}\right) - \left(\frac{2^4}{4} - \frac{2^3}{3}\right)$$

$$= 64 - \frac{64}{3} - 4 + \frac{8}{3}$$

$$= \frac{124}{3}$$

This example is shown in full to illustrate two things. First, as the functions get more complicated there are endless possibilities for students to become confused with the signs, so the use of parentheses is essential. Second, the facility with the manipulation of fractions on scientific calculators greatly reduces the potential for calculation error. Students are often tempted to use decimals, but by doing this the rounding errors can quickly compound, which is of importance when, for example, they calculate areas under curves.

Applications of integration

Areas under curves

The Mathematical Methods curriculum states that students should be able to 'use sums in the form of $\sum_i f(x)\delta x_i$ to estimate the area under the curve $y = f(x)$' (ACMMM124) and to 'interpret the integral $\int_a^b f(x)dx$ as the area under the curve $y = f(x)$ if $f(x) > 0$' (ACMMM125). It then continues to state that students should 'recognise the definite integral $\int_a^b f(x)dx$ as a limit of sums of the form $\sum_i f(x)\delta x_i$' (ACMMM126). These statements sum up the process of finding areas between curves and the x-axis.

The area between a curve and a straight line can be approximated by dividing the area into rectangles, as shown in Figure 13.11.

The height of each rectangle is found by finding the value of the function at each appropriate x-value, and the width of each rectangle can be represented by Δx. If there are k rectangles the total area is then approximated by:

$$A = \sum_i^k f(x)\Delta x$$

This formula calculates what is known as a Riemann sum.

Students should be able to see that by reducing the width of each rectangle, the error that is introduced by using the rectangles instead of the actual curve should be minimised until it approximates zero (Tarvainen, 2006). As $\Delta x \to 0$ the area between the curve and the x-axis over an interval $[a,b]$ is found by the definite integral:

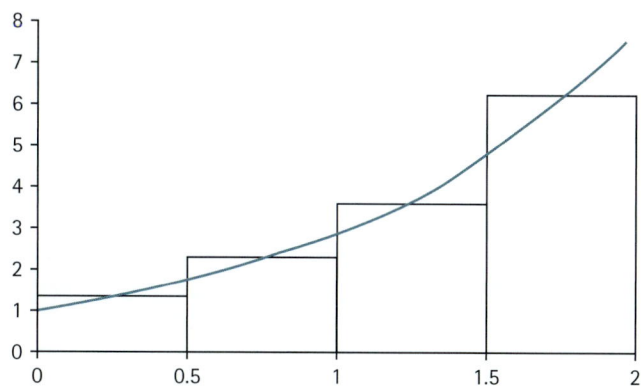

Figure 13.11 An approximation to the area under a curve using rectangles

$$A = \int_a^b f(x)dx$$

That is, the limit of the Riemann sum as the interval approaches zero is the same as the area between the curve and the x-axis. This process is easily simulated with the use of spreadsheets. As the intervals along the x-axis are reduced in size, the calculated area approaches a limiting value that is equal to the value obtained by integration. Students should also explore the result when part, or whole, of a graph is below the x-axis. For example, finding the value of the integration $\int_{-2\pi}^{2\pi} \sin(x)dx$ gives a very different answer to finding the area between the graph $f(x) = \sin x$ and the x-axis, $-2\pi \le x \le 2\pi$.

Being able to imagine the process of integration by using smaller and smaller rectangles is very useful for the students. A similar process of imagination is needed for students to follow the process of finding solids of revolution. If each rectangle is rotated around the x-axis, then the result is a series of cylinders, each with a radius equal to the value of the function and with the width equal to Δx. Once this is appreciated, the actual calculations are not difficult. Instructors may need a variety of representations for students to understand how the process works.

Students' misconceptions about integration

Students can have several problems in calculating the answers to integration and comprehending the process. Some errors arise from a poor understanding of fractions, fractional indices and factorisation. On a deeper level, they can have problems understanding notions such as 'positive' and 'negative' areas (Orton, 1983b). Sometimes, however, such area problems are merely owing to a reluctance of the student to make the effort to graph the function.

Orton (1983b) found that some students do not understand the idea of the Riemann sum. Students may treat the area under the curve as a literal representation of an area in nature (Thompson & Silverman, 2007) and get confused because a 'negative' area has no physical meaning (Tall & Rasslan, 1997, cited in Hall, 2010). Other problems arise from the way that the words 'definite' and 'indefinite' are used in general English. Therefore, for these students a 'definite' integral is one that is clear and accurate, while an 'indefinite' integral lasts for a prolonged period of time, or is unclearly defined (Hall, 2010).

Hall (2010) and Rasslan and Tall (1997) also report that the units that result from integration to solve real problems can be confusing for students. For example, the units that result in calculating an area are squared units, but if the question deals with velocity and acceleration, the units are in terms of distance and time.

IN PRACTICE

Generalising a rule beyond familiar cases

Maher (2019) described a Mathematics Methods lesson where the students were working on finding areas between curves and the x-axis. The class progressed smoothly until many asked for help when they were faced with the function $g(x) = (8 - x^2)$. 'Mr Jones', their teacher, led a discussion that guided the students to see that this function could be factorised by using the difference of two squares; 'that a particular "rule" could be generalised beyond familiar cases' (Maher, 2019, p. 153).

After the lesson Mr Jones stated that he had chosen this item deliberately because he wanted to challenge their ideas of the difference in two squares. He also expressed a dilemma often experienced by teachers – how long should he have left the students to struggle before intervening? Had he 'jumped the gun' too quickly? (Maher, 2019, p. 156).

Connecting it all: part 1 of the Fundamental Theorem of Calculus

Schwalbach and Dosemagen (2000) noted that the realisation that the derivative and antiderivative are opposite operations as a very important connection for calculus students. This is summarised by the Fundamental Theorem of Calculus (sometimes known as Part 1 of the Theorem).

If f is continuous on $[a,b]$, then $F(x) = \int_a^x f(t)dt$ has a derivative at every point of $[a,b]$ and:

$$\frac{dF}{dx} = \frac{d}{dx}\int_a^x f(t)dt = f(x), a \leq x \leq b$$

(Thomas & Finney, 1996, p. 333)

This equation states that the differential equation $\frac{dF}{dx} = f$ has a solution for every continuous function f. It also states that every continuous function f is the derivative of some other function, namely $\int_a^x f(t)dt$. In addition, it states that every continuous function has an antiderivative, and that integration and differentiation are the inverse processes of one another.

The problem of 'e' and the natural log (ln)

The Fundamental Theorem of Calculus allows us to deal with the problem of finding the indefinite integral of $\frac{1}{x}$.

If we follow the rules for polynomials:

$$f(x) = x^{-1}$$

then

$$\int x^{-1}dx = \frac{x^{-1+1}}{-1+1} = \frac{x^0}{0}$$

which is undefined.

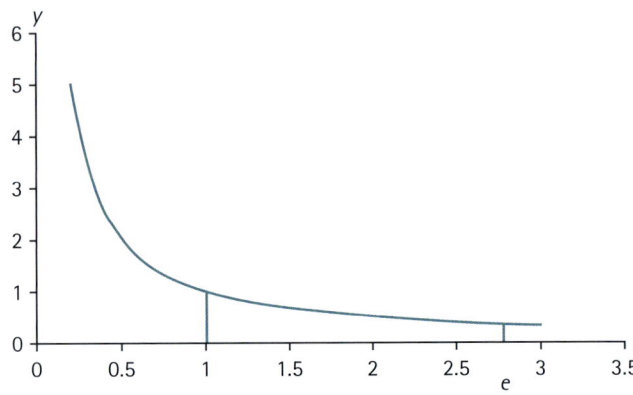

Figure 13.12 The graph of $f(x) = \frac{1}{x}$ showing the area from $x = 1$ to $x = e$

But we know from the Fundamental Theorem of Calculus that this integral must exist, and this is confirmed by looking at the graph of $f(x) = \frac{1}{x}$ (Figure 13.12).

In this figure the area under the curve from $x = 1$ to $x = e$ is 1 squared unit. From this we can deduce that:

$$\int_{1}^{e} \frac{1}{x} dx = \ln(e) = 1$$

Therefore, in general:

$$\int \frac{1}{x} dx = \ln(x)$$

And we can also find the converse, that is, if $f(x) = \ln(x)$, then $f'(x) = \frac{1}{x}$.

IN PRACTICE

Anticipating student errors

Knowing what mistakes are likely to be made and deciding what to do with these anticipated errors is an important part of being a teacher.

Maher (2019) describes a Mathematics Methods lesson where the students were asked to find this integral: $\int e^2 dx$. Several students tried to treat e^2 as a variable and not a number. In a discussion with the author after the lesson, 'Mr McLaren' stated that he had deliberately included this example knowing that students would make this error. He continued:

> It has probably only come through experience but knowing the pitfalls that they'll come across or that they'll fall into . . . I knew that this was a common mistake . . . you know how you learn out of confusion, so if you can be pointed or if you can be taken through it from confusion to understanding, then it is going to be more valuable . . . I wanted to let them make the mistake here and then talk about it. (pp. 182–3)

Conclusion

This chapter began with the concept of a mathematical function and then progressed to calculus. It should be remembered that even if students do not go on to study calculus that learning about functions is still very important because it is here they learn about mathematical relationships and covariational reasoning – important skills in many areas of the curriculum and life outside of school.

A deep understanding of calculus is not just learning the procedures but requires covariational reasoning and, contrary to what many people think about mathematics, imagination. Teachers are not only challenged by the complexity of some of the material but by the need to allow the students time to develop understanding through experience – this can be difficult when there is a set curriculum to be covered for examinations. Technology, when carefully used, can greatly assist in developing students' understanding.

BRINGING IT TOGETHER

1. A relationship that can be represented as a parabola with a vertical axis of symmetry is a function, but a relationship that can be represented as a circle is not a function. Explain why this is so. Why is it important for students to know the difference between functions and other algebraic relationships?

2. Compare these two functions:

$$f(x) = \begin{cases} \dfrac{x^2 - 9}{x - 3} & \text{if } x \neq 3 \\ 6 & \text{if } x = 3 \end{cases}$$

And

$$f(x) = x + 3$$

How are these the same, and how are they different? Do limits exist for all values of x for both functions?

3. Students can find it difficult to relate position, velocity and acceleration (McDermott, Rosenquist & van Zee, 1987). Possibly the hardest task in this topic is to work from acceleration versus time graphs. Here is an example.

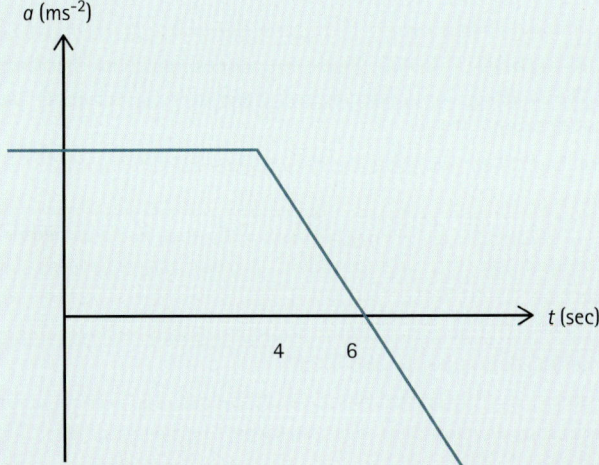

What is happening to the acceleration and velocity of the object between $t = 0$ and $t = 4$ seconds? What is happening to the acceleration and velocity of the object between $t = 4$ and $t = 6$ seconds? When the acceleration equals zero, is the object still moving? What is happening when the acceleration is negative?

4. You have given students a table of values of a function at various values of x so that they can calculate the Riemann sum. When faced with such problems, students may assume (1) that the largest value of the function is a maximum and (2) that a number is always

decreasing throughout an interval when the value of the function is less than the preceding value. What example can you draw to demonstrate that these assumptions are not always true?

5. A student comes to you and says that he has noticed something very interesting. If you work out the derivative of the formula for a sphere, the result is the formula to find the surface area of a sphere. If you work out the derivative of the formula for the area of a circle, the result is the formula to find the circumference of a circle. Why is this? What explanation/diagrams could you use to show why this happens?

Further resources

REFERENCES

Artigue, M. (1991). Analysis. In D. Tall (ed.), *Advanced mathematical thinking* (167–96). Dordrecht: Kluwer Academic Publishers.

Australian Curriculum, Assessment and Reporting Authority (ACARA) (2010 to present). *Essential Mathematics, General Mathematics, Mathematical Methods and Specialist Mathematics*. Sydney: ACARA. Retrieved from https://www.australiancurriculum.edu.au/senior-secondary-curriculum/mathematics/

Australian Institute for Teaching and School Leadership (AITSL) (2011). *Australian Professional Standards for Teachers*. Melbourne: AITSL.

Barnes, M. (1993). *Investigating change: An introduction to calculus for Australian schools (unit 5)*. Carlton: Curriculum Corporation.

Bezuidenhout, J. (2006). First-year university students' understanding of rate of change. *International Journal of Mathematical Education in Science and Technology*, 29(3), 389–99.

Carlson, M., Jacobs, S., Coe, E., Larsen, S. & Hsu, E. (2002). Applying covariational reasoning while modeling dynamic events: A framework and a study. *Journal for Research in Mathematics Education*, 33(5), 352–78.

Chance, B., del Mas, R. & Garfield, J. (2004). Reasoning about sampling distributions. In D. Ben-Zvi & J. Garfield (eds), *The challenges of developing statistical literacy, reasoning and thinking* (295–323). Dordrecht: Kluwer Academic Press.

Cornu, B. (1991). Limits. In D. Tall (ed.), *Advanced mathematical thinking* (153–66). Dordrecht: Kluwer Academic Publishers.

Cottrill, J., Dubinsky, E., Nichols, D., Schwingendorf, K., Thomas, K. & Vicakovic, D. (1996). Understanding the limit concept: Beginning with a coordinated process scheme. *Journal of Mathematical Behaviour*, 15, 167–92.

Davis, R. & Vinner, S. (1986). The notion of limit: Some seemingly unavoidable misconception stages. *Journal of Mathematical Behaviour*, 5, 281–303.

Ferrini-Mundy, J. & Graham, K. (1991). An overview of the calculus reform effort: Issues for learning, teaching, and curriculum development. *The American Mathematical Monthly*, 98(7), 627–35.

Hall, W. (2010). *Student misconceptions of the language of calculus: Definite and indefinite integrals. Proceedings of the 13th Annual Conference on Research in Undergraduate Mathematics Education*. Raleigh, NC: Mathematical Association of America.

Hughes Hallet, D. (1991). Vizualisation and calculus reform. In W. Zimmerman & S. Cunningham (eds), *Vizualisation in teaching and learning mathematics*, MAA Notes, 19, 121–6.

Kaplan, A., Ozturk, M. & Ocal, M. (2015). Relieving of misconceptions of derivative concept with derive. *International Journal of Research in Education and Science*, 1(1), 67–74.

Katz, V. (1993). *A history of mathematics: An introduction*. New York: Harper Collins.

Keene, K., Hall, W. & Duca, A. (2014). Sequence limits in calculus: Using design research and building on intuition to support instruction. *ZDM Mathematics Education*, 46, 561–74.

Maher, N. (2019). *Perspectives on pedagogical content knowledge in the senior secondary mathematics curriculum*. Doctoral thesis. University of Tasmania.

Mamona-Downs, J. (2001). Letting the intuitive bear on the formal: A didactical approach for the understanding of the limit of a sequence. *Educational Studies in Mathematics*, 48(2), 259–88.

McDermott, L.C., Rosenquist, M.L. & van Zee, E.H. (1987). Student difficulties in connecting graphs and physics: Examples from kinematics. *American Journal of Physics*, 55(6), 503–13.

Meyer, J. & Land, R. (2003). Threshold concepts and troublesome knowledge: Linkages to ways of thinking and practising within the disciplines. In C. Rust (ed.), *Improving student learning: Ten years on*. Oxford: Oxford Centre of Staff and Learning Development.

Motz, L. & Weaver, J. (1993). *The story of mathematics*. New York: Plenum Press.

Orton, A. (1983a). Students' understanding of differentiation. *Educational Studies in Mathematics*, 14, 235–50.

Orton, A. (1983b). Students' understanding of integration. *Educational Studies in Mathematics*, 14, 1–18.

Parameswaran, R. (2006). On understanding the notion of limits and infinitesimal quantities. *International Journal of Science and Mathematics Education*, 5, 193–216.

Rasslan, S. & Tall, D. (1997). Definitions and images for the definite integral concept. In A. Cockburn & E. Nardi (eds), *Proceedings of the 26th Psychology for Mathematics Education Conference (PME)*, (89–96). Norwich: PME.

Reaburn, R., Oates, G., Dharmadasa, K. & Brideson, M. (2018). In E.-J. Hsieh (ed.), *Relating flexibility of concept image and understanding of limits and derivatives. Proceedings of the 8th ICMI-East Asia Regional Conference on Mathematics Education (EARCOME)* (vol. 2, 239–47). Taipei: EARCOME.

Robert, A. & Schwarzenberger, R. (1991). Research in teaching and learning mathematics at an advanced level. In D. Tall (ed.), *Advanced mathematical thinking* (127–39). Dordrecht: Kluwer Academic Publishers.

Schwalbach, E. & Dosemagen, D. (2000). Developing students' understanding: Contextualising calculus concepts. *School Science and Mathematics*, 100(2), 90–8.

Szydlik, J. (2000). Mathematical beliefs and conceptual understanding of the limit of a function. *Journal for Research in Mathematics Education*, 31(3), 258–76.

Tall, D. (1992). The transition to advanced mathematical thinking: Functions, limits, infinity, and proof. In D. Grouws (ed.), *Handbook of research on mathematics teaching and learning* (495–511). New York: Macmillan Publishing Co.

Tall, D. & Vinner, S. (1981). Concept image and concept definition in mathematics with particular reference to limits and continuity. *Educational Studies in Mathematics*, 12(2), 151–69.

Tarvainen, K. (2006). How to make sure that the error in the f(x) dx term is insignificant when setting up definite integrals. *International Journal of Mathematical Education in Science and Technology*, 29(3), 359–70.

Thomas, G. & Finney, R. (1996). *Calculus and analytical geometry* (9th edn). Boston, MA: Addison-Wesley Publishing Co.

Thompson, P. & Silverman, J. (2007). The concept of accumulation in calculus. In M. Carlson & C. Rassmussen (eds), *Making the connection: Research in teaching in undergraduate mathematics* (117–31). Washington, DC: Mathematical Association of America.

Ubuz, B. (2007). Interpreting a graph and constructing its derivative graph: Stability and change in students' conceptions. *International Journal of Mathematical Education in Science and Technology*, 38(5), 609–37.

William, S. (1991). Model of limit held by college calculus students. *Journal for Research in Mathematics Education*, 22(3), 219–36.

Wolfram MathWorld (2020). *Function*. Retrieved from https://mathworld.wolfram.com/Function.html